Real Analysis

NORMAN B. HAASER
University of Notre Dame

JOSEPH A. SULLIVAN
Boston College

DOVER PUBLICATIONS, INC., New York

To Beatrice and Mary

This Dover edition, first published in 1991, is a revised and corrected republication of the work originally published in 1971 by Van Nostrand Reinhold Company, New York as part of The University Series in Mathematics. Certain passages have been deleted or replaced with new material, particularly the proof following Proposition 6 · 1 on page 172 and the material on pp. 302–303.

Manufactured in the United States of America
Dover Publications, Inc., 31 East 2nd Street, Mineola, New York 11501

Library of Congress Cataloging-in-Publication Data

Haaser, Norman B.
Real analysis / Norman B. Haaser, Joseph A. Sullivan. — Dover ed.
p. cm.
Originally published: New York : Van Nostrand Reinhold Co., 1971. (University series in mathematics)
Includes bibliographical references and index.
ISBN 0-486-66509-7
1. Mathematical analysis. 2. Functions of real variables. I. Sullivan, Joseph A. (Joseph Arthur), 1923– . II. Title.
QA300.H28 1990
515—dc20 90-45359
CIP

Preface

This book is a text for a first course in abstract analysis. Although the topics treated have been traditionally found in first year graduate courses, the presentation is such as to make them accessible to undergraduates with a good background in the calculus of functions of one and several variables. Preliminary versions of the text have been used both in courses for undergraduates and in courses for graduate students. In designing the course, due consideration has been given to the recommendations of the Committee on the Undergraduate Program in Mathematics of the Mathematical Association of America for a course in Real Analysis as described in their reports: *Pregraduate Preparation of Research Mathematicians* (May 1963) and *Preparation for Graduate Study in Mathematics* (November 1965).

As a text for a first course in abstract analysis, it is intended to provide a smooth transition from calculus to more advanced work in analysis. In outline the course consists of a study of the familiar concepts of calculus such as convergence, continuity, differentiation, and integration in a more general and abstract setting. This serves to reinforce and deepen the reader's understanding of the basic concepts of analysis and, at the same time, to provide a familiarity with the abstract approach to analysis which is valuable in many areas of applied mathematics and essential to the study of advanced analysis.

The first three chapters—Sets and Relations, The Real Number System, and Linear Spaces—are preliminary in nature and provide a foundation for what follows. Although they consist to a considerable extent in the introduction of notation and review of definitions and elementary results, there are a number of topics that are likely to be new to the reader. The introduction of the real numbers via Cauchy sequences of rationals not only establishes the basic properties of this number system, but also provides a model for the completion of an arbitrary metric space. The treatment of linear spaces emphasizes those aspects pertinent to analysis which might not be treated adequately in a Linear Algebra course.

The main body of the text starts with Chapter 4, Metric Spaces. In the context of metric spaces, we study the properties of completeness, compactness, and connectedness, as well as the continuity of functions. In the next chapter we illustrate the utility of metric spaces by proving a fixed-point theorem in this setting and then using it to obtain results on the existence of solutions of differential equations, integral equations, and systems of linear algebraic equations.

The study of integration begins in Chapter 6 with a review of the Riemann integral. Then the Lebesgue integral for functions from $\mathbf{R}^n$ to $\mathbf{R}$ is developed via the Daniell approach. This method of development is used because it makes good contact with the Riemann integral and, also, because it gets us to the central ideas more quickly than does the development via measure theory.

Returning to a study of abstract spaces we consider normed linear spaces which combine the properties of linear spaces and metric spaces. The primary model for these spaces is the Euclidean space $\mathbf{R}^n$ and important examples considered are the sequence spaces l_p and the function spaces $\mathscr{L}_p$. In the setting of normed linear spaces, we consider differentiation and some results on approximation including the Stone-Weierstrass Theorem.

In Chapter 9 the Fundamental Theorems of Calculus are obtained for the Lebesgue integral of functions from $\mathbf{R}$ to $\mathbf{R}$. Here a number of concepts are introduced which lead in a natural way to a discussion of the Stieltjes integral which is considered in Chapter 10.

In the final chapter we treat inner product spaces and orthogonal bases for such spaces. As examples of orthogonal bases for certain inner product spaces, we consider the complex exponential and trigonometric sequences, the Legendre polynomials, and the Hermite polynomials. Other examples occur in problems. Finally, pointwise convergence of Fourier series and Fourier integrals are discussed.

We wish to thank several anonymous reviewers who read an early set of class notes upon which part of this book was based and who made many helpful suggestions. We also express our gratitude to our students who aided us as we sought better methods of presenting the material. Finally, we wish to thank Mrs. Verna Osborne for her care and patience in producing an excellent typescript from the manuscript.

Norman B. Haaser
Joseph A. Sullivan

Contents

1 | *Sets and Relations*

1 · INTRODUCTION

The purpose of this chapter is to introduce terminology and to state some results which are used in our study of analysis. The discussion of sets is limited to topics needed for the work at hand and is informal in approach. Those interested in a more complete treatment should consult a book on set theory such as [19] or [37].

2 · SETS

By a set we mean a collection of objects; the objects are called the **elements** of the set. If A is a set, then "$a \in A$" denotes that a is an element of A. Sets will be denoted by braces, so that if P denotes some property, then $\{a : \text{P}\}$ is the set of all elements a such that a has the property P. Within the meaning of set we include a set which has no elements. This is called the **empty set** and is denoted by $\varnothing$.

A word now about notation for sets. If A is a set with only a few elements, we often denote the elements by different letters; for example, $A = \{a, b\}$. However, for a larger set it is natural to use indices; for example, $A = \{x_1, \cdots, x_n\}$ or $A = \{x_i : i \in S\}$ where $S = \{1, \cdots, n\}$. In general, if S is any set, then $\{x_s : s \in S\}$ is called an indexed set and S is called the index set. Note that any set A can be written as an indexed set as follows: $A = \{x_a : a \in A\}$ where $x_a = a$. Thus, this index notation can be used for any set. Observe that nothing we have said excludes the possibility that $x_s = x_t$ for $s \neq t$.

If every element of a set B is an element of the set A, then we say that B is a **subset** of A and we write $B \subset A$. If A is any set, then $\varnothing \subset A$. Two sets are **equal** if and only if they have the same elements. Thus, $A = B$ if and only if $A \subset B$ and $B \subset A$. If $B \subset A$ and $\varnothing \neq B \neq A$, then B is said to be a **proper subset** of A.

From a given set A we can form sets composed of any collection of elements of A. Thus, we obtain subsets of A. We also can form a set composed of all the subsets of A. This is called the **power set** of A and is denoted by $\mathscr{P}(A)$. Thus,

$$\mathscr{P}(A) = \{B : B \subset A\}.$$

If A and B are sets, then we can form their **union**, denoted by $A \cup B$. This is the set composed of all elements which are elements of A or B. Thus,

$$A \cup B = \{x : x \in A \text{ or } x \in B\}.$$

The **intersection** of the sets A and B, denoted by $A \cap B$, is the set composed of all elements which are elements of both A and B; that is,

$$A \cap B = \{x : x \in A \text{ and } x \in B\}.$$

Two sets are said to be **disjoint** if their intersection is empty.

Another set which can be formed from the sets A and B is the cartesian product $A \times B$. This depends on the concept of an ordered pair of elements. Intuitively we understand that an ordered pair (a, b) is composed of two elements one of which is considered to be the first element and the other the second element. An adequate approach would be to take ordered pair as an undefined notion with the understanding that $(a, b) = (c, d)$ if and only if $a = c$ and $b = d$. However, we will give a definition in terms of sets.

The set consisting of the single element a is called the **singleton** of a and is denoted by $\{a\}$. The (unordered) pair $\{a, b\}$ is $\{a\} \cup \{b\}$. The **ordered pair** (a, b) is defined to be the set $\{\{a\}, \{a, b\}\}$. Notice that if $a \in A$ and $b \in B$ then $(a, b) \in \mathscr{P}(\mathscr{P}(A \cup B))$. We now show that ordered pairs so defined have the basic property intuitively expected.

2 · 1 Proposition. *If (a, b) and (c, d) are ordered pairs of elements, then $(a, b) = (c, d)$ if and only if $a = c$ and $b = d$.*

PROOF. If $a = c$ and $b = d$, it is clear that $\{\{a\}, \{a, b\}\} = \{\{c\}, \{c, d\}\}$ and hence $(a, b) = (c, d)$. Now assume $(a, b) = (c, d)$; that is, $\{\{a\}, \{a, b\}\} = \{\{c\}, \{c, d\}\}$. Then there are two possibilities: either $\{a\} = \{c\}$ and $\{a, b\} = \{c, d\}$ or $\{a\} = \{c, d\}$ and $\{a, b\} = \{c\}$. Suppose $\{a\} = \{c\}$ and $\{a, b\} = \{c, d\}$. Then $a = c$ and $b = c$ or d. If $b = d$, the desired conclusion holds. If $b = c = a$, then $d = a = b$ and again the conclusion holds. Suppose now $\{a\} = \{c, d\}$ and $\{a, b\} = \{c\}$. Then $c = a$, $d = a$, and $b = c$. Thus, $a = c$ and $b = d$. ∎

The **cartesian** or **direct product** $A \times B$ of two sets A and B is the set of all ordered pairs (a, b) where $a \in A$ and $b \in B$. Thus,

$$A \times B = \{(a, b) : a \in A \text{ and } b \in B\}.$$

Assume now that we have a given fixed set X. The **complement** of a set $A \in \mathscr{P}(X)$, denoted by $\sim A$, is the set of all elements of X which are not elements of A; that is,

$$\sim A = \{x \in X : x \notin A\}.$$

If A and B are any sets in $\mathscr{P}(X)$, we define the set difference

$$B \sim A = \{x \in B : x \notin A\} = B \cap (\sim A).$$

Thus, $\sim A = X \sim A$.

Usually a set of sets is called a **collection** of sets. A nonempty collection $\mathscr{R}$ of sets in $\mathscr{P}(X)$ is called a **ring of sets** if $A \cup B \in \mathscr{R}$ and $A \sim B \in \mathscr{R}$ whenever $A, B \in \mathscr{R}$. If $\mathscr{R}$ has the additional property that $X \in \mathscr{R}$, then $\mathscr{R}$ is called an **algebra of sets**.

We now extend the definitions of union and intersection to collections of sets. An extension of the cartesian product is given in the next section. If $\mathscr{C}$ is a collection of sets in $\mathscr{P}(X)$, then its union and intersection are defined as follows:

$$\bigcup_{A \in \mathscr{C}} A = \{x \in X : x \in A \text{ for some } A \in \mathscr{C}\}$$

$$\bigcap_{A \in \mathscr{C}} A = \{x \in X : x \in A \text{ for all } A \in \mathscr{C}\}.$$

Note that if $\mathscr{C}$ is empty, then $\bigcup_{A \in \mathscr{C}} A = \varnothing$ and $\bigcap_{A \in \mathscr{C}} A = X$; if $\mathscr{C}$ consists of a single set A, then $\bigcup_{A \in \mathscr{C}} A = A$ and $\bigcap_{A \in \mathscr{C}} A = A$; and if $\mathscr{C}$ consists of the pair A and B then these definitions agree with those previously given for $A \cup B$ and $A \cap B$. A collection $\mathscr{C}$ of sets is said to be **pairwise disjoint** if every pair of distinct sets of the collection is disjoint.

If the collection $\mathscr{C}$ is denoted by the indexed set of sets $\{A_s : s \in S\}$, then the union and intersection of the collection are denoted by $\bigcup_{s \in S} A_s$ and $\bigcap_{s \in S} A_s$, respectively. In the cases where $S = \{1, \cdots, n\}$ and S is the set of all positive integers, the union is usually written as $\bigcup_{i=1}^{n} A_i$ and $\bigcup_{i=1}^{\infty} A_i$, respectively. Of course, similar remarks apply to the notation for intersection.

Important relations for the complement of the union and intersection of a collection $\{A_s : s \in S\}$ of sets in $\mathscr{P}(X)$ are given by the **DeMorgan laws**:

$$(2 \cdot 2) \qquad \sim\left(\bigcup_{s \in S} A_s\right) = \bigcap_{s \in S} (\sim A_s) \quad \text{and} \quad \sim\left(\bigcap_{s \in S} A_s\right) = \bigcup_{s \in S} (\sim A_s).$$

The proof of the DeMorgan laws is left to the reader (Problem 7).

PROBLEMS

1. If A, B, and C are sets, show that:

a. $A \cup B = B \cup A$; $A \cap B = B \cap A$

b. $A \cup (B \cup C) = (A \cup B) \cup C$; $A \cap (B \cap C) = (A \cap B) \cap C$

c. $A \cap (B \cup C) = (A \cap B) \cup (A \cap C)$

d. $A \cup (B \cap C) = (A \cup B) \cap (A \cup C)$

e. $A \cup \varnothing = A$; $A \cap \varnothing = \varnothing$

2. Determine $\bigcap_{n=1}^{\infty} I_n$ when I_n is the interval on the real line:

a. $I_n = \left(-\dfrac{1}{n}, \dfrac{1}{n}\right)$ **b.** $I_n = \left(0, \dfrac{1}{n}\right)$ **c.** $I_n = \left[0, \dfrac{1}{n}\right]$

3. Show that ("$\Leftrightarrow$" denotes "if and only if")

a. $A \subset B \Leftrightarrow A \cap B = A$

b. $A \subset B \Leftrightarrow A \cup B = B$

4. Describe $[1, 2] \times [2, 4]$ where $[1, 2]$ and $[2, 4]$ are closed intervals on the real line.

5. Show that ("$\Leftrightarrow$ "denotes" if and only if")
   a. $A \times B = \varnothing \Leftrightarrow A = \varnothing$ or $B = \varnothing$
   b. If $C \times D \neq \varnothing$, then $C \times D \subset A \times B \Leftrightarrow C \subset A$ and $D \subset B$.
   c. $(A \times B) \cup (C \times B) = (A \cup C) \times B$
   d. $(A \times B) \cap (C \times D) = (A \cap C) \times (B \cap D)$

6. If A, B and C are sets in $\mathscr{P}(X)$, show that
   a. $(A \sim B) \sim C = A \sim (B \cup C)$
   b. $A \sim (B \sim C) = (A \sim B) \cup (A \cap C)$
   c. $(A \cup B) \sim C = (A \sim C) \cup (B \sim C)$
   d. $A \sim (B \cup C) = (A \sim B) \cap (A \sim C)$
   e. If $B \subset A$, then $A \sim (A \sim B) = B$.

7. Prove the DeMorgan laws.

8. The **symmetric difference** of two sets A and B in $\mathscr{P}(X)$, denoted by $A \triangle B$, is defined by

$$A \triangle B = (A \sim B) \cup (B \sim A).$$

Show that:
   a. $A \triangle B = (A \cup B) \sim (A \cap B) = B \triangle A$
   b. $A \triangle (B \triangle C) = (A \triangle B) \triangle C$
   c. $A \triangle \varnothing = A$; $A \triangle A = \varnothing$
   d. $A \cap (B \triangle C) = (A \cap B) \triangle (A \cap C)$

9. If $\mathscr{R}$ is a ring of sets in $\mathscr{P}(X)$, show that if $A, B \in \mathscr{R}$, then
   a. $A \cap B \in \mathscr{R}$ b. $A \triangle B \in \mathscr{R}$ c. $\varnothing \in \mathscr{R}$

10. If $\mathscr{R}$ is a ring of sets in $\mathscr{P}(X)$, show that $\mathscr{R}$ is an algebra if and only if $\sim A \in \mathscr{R}$ whenever $A \in \mathscr{R}$.

11. If symmetric difference (Problem 8) is taken as addition and intersection is taken as multiplication, show that a ring of sets is a commutative ring in the algebraic sense (c.f. p. 23). What is the additive inverse of an element of this ring?

3 · RELATIONS, FUNCTIONS

For a fixed set X the relation "is an element of" determines the subset R of $X \times \mathscr{P}(X)$ such that $(a, A) \in R$ if and only if $a \in A$. Thus, this relation is completely characterized by the set R. This suggests that we can define a relation in terms of sets.

3 · 1 Definition. *Any subset R of $A \times B$ is called a* ***relation*** *from A to B.*

A relation from A to A will be called a relation in A.

Some remarks concerning this definition are in order. Note that the empty set is a relation from A to B for any sets A and B. Also, we should point out that what we have defined as a relation might in some treatments be called the graph of a relation. For example, the set R of points above the line $y = x$

in the cartesian plane is the "less than" relation in the set of real numbers; that is,

$$R = \{(x, y) : x < y\}.$$

We often write aRb to indicate that $(a, b) \in R$.

If R is a relation from A to B, then the relation R^{-1} from B to A given by

$$R^{-1} = \{(b, a) : (a, b) \in R\}$$

is called the **inverse** of R. Thus, if R is the "less than" relation in the set of real numbers, then R^{-1} is the "greater than" relation; "$(b, a) \in R^{-1}$ if and only if $(a, b) \in R$" becomes in this case "$b > a$ if and only if $a < b$."

Let R be a relation from A to B. If $A_1 \subset A$, then the set

$$R(A_1) = \{b \in B : (a, b) \in R \text{ for some } a \in A_1\}$$

is called the **image** of A_1 under R. If $B_1 \subset B$, then the image of B_1 under R^{-1} is called the **inverse image** of B_1 under R. Thus the inverse image of B_1 under R is the set

$$R^{-1}(B_1) = \{a \in A : (a, b) \in R \text{ for some } b \in B_1\}.$$

The inverse image of B is called the **domain** of R and is denoted by D_R. Therefore,

$$\mathrm{D}_R = R^{-1}(B) = \{a \in A : (a, b) \in R\}.$$

The image of A under R is called the **range** of R and is denoted by R_R. Thus,

$$\mathrm{R}_R = R(A) = \{b \in B : (a, b) \in R\}.$$

If R and S are relations such that $S \subset R$, then we say that S is a **restriction** of R or R is an **extension** of S. For example, the "less than" relation in the rational number system is the set of points above the line $y = x$ in the cartesian plane which have rational coordinates and is a restriction of the "less than" relation in the real number system.

If S is a relation from A to B and R is a relation from B to C, then their **composition**, $R \circ S$, is the relation from A to C such that

$$R \circ S = \{(a, c) : \text{for some } b, (a, b) \in S \text{ and } (b, c) \in R\}.$$

A function is a special type of relation.

3·2 Definition. *A **function** or **mapping** f from X into Y, denoted by $f: X \to Y$, is a relation from X to Y with the property that for each $x \in X$ there is a unique $y \in Y$ such that $(x, y) \in f$.*

Thus, the domain of f is X and if $(x, y_1) \in f$ and $(x, y_2) \in f$, then $y_1 = y_2$. Stated another way we can say that for any $x \in X$ the image of the singleton $\{x\}$ is a singleton. If $(x, y) \in f$, the element y is called the value of f at x and is written $f(x)$. Therefore, $f(\{x\}) = \{f(x)\}$. Notice that f is an element of

$\mathscr{P}(X \times Y)$ whereas $f(x)$ is an element of Y. If the range of f is Y, that is if $f(X) = Y$, then we say that f is a function from X **onto** Y.

Suppose f and g are functions with domains X and A, respectively. Then $g \subset f$ if and only if $A \subset X$ and $g(x) = f(x)$ for all $x \in A$. If such is the case, then g is called the **restriction** of f to A and is denoted by $f \mid A$.

Since a function is defined as a set, two functions f and g are equal if they are equal as sets; that is, $f \subset g$ and $g \subset f$. Thus, $f = g$ if and only if $D_f = D_g$ and $f(x) = g(x)$ for all $x \in D_f$. The usual way of specifying a function f is to give its domain D_f and some rule which determines $f(x)$ for each $x \in D_f$. For example, for any set X the function I_X such that $I_X(x) = x$ for all $x \in X$ is called the **identity** function on X. In a given discussion we may drop the subscript X and write simply I for the identity function on X if no confusion will arise as a result.

If f is a function from X into Y, then the inverse relation, $f^{-1} = \{(y, x) : (x, y) \in f\}$, is not, in general, a function. In fact, f^{-1} is a function if and only if $(y, x_1) \in f^{-1}$ and $(y, x_2) \in f^{-1}$ imply that $x_1 = x_2$. A function f is said to be **one-to-one** if $(x_1, y) \in f$ and $(x_2, y) \in f$ imply that $x_1 = x_2$ or, equivalently, if $f(x_1) = f(x_2)$ implies $x_1 = x_2$. Thus, we see that the inverse relation f^{-1} is a function if and only if f is one-to-one.

If f is a function from X into Y and g is a function from Y into Z, then the composition $g \circ f$ of g and f is the function from X into Z such that $g \circ f(x) = g(f(x))$. Composition is an associative operation; that is,

$$h \circ (g \circ f) = (h \circ g) \circ f.$$

In general, composition is not commutative; that is, it is not necessarily true that $g \circ f$ equals $f \circ g$. It is of interest to point out that the identity and inverse functions act as an identity and inverse with respect to the operation of composition in the following sense. If f is a function from X into Y, then $f \circ I_X = f$ and $I_Y \circ f = f$. Also, if $f : X \to Y$ and $g : Y \to X$ are such that $f \circ g = I_Y$ and $g \circ f = I_X$, then we can show that $g = f^{-1}$. For, if $(y, x) \in g$, then $x = g(y)$ and

$$f(x) = f(g(y)) = I_Y(y) = y.$$

Thus, $(y, x) \in g$ implies $(x, y) \in f$. Similarly, if $(x, y) \in f$, then $y = f(x)$ and

$$g(y) = g(f(x)) = I_X(x) = x.$$

Thus, $(x, y) \in f$ implies $(y, x) \in g$ and, therefore,

$$g = \{(y, x) : (x, y) \in f\} = f^{-1}$$

If $f : X \to Y$, $A \subset X$, and $B \subset Y$, then the image $f(A)$ of A under f and the inverse image $f^{-1}(B)$ of B under f are the sets:

$$f(A) = \{f(x) \in Y : x \in A\}$$

and

$$f^{-1}(B) = \{x \in X : f(x) \in B\}.$$

We can consider these expressions as rules defining a function f from $\mathscr{P}(X)$ into $\mathscr{P}(Y)$ and a function f^{-1} from $\mathscr{P}(Y)$ into $\mathscr{P}(X)$. These functions should perhaps have a designation different from the original function f and its inverse relation f^{-1} but this is not usually done. We now establish some properties of these functions defined on the power sets.

3 · 3 Proposition. *If $f: X \to Y$ and $\{B_s : s \in S\}$ is a collection of sets in $\mathscr{P}(Y)$, then*

$$f^{-1}\left(\bigcup_{s \in S} B_s\right) = \bigcup_{s \in S} f^{-1}(B_s)$$

and

$$f^{-1}\left(\bigcap_{s \in S} B_s\right) = \bigcap_{s \in S} f^{-1}(B_s).$$

PROOF. We prove only the first result. The proof of the second is similar and is left to the reader (Problem 5). We use the standard notation "$\Leftrightarrow$" to denote "if and only if" or "is equivalent to."

$$\begin{aligned} x \in f^{-1}\left(\bigcup_{s \in S} B_s\right) &\Leftrightarrow f(x) \in \bigcup_{s \in S} B_s \\ &\Leftrightarrow f(x) \in B_s \quad \text{for some } s \in S \\ &\Leftrightarrow x \in f^{-1}(B_s) \quad \text{for some } s \in S \\ &\Leftrightarrow x \in \bigcup_{s \in S} f^{-1}(B_s). \quad \blacksquare \end{aligned}$$

3 · 4 Proposition. *If $f: X \to Y$ and $\{A_s : s \in S\}$ is a collection of sets in $\mathscr{P}(X)$, then*

$$f\left(\bigcup_{s \in S} A_s\right) = \bigcup_{s \in S} f(A_s)$$

and

$$f\left(\bigcap_{s \in S} A_s\right) \subset \bigcap_{s \in S} f(A_s).$$

PROOF. We prove only the second result. The proof of the first is left to the reader (Problem 6). If $y \in f(\bigcap_{s \in S} A_s)$, then $y = f(x)$ for some $x \in \bigcap_{s \in S} A_s$. Then $x \in A_s$ for all $s \in S$ and thus $y \in f(A_s)$ for all $s \in S$; that is, $y \in \bigcap_{s \in S} f(A_s)$. $\blacksquare$

In Proposition 3 · 4 we cannot prove equality in the case of intersection. If $y \in \bigcap_{s \in S} f(A_s)$, then $y \in f(A_s)$ for all $s \in S$. This implies that for each $s \in S$ there exists an element $x_s \in A_s$ such that $y = f(x_s)$. But, it may happen that there is no element $x \in \bigcap_{s \in S} A_s$ for which $y = f(x)$. For example, if $f(x) = |x|$ for $x \in [-1, 1]$, $A_1 = [-1, 0]$, and $A_2 = [0, 1]$, then $f(A_1 \cap A_2) = f(\{0\}) = \{0\}$ whereas $f(A_1) \cap f(A_2) = [0, 1]$.

However, if f is a one-to-one function from X into Y, then we have the nicer result:

3 · 5 Proposition. *If f is a one-to-one function from X into Y and $\{A_s : s \in S\}$ is a collection of sets in $\mathscr{P}(X)$, then*

$$f\left(\bigcap_{s \in S} A_s\right) = \bigcap_{s \in S} f(A_s).$$

PROOF. If $y \in \bigcap_{s \in S} f(A_s)$, then $y \in f(A_s)$ for all $s \in S$ and, hence, for each $s \in S$ there is an element $x_s \in A_s$ such that $y = f(x_s)$. Since f is one-to-one and $f(x_s) = y$ for all $s \in S$, all of the x_s are equal, say $x_s = x$ for all $s \in S$. Then $x \in \bigcap_{s \in S} A_s$ and, therefore, $y \in f(\bigcap_{s \in S} A_s)$. Thus, $\bigcap_{s \in S} f(A_s) \subset f(\bigcap_{s \in S} A_s)$ and combining this with the result in Proposition 3 · 4 we have $\bigcap_{s \in S} f(A_s) = f(\bigcap_{s \in S} A_s)$. ▮

We consider now the behavior of f and f^{-1} with respect to complements and differences.

3 · 6 Proposition. *If $f: X \to Y$ and $B, C \in \mathscr{P}(Y)$, then*

$$f^{-1}(\sim B) = \sim(f^{-1}(B))$$

and

$$f^{-1}(B \sim C) = f^{-1}(B) \sim f^{-1}(C).$$

PROOF. The proof is left to the reader (Problem 7). ▮

To obtain corresponding results for the function f we again need to require that f be one-to-one and, in this case, f must also be onto.

3 · 7 Proposition. *If f is a one-to-one function from X onto Y and $A, C \in \mathscr{P}(X)$, then*

$$f(\sim A) = \sim(f(A))$$

and

$$f(A \sim C) = f(A) \sim f(C).$$

PROOF. If $y \in f(\sim A)$, then, since f is one-to-one, $y \notin f(A)$, that is, $y \in \sim(f(A))$. If $y \in \sim(f(A))$, then $y \notin f(A)$ and since f is onto, $y \in f(\sim A)$. Thus, $f(\sim A) = \sim(f(A))$. Then

$$f(A \sim C) = f(A \cap (\sim C)) = f(A) \cap f(\sim C) = f(A) \cap (\sim f(C)) = f(A) \sim f(C).$$

▮

Consider now two functions f and g from $\{1, 2\}$ into $A \cup B$ such that $f(1) = a \in A$, $f(2) = b \in B$, $g(1) = c \in A$, and $g(2) = d \in B$. Then $f = g$ if and only if $f(1) = g(1)$ and $f(2) = g(2)$; that is, if an only if $a = c$ and $b = d$. Recall that the characteristic property of ordered pairs in $A \times B$ is that $(a, b) = (c, d)$ if and only if $a = c$ and $b = d$. Thus, we may consider the ordered pair (a, b) to be the function f from $\{1, 2\}$ into $A \cup B$ such that $f(1) = a$ and $f(2) = b$. Then $A \times B$ becomes the set of all functions f from $\{1, 2\}$ into $A \cup B$ such that $f(1) \in A$ and $f(2) \in B$. This way of looking at ordered pairs provides a method for extending the concept of the cartesian product of sets.

If $\{A_s : s \in S\}$ is a collection of sets indexed by an arbitrary set S, then the product $\bigtimes_{s \in S} A_s$ is defined to be the set of all functions f from S into $\bigcup_{s \in S} A_s$ such that $f(s) \in A_s$. If $f(s) = x_s \in A_s$, then we can write $(x_s)_{s \in S}$ for f and

$$\bigtimes_{s \in S} A_s = \{(x_s)_{s \in S} : x_s \in A_s\}.$$

In the cases where $S = \{1, \cdots, n\}$ and S is the set of all positive integers the notation becomes:

$$\bigtimes_{i=1}^{n} A_i = A_1 \times \cdots \times A_n = \{(x_1, \cdots, x_n) : x_i \in A_i\}$$

and

$$\bigtimes_{i=1}^{\infty} A_i = \{(x_i) : x_i \in A_i\},$$

respectively.

If $A_s = A$ for all $s \in S$, then $\bigtimes_{s \in S} A_s$ is denoted by A^S and consists of all functions from S into A. Also, $(x_s)_{s \in S}$, which is the function f from S into A such that $f(s) = x_s$, is called a **family of points** in A. When $S = \{1, \cdots, n\}$, A^S is denoted by A^n and a family of points $(x_i)_{i \in S} = (x_1, \cdots, x_n)$ of A is called an **n-tuple** of points in A. When S is the set of all positive integers, a family of points $(x_i)_{i \in S}$ of A is called a **sequence of points** in A.

We now discuss an assumption concerning sets, called the Axiom of Choice, which is important in analysis. The **Axiom of Choice** states: *for any collection $\mathscr{C}$ of nonempty sets there exists a function f defined on $\mathscr{C}$ such that $f(A) \in A$ for all $A \in \mathscr{C}$.* We may consider that the function f, called a **choice function**, selects a particular element from each set $A \in \mathscr{C}$. We will use this assumption freely and usually without calling attention to the fact that we are using it. It will frequently occur in a guise such as: for each $A \in \mathscr{C}$ choose an element $a \in A$. Although this seems to be a very reasonable thing to be able to do, when no definite prescription can be given for the selection of the element a then the Axiom of Choice is required. It states that there is a function f which will select a particular element a of A.

It is easy to see that the Axiom of Choice can also be stated in the form: if $\{A_s : s \in S\}$ is a collection of nonempty sets, then the product $\bigtimes_{s \in S} A_s$ is nonempty.

PROBLEMS

1. Let f and g be functions defined on $\mathbf{R}$, the set of all real numbers, by the rules $f(x) = 3x + 2$ and $g(x) = x^2$.

- **a.** Show that f is one-to-one and g is not one-to-one.
- **b.** Determine the inverse relations f^{-1} and g^{-1} and show that f^{-1} is a function but that g^{-1} is not a function.
- **c.** Show that $g \circ f \neq f \circ g$.

2. If $f: X \to Y$, $g: Y \to Z$ and $h: Z \to W$, show that

$$h \circ (g \circ f) = (h \circ g) \circ f.$$

3. If $f: X \to Y$, $g: Y \to Z$, and f and g are one-to-one, show that $g \circ f$ is a one-to-one function from X into Z and that

$$(g \circ f)^{-1} = f^{-1} \circ g^{-1}.$$

4. Let (A_n) be a decreasing sequence of sets; that is, $A_n \supset A_{n+1}$ for each $n \in \mathbf{N}$, the set of positive integers. Define $B_0 = \bigcap_{n=1}^{\infty} A_n$ and $B_n = A_n \sim A_{n+1}$, $n \in \mathbf{N}$. Show that $\{B_k : k \in \{0\} \cup \mathbf{N}\}$ is a pairwise disjoint collection of sets and that

$$A_1 = \bigcup_{k=0}^{\infty} B_k.$$

5. If $f: X \to Y$ and $\{B_s : s \in S\}$ is a collection of sets in $\mathscr{P}(Y)$, prove

$$f^{-1}\left(\bigcap_{s \in S} B_s\right) = \bigcap_{s \in S} f^{-1}(B_s).$$

6. If $f: X \to Y$ and $\{A_s : s \in S\}$ is a collection of sets in $\mathscr{P}(X)$, prove

$$f\left(\bigcup_{s \in S} A_s\right) = \bigcup_{s \in S} f(A_s).$$

7. If $f: X \to Y$ and $B, C \in \mathscr{P}(Y)$, show that

$$f^{-1}(\sim B) = \sim(f^{-1}(B))$$

and

$$f^{-1}(B \sim C) = f^{-1}(B) \sim f^{-1}(C).$$

8. If $f: X \to Y$ and $B \in \mathscr{P}(Y)$, show that $f(f^{-1}(B)) \subset B$. Give an example where $f(f^{-1}(B)) \neq B$. What conditions on f insure that $f(f^{-1}(B)) = B$ for all $B \in \mathscr{P}(Y)$?

9. If $f: X \to Y$ and $A \in \mathscr{P}(X)$, show that $A \subset f^{-1}(f(A))$. Give an example where $A \neq f^{-1}(f(A))$. What condition on f insures that $A = f^{-1}(f(A))$ for all $A \in \mathscr{P}(X)$?

10. If $f: X \to Y$ and $A \in \mathscr{P}(X)$, show that $\sim f(A) \subset f(\sim A)$ if and only if f is onto Y.

11. If $(a_1, a_2, \cdots, a_n)$ and $(b_1, b_2, \cdots, b_n)$ are n-tuples in A^n, show that $(a_1, a_2, \cdots, a_n) = (b_1, b_2, \cdots, b_n)$ if and only if $a_i = b_i$, $i = 1, 2, \cdots, n$.

12. Give a one-to-one function from $\mathscr{P}(X)$ onto 2^X, where 2^X is the set of all functions from X into $\{0, 1\}$.

13. If the set A has m elements, how many elements are there in A^n?

4 · PARTIAL ORDERS AND EQUIVALENCE RELATIONS

Certain relations which are not functions but which have other special properties play an important rôle in mathematics. In this section we consider two special types of relations.

4 · 1 Definition. *A relation R in A is called a* ***partial order*** *in A if*

P_1. *aRa for all $a \in A$;*
P_2. *aRb and bRa imply $a = b$;*
P_3. *aRb and bRc imply aRc.*

A nonempty set A with a partial order defined in it is called a ***partially ordered set.***

Properties P_1 and P_3 are called, respectively, the reflexive and transitive properties of the relation. Frequently the notation $\prec$ is used for a partial order and $a \prec b$ is read "a precedes b." Note that any nonempty subset of a partially ordered set is itself a partially ordered set with respect to the partial order which is the restriction of the original partial order.

Some examples of partially ordered sets follow.

(1) The set $\mathbf{R}$ of all real numbers is partially ordered by the relation $\leq$. In this case P_1, P_2 and P_3 become:

$$a \leq a \text{ for all } a \in \mathbf{R};$$
$$a \leq b \text{ and } b \leq a \text{ imply } a = b;$$
$$a \leq b \text{ and } b \leq c \text{ imply } a \leq c.$$

(2) The power set $\mathscr{P}(X)$ of some fixed set X is partially ordered by $\subset$. This is called the partial ordering of $\mathscr{P}(X)$ by inclusion.

(3) Let $\mathbf{N}$ be the set of positive integers and let $a \mid b$ denote that a divides b. Then $\mathbf{N}$ is partially ordered by this relation.

(4) Let A be the set of all real-valued functions defined on a set X and let $f \leq g$ denote that $f(x) \leq g(x)$ for all $x \in X$. This relation $\leq$ is a partial order in A.

(5) Let A be the set of all functions f with domain $D_f \subset X$ and range $R_f \subset Y$. Let $g \prec f$ denote that f is an extension of g, that is, $g \subset f$. Then A is a partially ordered set and is said to be partially ordered by extension. Note that this example is closely related to example (2).

4 · 2 Definition. *Two elements a and b of a partially ordered set A are* ***comparable*** *if* $a \prec b$ *or* $b \prec a$.

4 · 3 Definition. *A partially ordered set is said to be* ***totally (completely, linearly) ordered*** *or to be a* ***chain*** *if every two elements of the set are comparable.*

The set of all real numbers, **R**, partially ordered by $\leq$ is a chain. In example (3) the subset $\{1, n, n^2, n^3, \cdots\}$, where n is any positive integer, is a chain.

If B is a subset of a partially ordered set A, then an element c in A is said to be an **upper bound** of B if $b \prec c$ for all $b \in B$. An element c in A is said to be the **least upper bound** of B if c is an upper bound of B and, if d is any other upper bound of B, then $c \prec d$. An element a_0 in the partially ordered set A is said to be a **maximal element** of A if $a \in A$ and $a_0 \prec a$ imply that $a = a_0$. In example (5) any function defined on X is a maximal element of A.

In many proofs a property of partially ordered sets, called Zorn's Lemma, seems to be more suitable to use than the Axiom of Choice. Though we will not do so, it can be shown that Zorn's Lemma is equivalent to the Axiom of Choice (c.f. [22] or [19]). We take Zorn's Lemma as another assumption.

4 · 4 Zorn's Lemma. *If A is a partially ordered set in which each chain has an upper bound, then A has a maximal element.*

Another important special type of relation is called an equivalence relation.

4 · 5 Definition. *A relation R in A is an* ***equivalence*** *relation if:*

E_1. *aRa for all* $a \in A$;
E_2. *aRb implies bRa*;
E_3. *aRb and bRc imply aRc.*

Notice that both an equivalence relation and a partial order have the reflexive and transitive properties. However, their second properties differ. Property E_2 is called the symmetry property.

Equality, as we use it in the sense of logical identity, is an example of an equivalence relation. Congruence and similarity of triangles in the plane are also equivalence relations. In the set of real numbers the relation R, defined by the rule: aRb if $a - b$ is an integral multiple of some fixed number c, is an equivalence relation.

Closely associated with the concept of an equivalence relation is that of a partition of a set. A **partition** of a set A is a pairwise disjoint collection of nonempty subsets of A whose union is A. We now show that an equivalence relation in A defines a partition of A and, conversely, a partition of A yields an equivalence relation in A.

Given an equivalence relation R in A, for each $a \in A$ let $R(a) = R(\{a\}) = \{x \in A : aRx\}$. We assert that the collection $\{R(a) : a \in A\}$ is a partition of A. For each $a \in A$, $a \in R(a)$ and, hence, $A = \bigcup_{a \in A} R(a)$. We complete the proof of the assertion by showing that if $R(a) \cap R(b) \neq \varnothing$, then $R(a) = R(b)$. Suppose $c \in R(a) \cap R(b)$; that is, aRc and bRc. Using the symmetry and transitivity of the equivalence relation, we obtain aRb. Also, if $x \in R(a)$, then aRx and, hence, bRx; that is, $x \in R(b)$. This shows that $R(a) \subset R(b)$. Then, by symmetry, $R(b) \subset R(a)$ and, therefore, $R(a) = R(b)$.

Conversely, if we have a partition of A then we define a relation R as follows: aRb if a and b belong to the same subset in the partition. It is easy to show that this is an equivalence relation in A.

The subsets $R(a)$ in a partition of A corresponding to an equivalence relation R are usually called **equivalence classes** and any element in an equivalence class is called a representative of that class. The set whose elements are the equivalence classes is called the **quotient set** of A with respect to R and is denoted by A/R (A modulo R). Thus, $R(a) \in A/R$. Often the equivalence class $R(a)$ is denoted by $[a]$.

For example, if $\mathbf{Z}$ is the set of integers, define a relation R on $\mathbf{Z} \times (\mathbf{Z} \sim \{0\})$ as follows:

$$(a, b)R(c, d) \quad \text{if} \quad ad = bc.$$

This is an equivalence relation and the associated equivalence classes are the rational numbers. This method of defining a rational number as an equivalence class of ordered pairs of integers is treated thoroughly in many books on modern abstract algebra. We will have more to say on this in the introduction to Chapter 2 and at that time will indicate how the expression of a rational number r as a quotient of two integers, $r = a/b$, corresponds to the fact that $(a, b) \in r$.

As another example, take a nonempty set X and let f be a function from X onto Y. Define a relation in X as follows: x_1Rx_2 if $f(x_1) = f(x_2)$. Clearly, this is an equivalence relation and the associated quotient set is $\{f^{-1}(y) : y \in Y\}$, where we have followed common practice in writing $f^{-1}(y)$ for $f^{-1}(\{y\})$. We can set up a one-to-one correspondence between X/R and Y as follows: define $f_* : X/R \to Y$ by the rule $f_*(f^{-1}(y)) = y$. Then f_* is a one-to-one function from X/R onto Y. If we define the natural quotient mapping $g : X \to X/R$ by the rule that $g(x)$ is the element of X/R which contains x, then

$$f = f_* \circ g.$$

PROBLEMS

1. Let A be the set of all functions f with domain $\mathrm{D}_f \subset X$ and range $\mathrm{R}_f \subset Y$ and let A be partially ordered by extension; that is, $g \prec f$ if and only if $g \subset f$. If $\{g_s : s \in S\}$ is a chain in A, show that the function g such that $\mathrm{D}_g = \bigcup_{s \in S} \mathrm{D}_{g_s}$ and $g(x) = g_s(x)$ for all $x \in \mathrm{D}_{g_s}$ is an upper bound for this chain in A.

2. Let **N** be the set of positive integers with a partial order defined by $a \prec b$ if $a \mid b$. Does every chain in this partially ordered set have an upper bound? Does every finite chain have an upper bound? Show that **N** is also partially ordered by the inverse of the given partial order. In this partially ordered set does every chain have an upper bound? Does this partially ordered set have a maximal element?

3. In $\mathbf{R} \times \mathbf{R}$, where $\mathbf{R}$ is the set of real numbers, define $(a, b) \prec (c, d)$ if either $a < c$ or $a = c$ and $b \leq d$. Verify that this defines a partial order in $\mathbf{R} \times \mathbf{R}$. This is called the lexicographical order of $\mathbf{R} \times \mathbf{R}$.

4. If **Z** is the set of all integers and m is a positive integer, define aRb if $m \mid b - a$. Show that R is an equivalence relation in **Z** and describe the associated partition of **Z**.

5. In the set of real numbers **R** define aRb if $a - b$ is a rational number. Show that R is an equivalence relation in **R** and describe the associated partition of **R**.

6. If r and s are rational numbers and $(a, b) \in r$ and $(c, d) \in s$, define

(1) $r + s$ to be the equivalence class containing $(ad + bc, bd)$,

(2) rs to be the equivalence class containing (ac, bd),

(3) assuming $b > 0$ and $d > 0$, $r < s$ if $ad < bc$.

Show that these definitions are independent of the particular representatives (a, b) of r and (c, d) of s.

7. If $\{A_s : s \in S\}$ is a partition of X, show that

$$R = \bigcup_{s \in S} (A_s \times A_s)$$

is an equivalence relation on X.

5 · COUNTABLE SETS

In everyday life we establish the number of elements in a set by counting. That is, we set up a one-to-one correspondence between elements of the set and elements of some set of positive integers $\{1, 2, \cdots, n\}$. This idea of the number of elements in a set can be generalized to the abstract sets considered in mathematics.

5 · 1 Definition. *Sets A and B are said to be* ***numerically equivalent*** *or to have the* ***same cardinal number*** *if there exists a one-to-one function from A onto B.*

If the set A is numerically equivalent to a set of positive integers $\{1, 2, \cdots, n\}$, then A is said to be **finite** and to have cardinal number n. The empty set is also considered to be finite and to have cardinal number 0. If the set A is numerically equivalent to the set **N** of all positive integers, then A is said to be **countably infinite** and to have cardinal number $\aleph_0$(aleph-naught).

A set which is either finite or countably infinite is called **countable** and a set which is not countable is called **uncountable.** Later we will show that the set of all real numbers is uncountable.

If A is a finite set with n elements and f is a one-to-one function from $\{1, 2, \cdots, n\}$ onto A, then, letting $f(i) = a_i$ for $i = 1, 2, \cdots, n$, the set A can be written as $\{a_i : i = 1, 2, \cdots, n\}$. Similarly, if A is countably infinite, then we can denote A as the indexed set $\{a_i : i \in \mathbf{N}\}$ where $a_i \neq a_j$ if $i \neq j$. This is called an **enumeration** of A.

In the rest of this section all sets mentioned are taken to be subsets of a set X which contains at least the real numbers. It is easy to see that numerical equivalence is an equivalence relation in $\mathscr{P}(X)$ and, hence, gives a partition of $\mathscr{P}(X)$ into equivalence classes. Since any two sets in the same equivalence class have the same cardinal number we may identify the cardinal number of a set with the equivalence class which contains the set.

Letting card A denote the cardinal number of the set A, we define a partial order in the set of all cardinal numbers of sets in $\mathscr{P}(X)$ as follows.

5 · 2 Definition. card $A \leq$ card B *if there is a one-to-one function from A into B.*

Since card A and card B are equivalence classes we should show that this definition does not depend on the particular representatives A and B of these classes. If $C \in$ card A and $D \in$ card B, then there are one-to-one mappings f from A onto C and g from B onto D. Then, if h is a one-to-one mapping from A into B, $g \circ h \circ f^{-1}$ is a one-to-one mapping from C into D. And, if k is a one-to-one mapping from C into D, then $g^{-1} \circ k \circ f$ is a one-to-one mapping from A into B. This shows that Definition 5 · 2 does not depend on the particular representatives A and B.

We define card $A <$ card B if card $A \leq$ card B and card $A \neq$ card B.

As we now show, if $A \subset B$, then card $A \leq$ card B. If A is empty, then the empty function is a one-to-one mapping from A into B. If A is nonempty, take the function $J : A \to B$ such that $J(a) = a$ for all $a \in A$. The function J is called the **inclusion mapping** from A into B; this is just the identity function I_A considered from a different point of view.

We now show that the relation defined by Definition 5 · 2 is a partial order, in fact, a linear order in the set of cardinal numbers of elements of $\mathscr{P}(X)$. Since $A \subset A$, card $A \leq$ card A for any $A \in \mathscr{P}(X)$. Also, it is easy to see that if card $A \leq$ card B card $B \leq$ card C, then card $A \leq$ card C; if f and g are respectively, one-to-one functions from A into B and B into C, then $g \circ f$ is a one-to-one function from A into C. The final requirement for a partial order, if card $A \leq$ card B and card $B \leq$ card A, then card A = card B, is more difficult to prove. First, we prove a lemma.

5 · 3 Lemma. *If* $A_1 \supset A_2 \supset A_3$ *and* card $A_1 =$ card A_3, *then* card $A_1 =$ card A_2.

PROOF. Let f be a one-to-one mapping from A_1 onto A_3; then $f(A_1) = A_3$. Define a collection $\{A_n\}$ of sets as follows:

$$A_n = f(A_{n-2}) \quad \text{for} \quad n \geq 4.$$

Since $A_3 \subset A_2 \subset A_1$, $A_{n+1} \subset A_n$ for $n \geq 3$. Thus $A_1 \supset A_2 \supset A_3 \supset A_4 \supset A_5 \supset \cdots$. Let $B_0 = \bigcap_{n=1}^{\infty} A_n$ and $B_k = A_k \sim A_{k+1}$ for $k \in \mathbf{N}$. Then $\{B_k : k = 0, 1, 2, \cdots\}$ is a pairwise disjoint collection of sets and

$$A_1 = \bigcup_{k=0}^{\infty} B_k = \bigcap_{n=1}^{\infty} A_n \cup (A_1 \sim A_2) \cup (A_2 \sim A_3) \cup (A_3 \sim A_4) \cup (A_4 \sim A_5) \cup \cdots$$

$$A_2 = \bigcup_{\substack{k=0 \\ k \neq 1}}^{\infty} B_k = \bigcap_{n=1}^{\infty} A_n \cup (A_2 \sim A_3) \cup (A_3 \sim A_4) \cup (A_4 \sim A_5) \cup (A_5 \sim A_6) \cup \cdots.$$

Since f is a one-to-one mapping of A_1 onto A_3,

$$f(A_k \sim A_{k+1}) = A_{k+2} \sim A_{k+3} \quad \text{for } k \in \mathbf{N};$$

that is, $f(B_k) = B_{k+2}$. Define the function g from A_1 into A_2 by

$$g(x) = \begin{cases} f(x) & \text{for } x \in \bigcup_{n=1}^{\infty} B_{2n-1} \\ x & \text{for } x \in \bigcup_{n=0}^{\infty} B_{2n}. \end{cases}$$

Then $g(B_{2n}) = B_{2n}$ and $g(B_{2n-1}) = B_{2n+1}$. Thus g is a one-to-one function from A_1 onto A_2 and, hence card $A_1 =$ card A_2. ▮

5 · 4 Theorem (Schröder-Bernstein). *If* card $A \leq$ card B *and* card $B \leq$ card A, *then* card $A =$ card B.

PROOF. Let f and g be one-to-one functions from A into B and from B into A, respectively. If we let $f(A) = B_1 \subset B$, $g(B) = A_2$, and $g(B_1) = A_3$, then $A \supset A_2 \supset A_3$. Also, $g(f(A)) = A_3$, that is, $g \circ f$ is a one-to-one function from A onto A_3. Hence, card $A =$ card A_3 and, therefore, by Lemma 5 · 3, card $A =$ card A_2. But since g maps B one-to-one onto A_2, card $A_2 =$ card B. ▮

With the proof of the Schröder-Bernstein Theorem we have completed the proof that the relation defined in Definition 5 · 2 is a partial order in the set of cardinal numbers of sets in $\mathscr{P}(X)$. We now show that this partial order is actually a total order.

5·5 Theorem. *For any sets A and B in $\mathscr{P}(X)$,* card $A \leq$ card B *or* card $B \leq$ card A.

PROOF. If A or B is empty, the result holds. Assume A and B are nonempty. Consider the set $\mathscr{F}$ of all one-to-one functions f with domain $D_f \subset A$ and range $R_f \subset B$. Partial order $\mathscr{F}$ by extension; that is, $g \prec f$ if and only if $g \subset f$. If $\{g_s : s \in S\}$ is a chain in $\mathscr{F}$, then the function g such that $D_g = \bigcup_{s \in S} D_{g_s}$ and $g(x) = g_s(x)$ for all $x \in D_{g_s}$ is an upper bound for this chain in $\mathscr{F}$. By Zorn's Lemma $\mathscr{F}$ has a maximal element, call it f_0. If $D_{f_0} \neq A$ and $R_{f_0} \neq B$, take $a_1 \in A \sim D_{f_0}$ and $b_1 \in B \sim R_{f_0}$ and let h be the extension of f_0 such that $h(a_1) = b_1$; since h is in $\mathscr{F}$, this contradicts the maximality of f_0 in $\mathscr{F}$ and, hence, $D_{f_0} = A$ or $R_{f_0} = B$; that is, card $A \leq$ card B or card $B \leq$ card A. ∎

The finite cardinal numbers satisfy the expected order relations. The empty function is a one-to-one mapping from $\varnothing$ into $\{1\}$ but there is no one-to-one mapping from $\varnothing$ onto $\{1\}$ and therefore $0 < 1$. The inclusion mapping is a one-to-one mapping from $\{1\}$ into $\{1, 2\}$. If f were a function from $\{1\}$ onto $\{1, 2\}$, we would have $f(1) = 1$ and $f(1) = 2$ which is impossible. Therefore $1 < 2$. In a similar manner we can see that $n < n + 1$ for any finite cardinal number n. If m is a nonzero finite cardinal number the inclusion mapping is a one-to-one mapping from $\{1, 2, \cdots, m\}$ into $\mathbf{N}$ and, therefore, $m \leq \aleph_0$. Since $n < n + 1 \leq \aleph_0$, we can conclude that $n < \aleph_0$ where n is any finite cardinal number. Thus, the cardinal numbers of countable sets satisfy the following relations:

$$0 < 1 < 2 < 3 < \cdots < \aleph_0.$$

We now show that $\aleph_0$ is the smallest infinite cardinal number.

5·6 Proposition. *If A is an infinite set, then* card $A \geq \aleph_0$.

PROOF. In A choose a countably infinite set $\{a_n : n \in \mathbf{N}\}$ as follows. Let a_1 be any element of A. If $a_1, \cdots, a_{n-1}$ have been chosen, let a_n be any element in $A \sim \{a_1, \cdots, a_{n-1}\}$, which is nonempty since A is infinite. Since $\{a_n : n \in \mathbf{N}\} \subset A$ and card $\{a_n : n \in \mathbf{N}\} = \aleph_0$, we have $\aleph_0 \leq$ card A. ∎

We might point out that the Axiom of Choice was needed in this proof. No specific prescription was or could be given for the selection of the set $\{a_n : n \in \mathbf{N}\}$.

Observe that we can now conclude that a set A is countable if and only if card $A \leq \aleph_0$.

5·7 Proposition. *Any subset of a countable set is countable.*

PROOF. Take $A \subset B$ where B is countable. Then card $A \leq$ card B and card $B \leq \aleph_0$. Therefore, card $A \leq \aleph_0$. ∎

5 · 8 Proposition. *If A and B are countable, then $A \times B$ is countable.*

PROOF. If A or B is empty, then the theorem is trivial. Assume A and B are nonempty. Since A and B are countable we can enumerate them:

$$A = \{a_i : i \in I\} \quad \text{and} \quad B = \{b_j : j \in J\}$$

where I and J are subsets of $\mathbf{N}$. Define the function f on $A \times B$ as follows:

$$f(a_i, b_j) = 2^i 3^j, \qquad i \in I, \quad j \in J.$$

This is a one-to-one mapping of $A \times B$ into $\mathbf{N}$ and, hence, card $(A \times B) \leq \aleph_0$. ∎

As a special case of Proposition 5 · 8 we see that $\mathbf{N}^2$ is countable. In fact, since the subset $\{1\} \times \mathbf{N}$ of $\mathbf{N}^2$ is countably infinite, it follows that $\mathbf{N}^2$ is countably infinite.

We now show that the set $\mathbf{Q}^+$ of positive rational numbers is countably infinite. If r is any positive rational number, then r has a unique representative (a, b) where a and b are positive integers with no common factors. The function f defined by $f(r) = (a, b)$ is a one-to-one mapping of $\mathbf{Q}^+$ into $\mathbf{N}^2$. Thus, card $\mathbf{Q}^+ \leq \aleph_0$. Since $f(\mathbf{Q}^+) \supset \{(n, 1) : n \in \mathbf{N}\}$, $\aleph_0 \leq$ card $\mathbf{Q}^+$ and, therefore, card $\mathbf{Q}^+ = \aleph_0$.

5 · 9 Proposition. *The union of a countable collection of countable sets is countable.*

PROOF. Let $\{A_i : i \in I \subset \mathbf{N}\}$ be a countable collection of countable sets. Then each A_i can be enumerated:

$$A_i = \{a_{ij} : j \in J_i \subset \mathbf{N}\}.$$

Assume $\bigcup_{i \in I} A_i$ is nonempty since otherwise the theorem is trivial, If $x \in \bigcup_{i \in I} A_i$, then $x = a_{ij}$ for at least one pair (i, j) with $i \in I$ and $j \in J_i$. Let $f(x) = (i, j)$ where i is the smallest positive integer such that $x = a_{ij}$. This gives a one-to-one mapping of $\bigcup_{i \in I} A_i$ into $\mathbf{N}^2$ and, therefore, $\bigcup_{i \in I} A_i$ is countable. ∎

Using Proposition 5 . 9 we can now state that the set $\mathbf{Q}$ of all rational numbers is countably infinite. Thus we may say that there are the same number of rational numbers as there are positive integers. In the next section we show that there are more real numbers than positive integers.

PROBLEMS

1. Prove that the set of all polynomials with rational coefficients is countable.

2. A complex number is said to be algebraic over the rationals if it is a zero of some polynomial with rational coefficients. Prove that the set of all algebraic numbers is countable.

3. Show that the set of all points in the plane $\mathbf{R}^2$ with rational coordinates is countable.

4. For any function f show that

$$\text{card } R_f \leq \text{card } D_f .$$

5. Show that $\mathscr{P}(X)$ and the set of all functions from X into $\{0, 1\}$ have the same cardinal number.

6. Show that for any set X

$$\text{card } X < \text{card } \mathscr{P}(X).$$

7. If f is an increasing real-valued function of a real variable, show that f has a countable number of discontinuities.

8. Show that every infinite set is numerically equivalent to a proper subset of itself.

6 · UNCOUNTABLE SETS

6 · 1 Proposition. *The set of real numbers in the open interval* (0, 1) *is uncountable.*

PROOF. Since $\{1/(n+1): n \in \mathbf{N}\} \subset (0, 1)$, card $(0, 1) \geq \aleph_0$. Suppose that card $(0, 1) = \aleph_0$. Each number in (0, 1) can be written uniquely as an infinite decimal which does not end in an infinite sequence of nines. Our supposition implies that all these decimals can be enumerated:

$$\begin{array}{l} a_1 = 0 \, . \, a_{11} a_{12} a_{13} \cdots a_{1n} \cdots \\ a_2 = 0 \, . \, a_{21} a_{22} a_{23} \cdots a_{2n} \cdots \\ a_3 = 0 \, . \, a_{31} a_{32} a_{33} \cdots a_{3n} \cdots \\ \cdots\cdots\cdots\cdots\cdots\cdots\cdots\cdots \\ a_n = 0 \, . \, a_{n1} a_{n2} a_{n3} \cdots a_{nn} \cdots \\ \cdots\cdots\cdots\cdots\cdots\cdots\cdots\cdots \end{array}$$

Let $b = 0 \, . \, b_1 b_2 b_3 \cdots b_n \cdots$ where $b_i = 3$ if $a_{ii} \neq 3$ and $b_i = 5$ if $a_{ii} = 3$. Then b is a real number in (0, 1) which is distinct from each a_i (they differ in the ith place). Thus, the set of all real numbers in (0, 1) cannot be countable. ∎

The function f defined by

$$f(x) = \begin{cases} \dfrac{2x-1}{x}, & x \in (0, \tfrac{1}{2}) \\[2ex] \dfrac{2x-1}{1-x}, & x \in [\tfrac{1}{2}, 1) \end{cases}$$

is a one-to-one mapping from (0, 1) onto the set $\mathbf{R}$ of all real numbers. Thus, (0, 1) has the same cardinal number as $\mathbf{R}$; this is called the cardinal number of the continuum and is denoted by c. From Propositions 5 · 6 and 6 · 1 we know that $\aleph_0 < c$.

6 · 2 Proposition. *Any subset A of* **R** *that contains an open interval has cardinal number c.*

PROOF. Suppose the open interval (a, b) is contained in A. Then $f(x) = a + (b - a)x$ defines a one-to-one mapping of the open interval $(0, 1)$ onto (a, b). Hence, card $(a, b) = c$. Since $(a, b) \subset A \subset \mathbf{R}$, card $A = c$. ∎

We now give an example of a subset of **R** which contains no open interval but which has cardinal number c. This set is called the **Cantor set**. The Cantor set may be described as follows. Delete the open middle third of the interval $[0, 1]$. Then delete the open middle third of each of the remaining intervals $[0, 1/3]$ and $[2/3, 1]$. Continue in this manner deleting the open middle third of each interval remaining from the previous step and what remains is called the Cantor set. Thus, if

$$C_1 = [0, \tfrac{1}{3}] \cup [\tfrac{2}{3}, 1]$$
$$C_2 = [0, \tfrac{1}{9}] \cup [\tfrac{2}{9}, \tfrac{1}{3}] \cup [\tfrac{2}{3}, \tfrac{7}{9}] \cup [\tfrac{8}{9}, 1]$$

and so on, then the Cantor set $C = \bigcap_{n=1}^{\infty} C_n$.

Another description of the Cantor set can be given by using ternary (base 3) notation for the real numbers in $[0, 1]$:

$$x = \sum_{k=1}^{\infty} \frac{x_k}{3^k} = 0.x_1 x_2 x_3 \cdots \qquad \text{where } x_k = 0, 1 \text{ or } 2.$$

The set C_1 consists of all numbers x in $[0, 1]$ having a ternary expansion with $x_1 = 0$ or 2. Note, for example, that

$$1/3 = 0.1000 \cdots = 0.0222 \cdots \text{ (ternary).}$$

Similarly, C_2 consists of all numbers x in $[0, 1]$ having a ternary expansion with $x_1 = 0$ or 2 and $x_2 = 0$ or 2. In general, C_n consists of all numbers x in $[0, 1]$ having a ternary expansion with each of $x_1, \ldots, x_n$ either a zero or a two. The Cantor set is the set of all real numbers in $[0, 1]$ which have a ternary expansion containing only zeros and twos.

Define a function f, called the Cantor function, from the Cantor set C into $[0, 1]$ as follows: If x is an element of C having ternary expansion $x = 0.x_1 x_2 x_3 \cdots$, where $x_k = 0$ or 2, then $f(x)$ is the number whose binary (base 2) expansion is $0.r_1 r_2 r_3 \cdots$ where $r_k = \frac{1}{2}x_k$. It is easy to see that f is a function from C onto $[0, 1]$ and, therefore, the Cantor set has cardinal number c.

Note that if (a, b) is an interval deleted at the nth stage in the geometric construction of the Cantor set, then in ternary notation

$$a = 0.a_1 a_2 \cdots a_{n-1} 0222 \cdots$$

and

$$b = 0.a_1a_2 \cdots a_{n-1}2000 \cdots$$

where $a_k = 0$ or 2 for $k \leq n - 1$. Thus, $f(a) = f(b)$.

Another example of an uncountable cardinal number can be obtained by considering the power set of the set **R** of all real numbers. The cardinal number of $\mathscr{P}(\mathbf{R})$ will be greater than c (Problem 6, p. 19).

PROBLEMS

1. Show that the set of all irrational numbers is uncountable.

2. Let $A = \{(x, y) : 0 < x < 1 \text{ and } 0 < y < 1\} \subset \mathbf{R}^2$. Using decimal representations, define $f(x, y) = 0.x_1y_1x_2y_2 \cdots$ when $x = 0.x_1x_2 \cdots$ and $y = 0.y_1y_2 \cdots$. Show that f is a one-to-one mapping of A into the interval $(0, 1)$ and, hence, conclude that card $A = c$.

3. If card $A =$ card C and card $B =$ card D, show that card $(A \times B) =$ card $(C \times D)$.

4. Show that $\mathbf{R}^2$ has cardinal number c.

5. A complex number is said to be transcendental over the rationals if it is not algebraic over the rationals (Problem 2, p. 18). Show that the set of transcendental numbers is uncountable.

6. Prove that the set of all real-valued functions defined on the interval $[0, 1]$ has cardinal number greater than c.

7. Show that the Cantor set contains no open interval.

8. Show that the number $\frac{1}{4}$ is in the Cantor set.

9. Show that the Cantor function is nondecreasing on C; that is, if $x < y$, then $f(x) \leq f(y)$.

10. Show that the set of all sequences of real numbers has cardinal number c.

2 | *The Real Number System*

1 · INTRODUCTION

A way of introducing the real number system, commonly used in introductory courses in analysis, is to give a set of axioms for the system. The real numbers are undefined objects with operations of addition and multiplication and an order relation which have properties specified by the given axioms. Also, one of the axioms states that the system is complete. This axiom may be given in various forms; for example, the least upper bound axiom states that a nonempty set of real numbers which is bounded above has a least upper bound.

Another approach to the real numbers is to introduce the natural numbers axiomatically and then to define successively the integers, rationals, and reals. This approach has the advantage of a relatively simple initial axiom system. However, the path to the real numbers is quite long and tedious [25]. The development as far as the rational number system can be found in many books on modern abstract algebra. A rational number is taken to be an equivalence class of ordered pairs of integers with second element nonzero where the equivalence is defined by: $(a, b)\ R\ (c, d)$ if $ad = bc$. Letting $[(a, b)]$ denote the equivalence class containing (a, b), if $r = [(a, b)]$ and $s = [(c, d)]$ one defines $r + s = [(ad + bc, bd)]$ and $rs = [(ac, bd)]$ (Problem 6, p. 14). Also, if b and d are positive, then

$$r < s \quad \text{if } ad < bc.$$

To obtain the representation of a rational number as a quotient of two integers identify the rational number $[(a, 1)]$ with the integer a. Note that $[(1, b)]$ is the multiplicative inverse of $[(b, 1)]$ since

$$[(b, 1)]\,[(1, b)] = [(b, b)] = [(1, 1)] = 1.$$

Then, for any rational number $r = [(a, b)]$, we have

$$r = [(a, b)] = [(a, 1)][(1, b)] = \frac{[(a, 1)]}{[(b, 1)]} = \frac{a}{b}.$$

In this chapter we will develop the real number system from the rational number system. This development may provide insights not gained by an axiomatic introduction of the real numbers. Also, the methods used in this development will be used later in a more general context.

2 · ORDERED RINGS AND FIELDS

The basic properties of the rational numbers are described by the statement that the rational number system is an archimedean-ordered field. These terms will be defined in this section. First we define a related algebraic structure called a ring.

2 · 1 Definition. A ***ring*** *is a set R with two functions from $R \times R$ into R, called addition and multiplication and denoted by the usual symbolism, which satisfy the following conditions:*

A_1. $(a + b) + c = a + (b + c)$ *for all* $a, b, c \in R$.
A_2. $a + b = b + a$ *for all* $a, b \in R$.
A_3. *There is an element* 0 *in* R *such that* $a + 0 = a$ *for all* $a \in R$.
A_4. *For each* $a \in R$ *there is an element* $-a \in R$ *such that* $a + (-a) = 0$.
M_1. $(ab)c = a(bc)$ *for all* $a, b, c \in R$.
D. $a(b + c) = ab + ac$ *and* $(b + c)a = ba + ca$ *for all* $a, b, c \in R$.

Usually the ring is denoted by R, the same letter used to denote its set of elements. Any function from $R \times R$ into R is called a **binary operation** on R. Thus, addition and multiplication are binary operations on R which are associative (A_1, M_1). Also, addition is commutative (A_2); there is an additive identity 0 in R (A_3); and each element a of R has an additive inverse $-a$ in R (A_4).

A ring R is called a **commutative ring** if it satisfies:

M_2. $ab = ba$ *for all* $a, b \in R$.

A ring R is said to be a **ring with unit** if it satisfies:

M_3. *There is an element* $1 \neq 0$ *in* R *such that* $a \cdot 1 = a = 1 \cdot a$ *for all* $a \in R$.

It is easy to show that the additive identity 0 and the unit 1 (if it exists) are unique. Also, each element $a \in R$ has a unique additive inverse $-a$. In a ring R the operation of subtraction is defined as follows:

$$a - b = a + (-b).$$

The following results are easily proved for a ring R:

(2 · 2) $\quad a + b = a + c$ implies $b = c$ (cancellation)

(2 · 3) $\quad -(-a) = a$ and $-(a + b) = -a + (-b)$

(2 · 4) $\quad a \cdot 0 = 0 = 0 \cdot a$ for all $a \in R$

(2 · 5) $\quad a(-b) = -(ab) = (-a)b$ and $(-a)(-b) = ab$.

If R has a unit, we also have:

(2 · 6) $\quad -a = (-1)a$.

The integers and the rational numbers provide examples of a commutative ring with unit. The rational numbers have an additional property not shared with the integers: each nonzero element has a multiplicative inverse.

2·7 Definition. *A* ***field*** **F** *is a commutative ring with unit which satisfies:*
M_4. *For each* $a \neq 0$ *in* **F** *there is an element* $a^{-1} \in \mathbf{F}$ *such that* $a \cdot a^{-1} = 1$.

The rational number system is a field. In a field **F** each nonzero element b has a unique multiplicative inverse b^{-1} and the operation of division is defined as follows:

$$\frac{a}{b} = ab^{-1} \quad \text{where } b \neq 0.$$

Notice that 2 · 4 implies that 0 cannot have a multiplicative inverse.

Since a field is a commutative ring with unit, the results 2 · 2–2 · 6 hold in a field. In a field **F** we also have:

(2 · 8) $ab = ac$ and $a \neq 0$ imply $b = c$ (cancellation)

(2 · 9) $(a^{-1})^{-1} = a$ and $(ab)^{-1} = a^{-1}b^{-1}$

(2 · 10) $\frac{a}{b} + \frac{c}{d} = \frac{ad + bc}{bd}$ and $\frac{a}{b} \cdot \frac{c}{d} = \frac{ac}{bd}$.

In the ring of integers there is a subset called the set of positive integers and there is a relation "less than" which can be defined in terms of this subset: $a < b$ if $b - a$ is positive. By recognizing the essential properties of the set of positive integers it is possible to generalize these notions so that they apply to other rings.

2·11 Definition. *A ring R is said to be an* ***ordered ring*** *if there exists a subset P of R such that*

(1) $0 \notin P$;
(2) *If* $a \neq 0$, *then one and only one of the elements* a *or* $-a$ *is in* P;
(3) *If* $a, b \in P$, *then* $a + b \in P$ *and* $ab \in P$.

The set P is called the ***positive cone*** *of R and the elements of P are said to be* ***positive.***

In an ordered ring the relation "less than" is defined as follows:

$$a < b \quad \text{if} \quad b - a \in P.$$

Observe that a is positive if and only if $0 < a$. If $a < 0$, then a is said to be **negative**. The relation "less than" (or "greater than") has the following properties in any ordered ring R:

(2 · 12) For any elements a and b in R one and only one of the following relations holds: $a < b$, $a = b$, $b < a$.

(2 · 13) $a < b$ and $b < c$ imply $a < c$

(2 · 14) $a < b$ implies $a + c < b + c$ for all $c \in R$.

(2 · 15) $a < b$ and $0 < c$ imply $ac < bc$

(2 · 16) $a < b$ and $c < 0$ imply $ac > bc$

(2 · 17) $a \neq 0$ implies $a^2 > 0$.

If R has a unit 1, then 2 · 17 implies that $1 > 0$.

We may assume that any ordered ring with unit contains the integers by identifying the ring elements $0, 1, -1, 1 + 1$, etc., with the integers $0, 1, -1, 2$, etc. An ordered ring R with unit is said to be an **archimedean-ordered ring** if for each pair of positive elements a and b in R there is a positive integer n such that $b < na$.

A field **F** is said to be (archimedean-) ordered if as a ring it is (archimedean-) ordered. In an ordered field the "less than" relation has the additional properties:

(2 · 18) $0 < a < b$ implies $0 < b^{-1} < a^{-1}$

(2 · 19) $a < b < 0$ implies $b^{-1} < a^{-1} < 0$.

The **absolute value** of an element a, denoted by $|a|$, is defined in an ordered ring as follows:

$$|a| = \begin{cases} a & \text{if } a \geq 0. \\ -a & \text{if } a < 0 \end{cases}$$

The basic properties of the absolute value of elements of an ordered ring are:

(1) $|a| \geq 0$; $|a| = 0$ if and only if $a = 0$
(2) $|ab| = |a|\,|b|$
(3) $|a + b| \leq |a| + |b|$ (Triangle Inequality).

These properties follow readily from the definition. In the proof of the Triangle Inequality the fact that $a \leq |a|$ and $-a \leq |a|$ can be helpful.

PROBLEMS

1. Prove: 2 · 2 – 2 · 6
2. Prove: 2 · 8 – 2 · 10
3. Prove: 2 · 12 – 2 · 17
4. Prove: 2 · 18 – 2 · 19
5. Prove the basic properties of absolute value.
6. Prove that $||a| - |b|| \leq |a - b|$.

7. If $\mathscr{F}$ is the set of all functions from a set X into a ring R, taking the usual definitions of addition and multiplication of functions, show that $\mathscr{F}$ is a ring.

8. A subset S of a ring R is called a **subring** if it is itself a ring under the operations in R restricted to S. Give minimal requirements for a subset of a ring to be a subring.

9. A ring R is isomorphic to a ring R' if there is a one-to-one mapping φ from R onto R' such that

$$\varphi(a+b) = \varphi(a) + \varphi(b) \quad \text{and} \quad \varphi(ab) = \varphi(a)\varphi(b).$$

If φ is an isomorphism of the ring R onto the ring R', show that $\varphi(0) = 0$, $\varphi(-a) = -\varphi(a)$, $\varphi(1) = 1$.

10. If R is the set of polynomials with rational coefficients, show that R is a ring when we use the usual definitions of addition and multiplication of polynomials. Let P be the set of polynomials in R with leading coefficient positive. Show P is a positive cone of R.

11. Show that an ordered field must be infinite.

12. Let a and b be elements of an ordered field $\mathbf{F}$ with $a < b$. Show that there exists a $c \in \mathbf{F}$ such that $a < c < b$.

3 · CAUCHY SEQUENCES

In the development of the real number system from the rational number system the concept of a Cauchy sequence will play a central rôle. First we recall the definitions of a sequence and a convergent sequence. A sequence of elements of a field $\mathbf{F}$ is a function from the set $\mathbf{N}$ of positive integers into $\mathbf{F}$.

3·1 Definition. *A sequence (a_n) of elements of an ordered field $\mathbf{F}$ is said to be **convergent** if there exists an element $a \in \mathbf{F}$ such that for each positive element ε in $\mathbf{F}$ there exists a postive integer n_0 such that*

$$|a_n - a| < \varepsilon \quad \textit{whenever} \quad n \geq n_0.$$

In this case we say that (a_n) converges to a and write $\lim a_n = a$.

As an example we show that the sequence $(1/n)$ of rational numbers converges to 0 in the rational field. Let ε be a positive rational number. Since the rational field is archimedean-ordered there exists a positive integer $n_0 > 1/\varepsilon$. Then

$$\left|\frac{1}{n} - 0\right| = \frac{1}{n} < \varepsilon \quad \text{whenever } n \geq n_0.$$

3·2 Definition. *A sequence (a_n) of elements of an ordered field $\mathbf{F}$ is called a **Cauchy sequence** if for each $\varepsilon > 0$ in $\mathbf{F}$ there exists a positive integer n_0 such that*

$$|a_n - a_m| < \varepsilon \quad \textit{whenever } m, n \geq n_0.$$

The definition of a Cauchy sequence differs from that of a convergent sequence in that the terms of a Cauchy sequence get close to each other whereas the terms of a convergent sequence get close to some fixed element in the ordered field. These may seem to be equivalent but such is not the case in every ordered field. A convergent sequence is always a Cauchy sequence but in some ordered fields a Cauchy sequence may fail to converge.

3·3 Proposition. *A convergent sequence in an ordered field* **F** *is a Cauchy sequence in* **F**.

PROOF. Let (a_n) be a sequence of elements in **F** such that $\lim a_n = a$. Then for each $\varepsilon > 0$ in **F** there exists a positive integer n_0 such that

$$|a_n - a| < \varepsilon/2 \quad \text{whenever } n \geq n_0 .$$

Thus

$$|a_n - a_m| \leq |a_n - a| + |a - a_m| < \varepsilon \quad \text{whenever } m, n \geq n_0 . \blacksquare$$

To show that the converse of this proposition does not hold, consider the sequence of rational numbers:

$$1, 1.4, 1.41, 1.414, 1.4142, \cdots$$

obtained by the familiar algorithm for "finding" the square root of 2. This is a Cauchy sequence but it does not converge to any element in the rational field.

In some ordered fields it is true that every Cauchy sequence converges.

3·4 Definition. *An ordered field* **F** *is said to be* **complete** *if every Cauchy sequence of elements in* **F** *converges to an element in* **F**.

The discussion above shows that the rational field is not complete. We show later that the real field is complete. In fact, the extension of the rational field to the real field is basically the extension of an incomplete ordered field to a complete ordered field.

We now prove some properties of Cauchy sequences in an arbitrary ordered field. A sequence (a_n) of elements in an ordered field **F** is said to be **bounded** if there exists an element M in **F** such that $|a_n| \leq M$ for all n.

3·5 Proposition. *A Cauchy sequence* (a_n) *of elements in an ordered field* **F** *is bounded.*

PROOF. Corresponding to the unit 1 in **F** there exists a positive integer n_0 such that

$$|a_n - a_m| < 1 \quad \text{whenever } m, n \geq n_0 .$$

Thus, whenever $n \geq n_0$,

$$|a_n| - |a_{n_0}| \leq |a_n - a_{n_0}| < 1;$$

that is,

$$|a_n| < |a_{n_0}| + 1.$$

If we let $M = \max\{|a_{n_0}| + 1, |a_k| : k = 1, \cdots, n_0 - 1\}$, then

$$|a_n| \leq M \quad \text{for all } n. \blacksquare$$

If (a_n) is a sequence of elements in an ordered field $\mathbf{F}$ and (n_k) is an increasing sequence of positive integers, i.e., $n_k < n_{k+1}$, then $(a_{n_k}) = (a_n) \circ (n_k)$ is called a **subsequence** of (a_n). For example, $(1/2k)$ is a subsequence of $(1/n)$. In this example $a_n = (1/n)$ and $n_k = 2k$.

3·6 Proposition. *If a Cauchy sequence (a_n) of elements in an ordered field $\mathbf{F}$ has a subsequence (a_{n_k}) which converges to a, then (a_n) converges to a.*

Proof. Take $\varepsilon > 0$ in $\mathbf{F}$. There exists a positive integer n_0 such that

$$|a_n - a_m| < \varepsilon/2 \quad \text{whenever } m, n \geq n_0.$$

Also, there exists a positive integer k_0 such that

$$|a_{n_k} - a| < \varepsilon/2 \quad \text{whenever } k \geq k_0.$$

If we take $k \geq k_0$ such that $n_k \geq n_0$, then

$$|a_n - a| \leq |a_n - a_{n_k}| + |a_{n_k} - a| < \varepsilon \quad \text{whenever } n \geq n_0. \blacksquare$$

Consider now Cauchy sequences of rational numbers. Let $\mathscr{C}$ denote the set of all Cauchy sequences of rational numbers. We can define binary operations of addition and multiplication in $\mathscr{C}$ so that $\mathscr{C}$ with these operations is a commutative ring with unit. If (a_n) and (b_n) are in $\mathscr{C}$, define

$$(a_n) + (b_n) = (a_n + b_n)$$

$$(a_n)(b_n) = (a_n b_n).$$

The fact that these definitions yield binary operations in $\mathscr{C}$, that is, $(a_n + b_n)$ and $(a_n b_n)$ are in $\mathscr{C}$ whenever (a_n) and (b_n) are in $\mathscr{C}$, follows readily from the inequalities:

$$|(a_n + b_n) - (a_m + b_m)| \leq |a_n - a_m| + |b_n - b_m|$$

and

$$\begin{aligned} |a_n b_n - a_m b_m| &\leq |a_n b_n - a_m b_n| + |a_m b_n - a_m b_m| \\ &\leq M_2 |a_n - a_m| + M_1 |b_n - b_m| \end{aligned}$$

where $|a_n| \leq M_1$ and $|b_n| \leq M_2$ for all n.

It is a routine matter to check that $\mathscr{C}$ with these operations satisfies the conditions for a commutative ring with unit. For example, the verification of A_2 is:

$$(a_n) + (b_n) = (a_n + b_n) = (b_n + a_n) = (b_n) + (a_n).$$

If we let $(\bar{a})$ denote the Cauchy sequence all of whose terms are the rational number a, then the element 0 in $\mathscr{C}$ is the Cauchy sequence $(\bar{0})$. The additive inverse of (a_n) is $-(a_n) = (-a_n)$. And, the unit element 1 in $\mathscr{C}$ is the Cauchy sequence $(\bar{1})$.

$\mathscr{C}$ is not a field however since a nonzero element in $\mathscr{C}$ may fail to have a multiplicative inverse. A sequence (b_n) such that $\lim b_n = 0$ is called a **null sequence**. The following proposition shows that any null sequence in $\mathscr{C}$, e.g., $(1/n)$, cannot have a multiplicative inverse in $\mathscr{C}$ and, hence, $\mathscr{C}$ is not a field.

3·7 Proposition. *If (a_n) and (b_n) are in $\mathscr{C}$ and $\lim b_n = 0$, then $\lim (a_n b_n) = 0$.*

PROOF. Since any Cauchy sequence is bounded, there exists a positive rational number M such that $|a_n| \leq M$ for all n. Let ε be any positive rational number. Then there is a positive integer n_0 such that

$$|b_n| < \varepsilon/M \quad \text{whenever } n \geq n_0$$

and, hence,

$$|a_n b_n| < \varepsilon \quad \text{whenever } n \geq n_0 . \blacksquare$$

PROBLEMS

1. If (a_n) and (b_n) are sequences of elements in an ordered field **F** such that $\lim a_n = a$ and $\lim b_n = b$, show that

$$\lim(a_n + b_n) = a + b \quad \text{and} \quad \lim(a_n b_n) = ab.$$

2. If in an ordered field **F** the sequence (a_n) converges to a, show that any subsequence (a_{n_k}) of (a_n) converges to a.

3. Prove that $\mathscr{C}$ is a commutative ring with unit.

4. Show that the sequence $(\bar{a})$ converges to a.

5. If $(b_n) \in \mathscr{C}$, $b_n \neq 0$ for all n, and (b_n) does not converge to zero, show that (b_n) has a multiplicative inverse in $\mathscr{C}$.

6. If (a_n) is a Cauchy sequence of integers in the rational field, does (a_n) converge?

7. In an archimedean-ordered field **F** show that

$$\lim \frac{1}{2^n} = \lim \frac{1}{n} = 0.$$

4·THE REAL NUMBERS

Let $\mathscr{C}$ be the set of all Cauchy sequences of rational numbers with

$$(a_n) + (b_n) = (a_n + b_n) \quad \text{and} \quad (a_n)(b_n) = (a_n b_n),$$

and let $\mathcal{N}$ be the subset of $\mathscr{C}$ consisting of the null sequences. Define a relation, denoted by R, in $\mathscr{C}$ as follows:

$$(a_n)\; R\; (b_n) \text{ if } (a_n) - (b_n) = (a_n - b_n) \in \mathcal{N}.$$

It is easy to check that this relation is an equivalence relation and, hence, determines a partition of $\mathscr{C}$ into equivalence classes which we denote by $[(a_n)]$.

4·1 Definition. *The real number system* **R** *is the quotient set* $\mathscr{C}/R$ *with the operations of addition and multiplication defined as follows:*

$$[(a_n)] + [(b_n)] = [(a_n) + (b_n)] \quad \textit{and} \quad [(a_n)][(b_n)] = [(a_n)(b_n)].$$

As usual when an operation on equivalence classes is defined in terms of representatives of the equivalence classes one must verify that it really is an operation on the equivalence classes; that is, that it is independent of the particular representative used. Take $(a_n)\; R\; (a_n')$ and $(b_n)\; R\; (b_n')$. Then $(a_n - a_n')$ and $(b_n - b_n')$ are null sequences and, hence,

$$((a_n + b_n) - (a_n' + b_n')) = ((a_n - a_n') + (b_n - b_n'))$$

is a null sequence; that is, $((a_n) + (b_n))\; R\; ((a_n') + (b_n'))$. Also,

$$(a_n b_n - a_n' b_n') = (b_n(a_n - a_n') + a_n'(b_n - b_n'))$$

is a null sequence; that is, $((a_n)(b_n))\; R\; ((a_n')(b_n'))$.

In the proof that **R** is a field we will need the following proposition.

4·2 Proposition. *If* $(a_n) \in \mathscr{C} \sim \mathcal{N}$, *then there exists a positive rational* ε *and a positive integer* n_0 *such that either* $a_n > \varepsilon$ *for all* $n \geq n_0$ *or* $a_n < -\varepsilon$ *for all* $n \geq n_0$.

PROOF. Since $(a_n) \notin \mathcal{N}$ there exists a positive rational ε such that for each n there exists an $m \geq n$ such that $|a_m| \geq 2\varepsilon$. Since $(a_n) \in \mathscr{C}$ there exists a positive integer n_0 such that

$$|a_n - a_m| < \varepsilon \quad \text{whenever } m, n \geq n_0.$$

Take $m \geq n_0$ such that $|a_m| \geq 2\varepsilon$, say $a_m \geq 2\varepsilon$. Then, for all $n \geq n_0$,

$$a_n > a_m - \varepsilon \geq \varepsilon.$$

If $a_m \leq -2\varepsilon$, then $a_n < -\varepsilon$ whenever $n \geq n_0$. ∎

4·3 Theorem. *The real number system* **R** *is a field.*

PROOF. The fact that **R** is a commutative ring with unit follows easily from the fact that $\mathscr{C}$ is a commutative ring with unit. For example, the verification of A_2 is:

$$[(a_n)] + [(b_n)] = [(a_n) + (b_n)] = [(b_n) + (a_n)] = [(b_n)] + [(a_n)].$$

The element 0 in $\mathbf{R}$ is $[(\bar{0})] = \mathcal{N}$. Each $[(a_n)]$ in $\mathbf{R}$ has $[-(a_n)]$ as its additive inverse. The unit or multiplicative identity 1 in $\mathbf{R}$ is $[(\bar{1})]$.

To complete the proof that $\mathbf{R}$ is a field we must verify M_4: if $[(a_n)] \neq 0$, then there exists an element $[(b_n)]$ in $\mathbf{R}$ such that $[(a_n)][(b_n)] = [(\bar{1})] = 1$. If $[(a_n)] \neq 0$, then $(a_n) \in \mathscr{C} \sim \mathcal{N}$ and, hence, by Proposition 4 · 2 there exists a positive rational ε and a positive integer n_0 such that $|a_n| > \varepsilon$ whenever $n \geq n_0$. Define (b_n) as follows:

$$b_n = \begin{cases} 0 & \text{for } n < n_0 \\ a_n^{-1} & \text{for } n \geq n_0 \end{cases}.$$

To show that $(b_n) \in \mathscr{C}$ take a positive rational number δ. Then for $m, n \geq n_0$

$$|b_n - b_m| = |a_n^{-1} - a_m^{-1}| = |a_n^{-1} a_m^{-1}|\,|a_m - a_n| < \varepsilon^{-2}\,|a_m - a_n|\,.$$

Since $(a_n) \in \mathscr{C}$ there exists a positive integer $m_0 \geq n_0$ such that

$$|a_m - a_n| < \varepsilon^2 \delta \quad \text{whenever } m, n \geq m_0$$

and, hence,

$$|b_n - b_m| < \delta \quad \text{whenever } m, n \geq m_0\,.$$

Finally, to show that $[(a_n)][(b_n)] = 1$ we must show that $(a_n b_n)\ R\ (\bar{1})$ or, equivalently, $(a_n b_n - 1) \in \mathcal{N}$. Clearly, for any positive rational number δ,

$$|a_n b_n - 1| = 0 < \delta \quad \text{whenever } n \geq n_0\,.$$

Thus, $[(a_n)][(b_n)] = 1$ and M_4 is verified. ▮

We now make $\mathbf{R}$ an ordered field by the introduction of a positive cone which we will denote by $\mathbf{R}^+$. Proposition 4 · 2 states that if $[(a_n)] \neq 0$, then, for some positive rational ε and some positive integer n_0, $a_n > \varepsilon$ for all $n \geq n_0$ or $a_n < -\varepsilon$ for all $n \geq n_0$. We define the positive cone $\mathbf{R}^+$ to be those real numbers $[(a_n)]$ such that, for some positive rational ε and some positive integer n_0, $a_n > \varepsilon$ for all $n \geq n_0$. It is not difficult to show that this definition is independent of the particular representative (a_n).

That $\mathbf{R}^+$ qualifies as a positive cone of a field will now be shown:

(1) Since $0 = [(\bar{0})]$, $0 \notin \mathbf{R}^+$.

(2) Proposition 4 · 2 implies that if $[(a_n)] \neq 0$, then one and only one of the elements $[(a_n)]$ or $-[(a_n)]$ is in $\mathbf{R}^+$.

(3) If $[(a_n)]$ and $[(b_n)]$ are in $\mathbf{R}^+$, then for some positive rationals ε and δ and for some positive integers n_0 and m_0 we have:

$$\begin{aligned} a_n &> \varepsilon \quad \text{whenever } n \geq n_0 \\ b_n &> \delta \quad \text{whenever } n \geq m_0\,. \end{aligned}$$

Therefore, if $n \geq \max\,\{n_0, m_0\}$,

$$a_n + b_n > \varepsilon + \delta > 0 \quad \text{and} \quad a_n b_n > \varepsilon\,\delta > 0.$$

Thus, $[(a_n)] + [(b_n)] \in \mathbf{R}^+$ and $[(a_n)][(b_n)] \in \mathbf{R}^+$.

In the ordered field **R** we define the "less than" relation in the standard way:

$$[(a_n)] < [(b_n)] \text{ if } [(b_n)] - [(a_n)] \in \mathbf{R}^+,$$

that is, $[(a_n)] < [(b_n)]$ if for some positive rational ε and some positive integer n_0, $b_n - a_n > \varepsilon$ whenever $n \geq n_0$.

Identifying the real numbers $[(\bar{0})]$, $[(\bar{1})]$, $[(-\bar{1})]$, $[(\bar{2})]$, etc., with the integers 0, 1, -1, 2, etc., we can show that **R** is an archimedean-ordered field.

4·4 Theorem. *The real number system* **R** *is an archimedean-ordered field.*

PROOF. Let $[(a_n)]$ and $[(b_n)]$ be positive real numbers. We wish to show that there is a positive integer m such that $m[(a_n)] > [(b_n)]$. There exists a positive rational ε and a positive integer n_0 such that

$$a_n > \varepsilon \quad \text{whenever } n \geq n_0.$$

Since $(b_n) \in \mathscr{C}$, it is bounded; that is, there exists a positive rational M such that

$$b_n \leq M \quad \text{for all } n.$$

Using the fact that the rational field is archimedean-ordered, we know that there exists a positive integer m such that

$$m\varepsilon > M.$$

Then

$$ma_n - b_n > m\varepsilon - M > 0 \quad \text{whenever } n \geq n_0.$$

Therefore,

$$[(b_n)] < [(ma_n)] = [(\bar{m})][(a_n)] = m[(a_n)]. \ \blacksquare$$

In our discussion we have identified the integer m with the rational number m and then with the real number $[(\bar{m})]$. We would like now to show that any rational number a can be identified with the real number $[(\bar{a})]$. In order to justify this identification we must show that elements a in the rational field behave the same as the corresponding elements $[(\bar{a})]$ in the real field. By "behave the same" we mean with respect to the operations of addition and multiplication and also with respect to the order. In more formal language we wish to show that the rational field is isomorphic to a subset of the real field.

4·5 Definition. *If* **F** *and* **F′** *are fields, then* **F** *is* ***isomorphic*** *to a subset* **S′** *of* **F′** *if there is a one-to-one function f from* **F** *onto* **S′** *such that for all a and b in* **F**

$$f(a+b) = f(a) + f(b) \quad \text{and} \quad f(ab) = f(a)f(b).$$

4·6. Theorem. *The rational field is isomorphic to a subset of the real field.*

Proof. For any rational number a, let $f(a) = [(\bar{a})]$. This is clearly a one-to-one function from the rational field $\mathbf{F}$ onto the subset $\mathbf{S}' = \{[(\bar{a})] : a \text{ rational}\}$ of $\mathbf{R}$. For, if $f(a) = f(b)$, that is, $[(\bar{a})] = [(\bar{b})]$, then $(\overline{a - b})$ is a null sequence and hence $a = b$. Also

$$f(a + b) = [(\overline{a + b})] = [(\bar{a})] + [(\bar{b})] = f(a) + f(b)$$

and

$$f(ab) = [(\overline{ab})] = [(\bar{a})][(\bar{b})] = f(a)f(b). \blacksquare$$

The isomorphism f also preserves the order relation. If a is a positive rational number, then $f(a) = [(\bar{a})]$ is a positive real number. This implies that if $a < b$ then $[(\bar{a})] < [(\bar{b})]$.

Although a rational number may be considered to be a real number, it will sometimes be convenient to use a notational device to indicate whether a rational number is being considered as an element of the rational field or the real field. When we wish to emphasize that the rational number a is being considered as an element of the real field we will write a^*. Note that $a^* = [(\bar{a})]$.

4·7 Theorem. *Between any two distinct real numbers there is a rational number.*

Proof. If x and y are real numbers such that $x < y$, we wish to show that there is a rational number c such that

$$x < c^* < y.$$

Let $x = [(a_n)]$ and $y = [(b_n)]$. Then $[(a_n)] < [(b_n)]$ implies that for some positive rational ε and some positive integer n_0

$$b_n - a_n > \varepsilon \quad \text{whenever } n \geq n_0.$$

Since (a_n) and (b_n) are in $\mathscr{C}$ there exists a positive integer $m_0 \geq n_0$ such that

$$|a_n - a_m| < \varepsilon/4 \quad \text{and} \quad |b_n - b_m| < \varepsilon/4 \quad \text{whenever } m, n \geq m_0.$$

Let $c = \frac{1}{2}(a_{m_0} + b_{m_0})$. Then, for all $n \geq m_0$,

$$c - a_n > \tfrac{1}{2}(a_{m_0} + b_{m_0}) - a_{m_0} - \varepsilon/4 = \tfrac{1}{2}(b_{m_0} - a_{m_0}) - \varepsilon/4 > \varepsilon/4$$

and

$$b_n - c > b_{m_0} - \varepsilon/4 - \tfrac{1}{2}(a_{m_0} + b_{m_0}) = \tfrac{1}{2}(b_{m_0} - a_{m_0}) - \varepsilon/4 > \varepsilon/4.$$

Thus,

$$[(a_n)] < [(\bar{c})] < [(b_n)]$$

or equivalently

$$x < c^* < y. \blacksquare$$

The following proposition should provide the hindsight to make our development of the real numbers from the rational numbers seem natural. We will show that if $x = [(a_n)]$ then $\lim a_n^* = x$. Thus, we have represented a real number x by a sequence (a_n) of rationals which as a sequence of real numbers converges to x. Tracing our way back, for any real number x, we find that Theorem 4 · 7 implies that there is a sequence (a_n) of rationals which converges to x in **R**. Since (a_n) converges in **R**, it is a Cauchy sequence in **R** and, hence, also in the rational field. Also, two sequences converge to the same number if and only if they differ by a null sequence. This suggests defining a real number as an equivalence class of Cauchy sequences of rationals where two sequences in the same equivalence class differ by a null sequence.

4·8 Proposition. *If* $x = [(a_n)]$, *then* $\lim a_n^* = x$.

PROOF. Take $\varepsilon > 0$ in **R** and let δ be a rational number such that $0 < \delta^* < \varepsilon/2$. Since $(a_n) \in \mathscr{C}$, there exists a positive integer n_0 such that

$$|a_n - a_m| < \delta \quad \text{whenever } m, n \geq n_0 .$$

Thus, for each $n \geq n_0$, we have

$$2\delta - (a_n - a_m) > \delta \quad \text{and} \quad 2\delta - (a_m - a_n) > \delta \quad \text{whenever } m \geq n_0 .$$

This implies that, for each $n \geq n_0$,

$$[(\bar{a}_n) - (a_m)] < [(\overline{2\delta})] \quad \text{and} \quad [(a_m) - (\bar{a}_n)] < [(\overline{2\delta})],$$

that is,

$$a_n^* - x < 2\delta^* < \varepsilon \quad \text{and} \quad x - a_n^* < 2\delta^* < \varepsilon.$$

Therefore,

$$|a_n^* - x| < \varepsilon \quad \text{whenever } n \geq n_0$$

and, hence,

$$\lim a_n^* = x. \blacksquare$$

PROBLEMS

1. Defining $(a_n)\, R\, (b_n)$ if $(a_n - b_n) \in \mathscr{N}$, show that this is an equivalence relation.

2. Verify that the real number system, **R**, is a commutative ring with unit.

3. If the field **F** is isomorphic to the subset **S**′ of the field **F**′, show that **S**′ is a subfield of **F**′.

4. Show that between any two distinct real numbers there is an infinite number of rational numbers.

5. If A is a ring with unit, a subset I of A is called an **ideal** if

$$(1)\quad a, b \in I \Rightarrow a - b \in I$$
$$(2)\quad a \in A \text{ and } b \in I \Rightarrow ab \in I.$$

a. If we define a relation in A by aRb if $a - b \in I$, show that this is an equivalence relation.

b. Show that $\mathcal{N}$ is an ideal in $\mathscr{C}$.

6. If x is any real number, show that there is a real number greater than x.

7. If x is a real number, show that there is an integer n such that

$$n \leq x < n + 1.$$

5 · COMPLETENESS OF THE REAL NUMBER SYSTEM

Our objective in the development of the real number system is to obtain a complete ordered field by extending the incomplete ordered field of rationals. In this section we will show that we have gained our objective.

5·1 Theorem. *The real field* **R** *is complete.*

PROOF. Let (x_n) be a Cauchy sequence of real numbers. For each positive integer n take a rational number a_n such that $x_n < a_n^* < x_n + 1/n$. To show that $(a_n) \in \mathscr{C}$, let ε be any positive rational number. There exists a positive integer $n_0 > 3/\varepsilon$ such that

$$|x_n - x_m| < \varepsilon^*/3 \quad \text{whenever } m, n \geq n_0.$$

Then

$$|a_n^* - a_m^*| \leq |a_n^* - x_n| + |x_n - x_m| + |x_m - a_m^*| < \varepsilon^* \quad \text{whenever } m, n \geq n_0,$$

and, therefore,

$$|a_n - a_m| < \varepsilon \quad \text{whenever } m, n \geq n_0.$$

Letting $x = [(a_n)]$, we can show that $\lim x_n = x$. Take $\delta > 0$ in **R**. Since $\lim a_n^* = x$, there exists a positive integer $m_0 > 2/\delta$ such that

$$|a_n^* - x| < \delta/2 \quad \text{whenever } n \geq m_0.$$

Then

$$|x_n - x| \leq |x_n - a_n^*| + |a_n^* - x| < \delta \quad \text{whenever } n \geq m_0$$

and, therefore, $\lim x_n = x$. ∎

We now prove some other properties of the real number system which are equivalent to completeness and which in some instances are more convenient to use. We call a sequence (x_n) of real numbers **nondecreasing** if $x_n \leq x_{n+1}$ for all n.

5·2 Theorem. *A bounded nondecreasing sequence of real numbers converges.*

PROOF. Let (x_n) be a bounded nondecreasing sequence of real numbers. We will show that (x_n) is a Cauchy sequence. Suppose (x_n) is not a Cauchy sequence. Then there is a positive number ε such that for every positive integer n there exist positive integers $r, s \geq n$ such that

$$|x_r - x_s| \geq \varepsilon.$$

Since (x_n) is nondecreasing this implies that for every positive integer n there exists a positive integer $r > n$ such that

$$x_r - x_n \geq \varepsilon.$$

Thus, there exists a positive integer $n_1 > 1$ such that

$$x_{n_1} \geq x_1 + \varepsilon.$$

Moreover, there exists a positive integer $n_2 > n_1$ such that

$$x_{n_2} \geq x_{n_1} + \varepsilon \geq x_1 + 2\varepsilon.$$

Continuing in this manner, we obtain a subsequence (x_{n_k}) such that $x_{n_k} \geq x_1 + k\varepsilon$. This contradicts the fact that (x_n) is bounded. Hence, (x_n) is a Cauchy sequence of real numbers and therefore converges. ∎

If we define a sequence (x_n) of real numbers to be **nonincreasing** if $x_n \geq x_{n+1}$ for all n, then we easily obtain the result: *a bounded nonincreasing sequence of real numbers converges.*

A subset S of $\mathbf{R}$ is said to be **bounded above** if there exists an element $b \in \mathbf{R}$ such that $x \leq b$ for all $x \in S$; in such a case b is called an **upper bound** of S. An element $c \in \mathbf{R}$ is said to be the **least upper bound** of a subset S of $\mathbf{R}$ if c is an upper bound of S and no element smaller than c is an upper bound of S. The least upper bound of S is denoted by sup S (supremum S) or lub S. Thus, $c = \sup S$ if and only if it satisfies:

(1) $x \leq c$ for all $x \in S$

(2) for each $\varepsilon > 0$ there exists an $x \in S$ such that $x > c - \varepsilon$.

The greatest lower bound of a set is defined in a similar way and is denoted by inf S (infinum S) or glb S.

5·3 Theorem. *If S is a nonempty set of real numbers which is bounded above, then S has a least upper bound.*

PROOF. Let A and B be the sets of real numbers which are not and are, respectively, upper bounds of S. Since S is nonempty and is bounded above, A and B are nonempty. Also, any element of A is less than any element of B. Take $a_1 \in A$ and $b_1 \in B$. Construct a nondecreasing sequence (a_n) and a nonincreasing sequence (b_n) as follows: if $\frac{1}{2}(a_n + b_n) \in A$, let $a_{n+1} = \frac{1}{2}(a_n + b_n)$ and $b_{n+1} = b_n$; if $\frac{1}{2}(a_n + b_n) \in B$, let $a_{n+1} = a_n$ and $b_{n+1} = \frac{1}{2}(a_n + b_n)$. Then,

(a_n) and (b_n) are bounded nondecreasing and nonincreasing sequences, respectively, and therefore they converge. Let $a = \lim a_n$ and $b = \lim b_n$. Since $b_n - a_n = (\frac{1}{2})^{n-1}(b_1 - a_1)$, $b = a$. Let $c = a = b$.

We now show that $c = \sup S$. For any $x \in S$, $x \leq b_n$ for all n. Therefore, $x \leq b = c$ and c is an upper bound of S. To show that c is the least upper bound of S take $y < c$. Then, for some n, $y < a_n$. Since a_n is not an upper bound of S, y is not an upper bound of S. Thus, $c = \sup S$. ∎

PROBLEMS

1. Prove that a bounded nonincreasing sequence of real numbers converges.

2. If $\lim b_n = b$ and $a \leq b_n$ for all n, show that $a \leq b$.

3. If S is a nonempty set of real numbers which is bounded below, show that S has a greatest lower bound.

4. Show that **R** has the Dedekind property: If **R** is partitioned into two sets L and U such that each element of L is less than each element of U, then L has a greatest element or U has a least element.

5. A point $c \in \mathbf{R}$ is said to be a **limit point** of the sequence (x_n) of real numbers if for each $\varepsilon > 0$ and each positive integer m there is a positive integer $n \geq m$ such that $|x_n - c| < \varepsilon$. Show that a bounded sequence (x_n) of real numbers has at least one limit point.

6. Show that a bounded sequence of real numbers has a convergent subsequence.

7. If $\lim a_n = a$, $\lim b_n = b$, and $a_n \leq b_n$ for all n, show that $a \leq b$.

8. If (x_n) is a bounded sequence of real numbers, let

$$y_k = \inf\{x_n : n \geq k\} \quad \text{and} \quad z_k = \sup\{x_n : n \geq k\}.$$

a. Show that (y_k) and (z_k) are, respectively, nondecreasing and nonincreasing sequences and

$$\lim y_k = \sup\{y_k\} \quad \text{and} \quad \lim z_k = \inf\{z_k\}.$$

b. Define the **lower limit** and **upper limit** of (x_n) as follows:

$$\underline{\lim}\, x_n = \lim_{k\to\infty} \inf\{x_n : n \geq k\} = \sup_k \inf\{x_n : n \geq k\}$$

$$\overline{\lim}\, x_n = \lim_{k\to\infty} \sup\{x_n : n \geq k\} = \inf_k \sup\{x_n : n \geq k\}.$$

Show that

$$\underline{\lim}\, x_n \leq \overline{\lim}\, x_n.$$

c. Show that $\underline{\lim}\, x_n$ and $\overline{\lim}\, x_n$ are limit points of (x_n).

d. Show that $\lim x_n = x \Leftrightarrow \underline{\lim}\, x_n = x = \overline{\lim}\, x_n$.

6 · THE COMPLEX NUMBERS

Another important number system in analysis is the complex number system. The property of the complex numbers which makes them useful in many instances is that every nonconstant polynomial over the complex field can be factored into a product of linear factors. This fact, known as the Fundamental Theorem of Algebra, will be proved in Chapter 4. We give only a brief discussion of complex numbers here, assuming that the reader is versed in routine manipulations with these numbers.

6·1 Definition. *The complex number system* **C** *is the set* $\mathbf{R} \times \mathbf{R}$ *with operations of addition and multiplication defined as follows:*

$$(a, b) + (c, d) = (a + c, b + d)$$

$$(a, b) \cdot (c, d) = (ac - bd, ad + bc).$$

The verification that **C** is a field is straightforward and is left to the reader (Problem 1). The zero of **C** is (0, 0) and the unit of **C** is (1, 0).

Let **S** be the subfield of **C** consisting of those elements of **C** with second element zero; that is,

$$\mathbf{S} = \{(x, 0) : x \in \mathbf{R}\}.$$

Then the function φ from **R** onto **S** defined by

$$\varphi(x) = (x, 0)$$

is an isomorphism of **R** onto **S** (Problem 6). Thus, **R** and **S** may be identified; that is, the real number field may be considered as a subfield of the complex field.

If $(x, y) \in \mathbf{C}$, then, letting $i = (0, 1)$ and identifying complex numbers having second element zero with the corresponding real numbers, we have

$$(x, y) = (x, 0) + (0, y) = (x, 0) + (y, 0) \cdot (0, 1) = x + yi.$$

This gives the more familiar notation for complex numbers. If $z = (x, y)$, then x is called the **real part** of z, denoted by $\mathrm{Rl}(z)$, and y is called the **imaginary part** of z, denoted by $\mathrm{Im}(z)$.

Note that

$$i^2 = (0, 1)^2 = (-1, 0) = -1.$$

This shows that **C** cannot be an ordered field. For, in an ordered field, the square of any nonzero element is positive. We would then have $(-1, 0) = -(1, 0)$ and $(1, 0)$ both positive which contradicts a basic property of the positive cone of an ordered field.

Although **C** is not an ordered field we can define an absolute value in this system which has the same basic properties as the absolute value in the real number system. Define the **absolute value** of the complex number $z = (x, y)$ as follows:

$$|z| = \sqrt{x^2 + y^2}.$$

This absolute value has the following properties:

(1) $|z| \geq 0$; $|z| = 0$ if and only if $z = 0$

(2) $|z_1 z_2| = |z_1|\,|z_2|$

(3) $|z_1 + z_2| \leq |z_1| + |z_2|$.

The definition of convergence of a sequence (z_n) of complex numbers has the same form as that for convergence of a sequence of real numbers: *the sequence (z_n) converges to z if for each positive real number ε there is a positive integer n_0 such that*

$$|z_n - z| < \varepsilon \quad \textit{whenever } n \geq n_0.$$

Similarly, the definition of a Cauchy sequence of complex numbers corresponds to that for real numbers.

It is easy to show that the complex number system **C** is complete. If (z_n) is a Cauchy sequence of complex numbers and $z_n = (x_n, y_n)$, then, since

$$|x_n - x_m| \leq |z_n - z_m| \quad \text{and} \quad |y_n - y_m| \leq |z_n - z_m|,$$

(x_n) and (y_n) are Cauchy sequences of real numbers. Thus, these sequences converge, say $\lim x_n = x$ and $\lim y_n = y$. Then, if $z = (x, y)$, $\lim z_n = z$ (Problem 7).

If we consider the complex number $z = (x, y)$ to be the point in the Euclidean plane with coordinates x and y, then $|z_1 - z_2|$ is the distance between the points z_1 and z_2.

PROBLEMS

1. Show that **C** is a field.

2. If $z = (x, y)$, define the **conjugate** of z by: $\bar{z} = (x, -y)$. Show that:

a. $z\bar{z} = |z|^2$

b. $z + \bar{z} = 2\text{Rl}(z)$

c. $|\bar{z}| = |z|$

d. $\overline{z_1 z_2} = \bar{z}_1 \bar{z}_2$

e. $\overline{z_1 + z_2} = \bar{z}_1 + \bar{z}_2$

3. If $f(z) = \bar{z}$, show that f is continuous on **C**. That is, the operation of taking the complex conjugate is continuous.

4. Show that:

a. $|z_1 z_2| = |z_1|\,|z_2|$

b. $|z_1 + z_2| \leq |z_1| + |z_2|$.

5. If $e^{it} = \cos t + i \sin t$, show that:

a. $|e^{it}| = 1$

b. This function is a one-to-one mapping of $[0, 2\pi)$ onto the unit circle in **C** (or $\mathbf{R}^2$).

c. Any complex number z can be written in the form re^{it} where $r = |z|$.

d. $r_1 e^{it_1} \cdot r_2 e^{it_2} = r_1 r_2 e^{i(t_1 + t_2)}$.

6. Verify that the function φ defined by $\varphi(x) = (x, 0)$ is an isomorphism of **R** into **C**.

7. Let (x_n) and (y_n) be sequences of real numbers such that $\lim x_n = x$ and $\lim y_n = y$. If $z_n = (x_n, y_n)$, show that $\lim z_n = (x, y)$.

8. If $z_1 = r_1 e^{i\theta_1}$ is a nonzero complex number and n is a positive integer, determine all complex numbers z satisfying the equation

$$z^n = z_1.$$

3 | *Linear Spaces*

1 · INTRODUCTION

An important aspect of modern analysis is the recognition and exploitation of the fact that certain sets of functions form linear spaces. The basic concepts of linear or vector spaces are no doubt quite familiar to the reader. Our purpose in this chapter is to review these concepts briefly and to emphasize certain aspects that will be of particular importance for our later work. For example, many of the interesting spaces considered in analysis are of infinite dimension.

2 · LINEAR SPACES

2 · 1 Definition. *A* ***linear (vector) space*** *X over a field* **F** *is a set of elements together with a function, called addition, from $X \times X$ into X and a function, called scalar multiplication, from* **F** $\times X$ *into X which satisfy the following conditions for all $x, y, z \in X$ and $\alpha, \beta \in$* **F**:

A_1. $(x + y) + z = x + (y + z)$
A_2. $x + y = y + x$
A_3. *There is an element* 0 *in X such that $x + 0 = x$ for all $x \in X$.*
A_4. *For each $x \in X$ there is an element $-x \in X$ such that $x + (-x) = 0$.*
S_1. $\alpha(x + y) = \alpha x + \alpha y$
S_2. $(\alpha + \beta)x = \alpha x + \beta x$
S_3. $\alpha(\beta x) = (\alpha\beta)x$
S_4. $1 \cdot x = x$

Properties $A_1 - A_4$ imply that X is an abelian group under addition and Properties $S_1 - S_4$ relate the operation of scalar multiplication to addition in X and to addition and multiplication in **F**.

The elements of X are called **points** or **vectors** and the elements of **F** are called **scalars**. In our discussions the field **F** will always be the real field **R** or the complex field **C**. Note that a linear space over **C** can also be considered as a linear space over **R** since **R** is a subset of **C**.

In general we will abide by the notational convention that Greek letters denote scalars and Roman letters denote vectors. Frequent use will be made of superscripts as well as subscripts

A familiar example of a linear space is the space $V_n(\mathbf{R})$ where the vectors are n-tuples of real numbers and the scalars are real numbers with addition and scalar multiplication defined by

$$(\alpha^1, \cdots, \alpha^n) + (\beta^1, \cdots, \beta^n) = (\alpha^1 + \beta^1, \cdots, \alpha^n + \beta^n)$$
$$\gamma(\alpha^1, \cdots, \alpha^n) = (\gamma\alpha^1, \cdots, \gamma\alpha^n).$$

That $V_n(\mathbf{R})$ satisfies the requirements for a linear space over $\mathbf{R}$ is easily verified. Similarly, the set of all n-tuples of complex numbers with the above definitions of addition and scalar multiplication is a linear space over $\mathbf{C}$ and is denoted by $V_n(\mathbf{C})$. In the case where $n = 1$, we understand that $V_1(\mathbf{F})$ is the same as $\mathbf{F}$ and, hence, $\mathbf{F}$ is a linear space over $\mathbf{F}$.

Before giving more examples of linear spaces we discuss some elementary properties which can be derived directly from the definition of a linear space. For example, by adding $-x$ to both sides of $x + y = x$, we can show that if $x + y = x$ for some $x \in X$ then $y = 0$. This implies that the zero element specified in A_3 is unique. This result can also be used to show that $0x = 0$ for all $x \in X$ since

$$0x = (0 + 0)x = 0x + 0x.$$

Similarly, we have: $\alpha 0 = 0$ for all $\alpha \in \mathbf{F}$. Observe that we use the symbol 0 to denote both the zero vector and the zero scalar. The context indicates which meaning is intended. It is also easy to show that for each $x \in X$ the element $-x$ specified in A_4 is unique. It then follows that $-x = (-1)x$. Subtraction in X is defined by the rule: $x - y = x + (-y)$.

Many of the linear spaces of interest to us are special cases of the linear space introduced in the following proposition.

2 · 2 Proposition. *The set X of all functions from a nonempty set T into a field $\mathbf{F}$ with addition and scalar multiplication defined by*

$$[f + g](t) = f(t) + g(t) \quad \textit{and} \quad [\alpha f](t) = \alpha f(t);\ f, g \in X, \quad t \in T, \quad \alpha \in \mathbf{F}$$

is a linear space over $\mathbf{F}$.

Proof. Conditions A_1, A_2, and $S_1 - S_4$ of Definition 2 · 1 follow immediately from corresponding properties of $\mathbf{F}$. The zero element in X is the function whose value is zero at all points in T. Also, for each $f \in X$ the function $-f$ is defined by $[-f](t) = -f(t)$. ∎

If in Proposition 2 · 2 we let $T = \{1, \cdots, n\}$, then we conclude that $V_n(\mathbf{R})$ and $V_n(\mathbf{C})$ are linear spaces over $\mathbf{R}$ and $\mathbf{C}$, respectively. If $T = \mathbf{N}$, the set of all positive integers, then X is the set of all sequences of elements of $\mathbf{F}$ with addition and scalar multiplication defined by

$$(\alpha^n) + (\beta^n) = (\alpha^n + \beta^n) \quad \text{and} \quad \gamma(\alpha^n) = (\gamma\alpha^n).$$

Thus, these sequences with the given operations form a linear space over $\mathbf{F}$ which we denote by $V_\infty(\mathbf{F})$.

If, in Proposition 2 · 2, $T = \{(i, j): i = 1, \cdots, m; j = 1, \cdots, n\}$, then X is the set of all $m \times n$ matrices with entries in **F**. If we denote a matrix by

$$A = [\alpha_j^{\,i}] = \begin{bmatrix} \alpha_1^{\,1} & \alpha_2^{\,1} \cdots \alpha_n^{\,1} \\ \alpha_1^{\,2} & \alpha_2^{\,2} \cdots \alpha_n^{\,2} \\ \cdots & \cdots \\ \alpha_1^{\,m} & \alpha_2^{\,m} \cdots \alpha_n^{\,m} \end{bmatrix} \qquad \text{where } \alpha_j^{\,i} \in \mathbf{F},$$

then the operations in X can be written

$$A + B = [\alpha_j^{\,i}] + [\beta_j^{\,i}] = \begin{bmatrix} \alpha_1^{\,1} + \beta_1^{\,1} & \alpha_2^{\,1} + \beta_2^{\,1} \cdots \alpha_n^{\,1} + \beta_n^{\,1} \\ \alpha_1^{\,2} + \beta_1^{\,2} & \alpha_2^{\,2} + \beta_2^{\,2} \cdots \alpha_n^{\,2} + \beta_n^{\,2} \\ \cdots & \cdots \\ \alpha_1^{\,m} + \beta_1^{\,m} & \alpha_2^{\,m} + \beta_2^{\,m} \cdots \alpha_n^{\,m} + \beta_n^{\,m} \end{bmatrix} = [\alpha_j^{\,i} + \beta_j^{\,i}]$$

and

$$\gamma A = \gamma[\alpha_j^{\,i}] = \begin{bmatrix} \gamma\alpha_1^{\,1} & \gamma\alpha_2^{\,1} \cdots \gamma\alpha_n^{\,1} \\ \gamma\alpha_1^{\,2} & \gamma\alpha_2^{\,2} \cdots \gamma\alpha_n^{\,2} \\ \cdots & \cdots \\ \gamma\alpha_1^{\,m} & \gamma\alpha_2^{\,m} \cdots \gamma\alpha_n^{\,m} \end{bmatrix} = [\gamma\alpha_j^{\,i}].$$

Thus, the set of all $m \times n$ matrices with these operations is a linear space over **F** which we denote by $\mathbf{F}_n^{\,m}$.

A subset E of a linear space X over a field **F** is called a (linear) **subspace** if $\alpha x + \beta y \in E$ whenever $x, y \in E$ and $\alpha, \beta \in \mathbf{F}$. The condition that $\alpha x + \beta y \in E$ for all $x, y \in E$ and $\alpha, \beta \in \mathbf{F}$ implies that the addition and scalar multiplication functions when restricted to $E \times E$ and $\mathbf{F} \times E$, respectively, map into E. It also implies that $0 \in E$ and $-x \in E$ whenever $x \in E$. Since all the other conditions for a linear space will hold in E because they hold in X, we see that a subspace E is itself a linear space over **F**. Note that the singleton $\{0\}$ is a subspace of any linear space; it is called the trivial linear space. A subspace is also called a **linear manifold.**

Consider now the space X of all real-valued functions defined on the closed interval $[a, b]$ with the usual definitions of addition and scalar multiplication. This is a linear space over **R** by Proposition 2 · 2. Let E be the subset of X consisting of all continuous functions in X. Since $\alpha f + \beta g$ is continuous if f and g are continuous and α and β are any real numbers, E is a subspace of X. This space will be denoted by $C[a, b]$.

Given a linear space X and a subspace E of X, we can form another linear space as follows. In X define a relation $a \equiv b$ if $a - b \in E$. It is easy to check that this is an equivalence relation. Also, this equivalence relation has the following properties: if $a \equiv b$ and $c \equiv d$, then $a + c \equiv b + d$, and if $a \equiv b$, then $\alpha a \equiv \alpha b$ for all $\alpha \in \mathbf{F}$. Let X/E denote the quotient set with respect to this equivalence relation and $[a]$ denote an equivalence class with a representative a. Then X/E is a linear space when the operations of addition and scalar multiplication are defined by

$$[a] + [b] = [a + b] \quad \text{and} \quad \alpha[a] = [\alpha a].$$

As usual in such a situation one must show that these definitions are independent of the representative used; e.g. one must show that if $[a] = [a']$ and $[b] = [b']$, then

$$[a] + [b] = [a + b] = [a' + b'] = [a'] + [b'].$$

The linear space X/E is called the **quotient space** of X modulo E.

PROBLEMS

1. Verify that $V_n(\mathbf{R})$ is a linear space.

2. If X is a linear space over a field $\mathbf{F}$, show that

a. $-x = (-1)x$

b. If $x, y \in X$, then there is a unique $z \in X$ such that

$$x + z = y.$$

c. If $\alpha x = 0$, then $\alpha = 0$ or $x = 0$.

3. Let X be a linear space over the field $\mathbf{F}$ and take $x, y \in X$. Show that

$$S = \{\alpha x + \beta y : \alpha, \beta \in \mathbf{F}\}$$

is a subspace.

4. Let X be the set of all polynomials of degree less than n with real coefficients. With the usual definitions of addition of polynomials and multiplication of a polynomial by a real number show that X is a linear space over $\mathbf{R}$.

5. Let X be the set of real-valued functions which are Riemann integrable over $[a, b]$. With the usual definitions of addition and multiplication by a real number, show that X is a linear space over $\mathbf{R}$.

6. Let X be the set of real-valued functions which are differentiable on $[a, b]$. Show that X is a subspace of the space of continuous functions on $[a, b]$.

7. Let X be the set of all functions from a nonempty set T into a linear space Y over the field $\mathbf{F}$. If in X we define

$$[f + g](t) = f(t) + g(t) \quad \text{and} \quad [\alpha f](t) = \alpha f(t);\ f, g \in X, \quad t \in T, \quad \alpha \in \mathbf{F},$$

show that X is a linear space over $\mathbf{F}$.

8. If X is a linear space and E is a subspace of X, verify that X/E is a linear space.

9. If X is the linear space of real-valued functions defined on $[a, b]$, let $E = \{f : \int_a^b f = 0\}$. Show that E is a subspace of X. Describe the elements of the space X/E.

3 · HAMEL BASES

In $V_3(\mathbf{R})$ any vector $x = (\alpha^1, \alpha^2, \alpha^3)$ can be written in the form $x = \alpha^1 i + \alpha^2 j + \alpha^3 k$ where $i = (1, 0, 0)$, $j = (0, 1, 0)$, and $k = (0, 0, 1)$. The triple (i, j, k) is called a basis for $V_3(\mathbf{R})$. In this section we generalize this notion and discuss the concept of a Hamel basis for an arbitrary linear space, including the case where the Hamel basis is not finite. A finite Hamel basis is usually called simply a basis.

Let X be a linear space over the field $\mathbf{F}$. Recall that a family $(x_s)_{s \in S}$ of points in X is a function A from an index set S into X such that $A(s) = x_s$. A subfamily $(x_s)_{s \in S'}$ of $(x_s)_{s \in S}$ is the restriction of A to a subset S' of S. The subfamily is called finite if S' has a finite number of elements.

3 · 1 Definition. *A family $(x_s)_{s \in S}$ of points in X is said to be **linearly independent** if for each finite subfamily $(x_s)_{s \in S'}$ of $(x_s)_{s \in S}$*

$$\sum_{s \in S'} \alpha^s x_s = 0 \quad \textit{implies} \quad \alpha^s = 0 \quad \textit{for all } s \in S'.$$

A family of points in X is said to be **linearly dependent** if it is not linearly independent. Thus, $(x_s)_{s \in S}$ is linearly dependent if for some finite subfamily $(x_s)_{s \in S'}$ and for some scalars α^s $(s \in S')$, not all zero, we have $\sum_{s \in S'} \alpha^s x_s = 0$.

For example, the triple (i, j, k) of vectors in $V_3(\mathbf{R})$ is linearly independent since

$$\begin{aligned} \alpha^1 i + \alpha^2 j + \alpha^3 k &= \alpha^1(1, 0, 0) + \alpha^2(0, 1, 0) + \alpha^3(0, 0, 1) \\ &= (\alpha^1, \alpha^2, \alpha^3) = (0, 0, 0) \end{aligned}$$

implies $\alpha^1 = \alpha^2 = \alpha^3 = 0$.

It is clear from the definition that any subfamily of a linearly independent family is also linearly independent.

The following result shows that what the elements of the index set S are is of no importance in questions concerning linear independence of a family. As we might expect, it is only how many elements there are in S and how these indices correspond to points of X that are of any consequence.

3 · 2 Proposition. *Let $A = (x_s)_{s \in S}$ and $B = (y_t)_{t \in T}$ be two families of points in X for which there is a one-to-one function f from S onto T such that $A = B \circ f$. Then A and B are both linearly independent or both linearly dependent.*

PROOF. Suppose A is linearly independent. Take a finite subfamily $(y_t)_{t \in T'}$ of B and let $S' = f^{-1}(T')$. Then, for each $t \in T'$ there is a unique $s \in S'$ such that $f(s) = t$ and, hence,

$$y_t = B(t) = B(f(s)) = A(s) = x_s.$$

Thus,

$$\sum_{t\in T'} \alpha^t y_t = 0 \;\Rightarrow \sum_{s\in S'} \alpha^{f(s)} x_s = 0 \;\;\Rightarrow\;\; \alpha^{f(s)} = 0 \quad \text{for all } s \in S';$$

that is, $\alpha^t = 0$ for all $t \in T'$. This shows that B is linearly independent. Similarly, if B is linearly independent, then A is linearly independent. ▌

3·3 Proposition. *If $A = (x_s)_{s\in S}$ is linearly independent, then A is one-to-one.*

PROOF. If s and t are distinct elements in S such that $x_s = x_t$, then $x_s - x_t = 0$. This implies that A is linearly dependent. ▌

Proposition 3 · 3 implies that the number of points in the range $\{x_s : s \in S\}$ of a linearly independent family $A = (x_s)_{s\in S}$ is the same as the number of elements in the index set S. That is, any two distinct terms x_s and x_t in A are distinct points of X. Also, if $A = (x_s)_{s\in S}$ is linearly independent and $B = A \circ A^{-1}$, that is, B is the family with index set $\{x_s : s \in S\}$ and $B(x_s) = x_s$, then Proposition 3 · 2 implies that B is linearly independent.

A sum $\sum_{i=1}^{n} \alpha^i x_i$, where $\alpha^i \in \mathbf{F}$, is called a **linear combination** of the vectors $x_1, \cdots, x_n$ in X; α^i is called the **coefficient** of x_i. We say that a subset E of X **spans** X if every element of X can be written as a linear combination of a finite number of elements of E.

3·4 Definition. *Any linearly independent family of points in X whose range spans X is called a **Hamel basis** of X.*

In the space $V_n(\mathbf{R})$ the family $(e_1, \cdots, e_n)$, where $e_i = (\delta_i^{\,1}, \cdots, \delta_i^{\,n})$ and $\delta_i^{\,j} = 0$ for $j \neq i$ and $\delta_i^{\,i} = 1$, is linearly independent. Also, the range of this family spans $V_n(\mathbf{R})$ since any vector $(\alpha^1, \cdots, \alpha^n)$ in $V_n(\mathbf{R})$ can be written

$$(\alpha^1, \cdots, \alpha^n) = \sum_{i=1}^{n} \alpha^i e_i .$$

Thus, $(e_1, \cdots, e_n)$ is a basis of $V_n(\mathbf{R})$.

3·5 Theorem. *If B is a Hamel basis of X, then each nonzero vector in X has a unique representation as a linear combination, with nonzero coefficients, of a finite number of vectors in the range of B.*

PROOF. Suppose $B = (x_s)_{s\in S}$ and let x be a nonzero vector in X. Since $\{x_s : s \in S\}$ spans X, then x can be written as a linear combination of a finite number of vectors in this set:

$$x = \sum_{s\in S_1} \alpha^s x_s$$

where S_1 is a finite subset of S and $\alpha^s \neq 0$ for all $s \in S_1$.

Suppose that we have another representation:

$$x = \sum_{s \in S_2} \beta^s x_s$$

where S_2 is a finite subset of S and $\beta^s \neq 0$ for all $s \in S_2$. Let $S_3 = S_1 \cup S_2$. Then, if we set $\alpha^s = 0$ for $s \in S_3 \sim S_1$ and $\beta^s = 0$ for $s \in S_3 \sim S_2$, we have

$$x = \sum_{s \in S_3} \alpha^s x_s = \sum_{s \in S_3} \beta^s x_s .$$

Therefore,

$$\sum_{s \in S_3} (\alpha^s - \beta^s) x_s = 0$$

and, hence, $\alpha^s = \beta^s$ for all $s \in S_3$. This shows that $S_1 = S_2$ and $\alpha^s = \beta^s$ for all $s \in S_1$; that is, the representation is unique. ▌

We will show that every nontrivial linear space has a Hamel basis and that the ranges or, equivalently, the index sets of any two Hamel bases of a space have the same cardinal number. This cardinal number is called the dimension of the space. The dimension of the trivial space $\{0\}$ is taken to be zero. Since we have shown that $(e_1, \cdots, e_n)$ is a basis of $V_n(\mathbf{R})$, the dimension of this space is n.

In the discussion following Proposition 3 · 3 we showed that to each linearly independent family $A = (x_s)_{s \in S}$ there corresponds a linearly independent family B with index set $\{x_s : s \in S\}$ such that $B(x_s) = x_s$. Furthermore, it is clear that if A is a Hamel basis of X, then B is also a Hamel basis of X. We now limit our attention to families of the same type as B; that is, we consider families I_A where A is a subset of X and I_A is the identity function on A.

Let $\mathscr{F}$ be the set of all linearly independent families I_A where $A \in \mathscr{P}(X)$ and let $\mathscr{F}$ be partially ordered by extension; that is, $I_A \prec I_B$ if $A \subset B$.

3 · 6 Proposition. *A family I_B in $\mathscr{F}$ is a Hamel basis if and only if it is a maximal element of $\mathscr{F}$.*

PROOF. Let I_B be a Hamel basis of X and suppose that I_A is a family in $\mathscr{F}$ such that $I_B \prec I_A$. If $A \sim B \neq \varnothing$, take $x \in A \sim B$. Since I_B is a Hamel basis, there exist vectors $b_1, \cdots, b_n$ in B and scalars $\alpha^1, \cdots, \alpha^n$ such that

$$x = \sum_{i=1}^{n} \alpha^i b_i .$$

This shows that I_A is not linearly independent. Therefore, $A = B$ and $I_A = I_B$. Hence I_B is maximal in $\mathscr{F}$.

Now suppose I_B is a maximal element of $\mathscr{F}$ and take $x \notin B$. Then, $I_{\{x\} \cup B}$ is not linearly independent; that is, there exist vectors $b_1, \cdots, b_n$ in B and scalars $\alpha, \alpha^1, \cdots, a^n$, not all zero, such that

$$\alpha x + \sum_{i=1}^{n} \alpha^i b_i = 0.$$

Since I_B is linearly independent, $\alpha \neq 0$ and, hence,

$$x = \sum_{i=1}^{n} \left(-\frac{\alpha^i}{\alpha}\right) b_i .$$

Thus, B spans X and I_B is a Hamel basis of X. ∎

3 · 7 Proposition. *If $I_A \in \mathscr{F}$, then there exists a Hamel basis I_B in $\mathscr{F}$ such that $I_A \prec I_B$.*

PROOF. Sometimes to avoid an excess of subscripts we use the notation $I \mid B$ for I_B. Let $\mathscr{H}$ be the partially ordered subset of $\mathscr{F}$ consisting of those families in $\mathscr{F}$ which are extensions of I_A. Any chain $\{I_C : C \in \mathscr{C}\}$ in $\mathscr{H}$ is bounded above by the family $I|(\bigcup_{C \in \mathscr{C}} C)$ in $\mathscr{H}$. ($I|(\bigcup_{C \in \mathscr{C}} C)$ is linearly independent since any finite subset of $\bigcup_{C \in \mathscr{C}} C$ will lie in some set $C \in \mathscr{C}$.) By Zorn's Lemma $\mathscr{H}$ has a maximal element I_B. Since I_B is also a maximal element of $\mathscr{F}$, I_B is a Hamel basis of X. ∎

3 · 8 Theorem. *Any nontrivial linear space X has a Hamel basis.*

PROOF. Let x be a nonzero vector in X. Then $I_{\{x\}}$ is a linearly independent family and, hence, by Proposition 3 · 7 there is a Hamel basis I_B of X such that $I_{\{x\}} \prec I_B$. ∎

In fact, Proposition 3 · 7 implies that a nontrivial linear space has an infinite number of Hamel bases. We now show that any two Hamel bases of X have the same cardinal number; that is, their index sets have the same cardinal number.

3 · 9 Theorem. *Any two Hamel bases for a linear space have the same cardinal number.*

PROOF. Since to any Hamel basis $(x_s)_{s \in S}$ of X there corresponds a Hamel basis I_B with $B \in \mathscr{P}(X)$ such that card S = card B, we need only consider Hamel bases of the form I_B.

Let $I \mid B$ and $I \mid C$ be Hamel bases of X and let $\mathscr{G}$ be the set of all one-to-one functions g from a subset of B onto a subset of C such that $I \mid (\mathrm{R}_g \cup (B \sim \mathrm{D}_g))$ is linearly independent. Partial order $\mathscr{G}$ by extension; that is, $g \prec h$ if $\mathrm{D}_g \subset \mathrm{D}_h$ and $g(x) = h(x)$ for all $x \in \mathrm{D}_g$.

Observe that $\mathscr{G}$ is nonempty. For if $c \in C$ then since $I \mid B$ is a Hamel basis of X, c has a unique representation

$$(3 \cdot 10) \qquad c = \sum_{i=1}^{n} \alpha^i b_i \qquad \text{where } b_i \in B \quad \text{and} \quad \alpha^i \neq 0.$$

Let g be the function with domain $\{b_1\}$ such that $g(b_1) = c$. Then, $I \mid (\mathrm{R}_g \cup (B \sim \mathrm{D}_g)) = I \mid (\{c\} \cup (B \sim \{b_1\})$ is linearly independent. Otherwise c would be a linear combination of a finite number of elements in $B \sim \{b_1\}$ and this would contradict the uniqueness of $3 \cdot 10$. Thus, $g \in \mathscr{G}$.

Let $\{g_t : t \in T\}$ be a chain in $\mathscr{G}$. If we define the function g by $D_g = \bigcup_{t \in T} D_{g_t}$ and $g(x) = g_t(x)$ if $x \in D_{g_t}$, then g is a one-to-one function from a subset of B onto a subset of C such that $I|(R_g \cup (B \sim D_g))$ is linearly independent. For, if S is a finite subset of $R_g \cup (B \sim D_g)$, assume $S = S_1 \cup S_2$ where $S_1 \subset R_g$ and $S_2 \subset B \sim D_g$. Then, for some $t \in T$, $S_1 \subset R_{g_t}$ and $S_2 \subset B \sim D_{g_t}$. Since $I|(R_{g_t} \cup (B \sim D_{g_t}))$ is linearly independent, $I|S$ must be linearly independent and, hence, $I|(R_g \cup (B \sim D_g))$ is linearly independent. Thus, $g \in \mathscr{G}$ and g is an upper bound for the given chain.

Zorn's Lemma implies that $\mathscr{G}$ has a maximal element, call it h. If we can show that $D_h = B$, then we will have card $B \leqq$ card C. By the symmetry of B and C in the hypothesis we would also have card $C \leqq$ card B and, hence, card $B =$ card C.

Now suppose $D_h \neq B$. Then $I|(C \cup (B \sim D_h))$ is not linearly independent but $I|(R_h \cup (B \sim D_h))$ is linearly independent and, therefore, $R_h \neq C$. Take $c_0 \in C \sim R_h$ and consider the two alternatives: $I|(\{c_0\} \cup R_h \cup (B \sim D_h))$ is linearly independent or it is not. If $I|(\{c_0\} \cup R_h \cup (B \sim D_h))$ is linearly independent, take any element $b_0 \in B \sim D_h$ and let h' be the extension of h to $D_h \cup \{b_0\}$ such that $h'(b_0) = c_0$. Then, $I|(R_{h'} \cup (B \sim D_{h'}))$ is linearly independent. If $I|(\{c_0\} \cup R_h \cup (B \sim D_h))$ is not linearly independent, then c_0 can be written as a linear combination of a finite number of elements of $R_h \cup (B \sim D_h)$ in which at least one element of $B \sim D_h$, call it b_0, will appear with non-zero coefficient. Let h' be the extension of h to $D_h \cup \{b_0\}$ such that $h'(b_0) = c_0$. Then, $I|(R_{h'} \cup (B \sim D_{h'}))$ is linearly independent. Thus, in either case, since h' is clearly one-to-one, we have obtained an element of $\mathscr{G}$ which is an extension of h. This contradicts the maximality of h in $\mathscr{G}$ and, hence, we can conclude that $D_h = B$. ∎

Since any two Hamel bases for a given linear space have the same cardinal number, this cardinal number is a property of the space itself and we have already noted that it is called the dimension of the space.

It is easy to show that the space $V_\infty(\mathbf{R})$ of all sequences of real numbers has infinite dimension. Let $e_i = (\delta_i^{\,j})$ where $\delta_i^{\,j} = 0$ for $i \neq j$ and $\delta_i^{\,i} = 1$. The family $(e_i)_{i=1}^{\infty}$ is clearly linearly independent. Since $V_\infty(\mathbf{R})$ has a Hamel basis which is an extension of this family, the dimension of $V_\infty(\mathbf{R})$ must be infinite. One might think at first that $(e_i)_{i=1}^{\infty}$ itself would be a Hamel basis for $V_\infty(\mathbf{R})$ but such is not the case. A sequence with an infinite number of nonzero terms cannot be expressed as a linear combination of a finite number of elements of $\{e_i : i = 1, 2, \cdots\}$.

PROBLEMS

1. Show that a linearly independent family cannot have the zero vector in its range.

2. Show that any subfamily of a linearly independent family is linearly independent.

3. If I_S is linearly independent but $I_{\{x\} \cup S}$ is linearly dependent, show that x can be written as a linear combination of a finite number of elements of S.

4. If $\mathbf{F}_n{}^m$ is the space of all $m \times n$ matrices with entries in $\mathbf{F}$, show that $\dim \mathbf{F}_n{}^m = mn$.

5. Show that the linear space $C[a, b]$ has infinite dimension.

6. If Y is a subspace of a linear space X, show that $\dim Y \leqq \dim X$. Give an example where $Y \neq X$ but $\dim Y = \dim X$.

7. If $f_n(x) = x^n$, show that the family $(f_n)_{n=0}^{\infty}$ is linearly independent in $C[0, 1]$.

8. Show that a polynomial is an even function if and only if it contains only even powers.

4 · LINEAR TRANSFORMATIONS

In this section we consider functions from one linear space into another which preserve the algebraic structure.

4 · 1 Definition. *If X and Y are linear spaces over the same field* $\mathbf{F}$, *then a function T from X into Y is called a* ***linear transformation*** *or* ***linear operator*** *if*

$$T(\alpha^1 x_1 + \alpha^2 x_2) = \alpha^1 T(x_1) + \alpha^2 T(x_2)$$

for all $x_1, x_2 \in X$ and $\alpha^1, \alpha^2 \in \mathbf{F}$.

4 · 2 Proposition. *If T is a linear transformation from X into Y, then $T(X)$ is a subspace of Y. If T is one-to-one, then T^{-1} is a linear transformation from $T(X)$ onto X.*

PROOF. For any $y_1, y_2 \in T(X)$ there are vectors $x_1, x_2 \in X$ such that $T(x_1) = y_1$ and $T(x_2) = y_2$. Since T is linear, for any $\alpha^1, \alpha^2 \in \mathbf{F}$,

$$\alpha^1 y_1 + \alpha^2 y_2 = \alpha^1 T(x_1) + \alpha^2 T(x_2) = T(\alpha^1 x_1 + \alpha^2 x_2).$$

Thus, $\alpha^1 y_1 + \alpha^2 y_2 \in T(X)$ and, hence, $T(X)$ is a subspace of Y. Also, if T^{-1} exists, then

$$T^{-1}(\alpha^1 y_1 + \alpha^2 y_2) = \alpha^1 x_1 + \alpha^2 x_2 = \alpha^1 T^{-1}(y_1) + \alpha^2 T^{-1}(y_2). \blacksquare$$

This proposition shows that a linear transformation preserves the linear space structure. Also, a linear transformation T from X into Y maps the zero in X into the zero in Y: for any $x \in X$.

$$T(0) = T(0 \cdot x) = 0 \cdot T(x) = 0.$$

We now show that if zero is the only point of X which is mapped into zero then T is one-to-one.

4 · 3 Proposition. *A linear transformation T is one-to-one if and only if $T(x) = 0$ implies $x = 0$.*

PROOF. First assume that T is one-to-one and that $T(x) = 0$. Since $T(0) = 0$, we have $T(x) = T(0)$ and, hence, $x = 0$. Now assume that $T(x) = 0$ implies $x = 0$. If $T(x_1) = T(x_2)$, then

$$T(x_1 - x_2) = T(x_1) - T(x_2) = 0$$

and therefore $x_1 - x_2 = 0$ or, equivalently, $x_1 = x_2$. ∎

4 · 4 Definition. *A linear transformation T from X into Y which is one-to-one is called an* ***isomorphism*** *of X into Y. Two linear spaces X and Y over the field* **F** *are said to be* ***isomorphic*** *if there is an isomorphism of X* onto *Y.*

Isomorphic linear spaces are essentially the same since there is a one-to-one correspondence between their points and also the algebraic operations in the two spaces correspond. That is, if y_1 and y_2 in Y correspond to x_1 and x_2 in X, then $y_1 + y_2$ corresponds to $x_1 + x_2$ and αy_1 corresponds to αx_1 for any $\alpha \in \mathbf{F}$.

We now show that any two linear spaces over the same field that have the same dimension are isomorphic.

4 · 5 Theorem. *Two linear spaces X and Y over the same field* **F** *are isomorphic if and only if they have the same dimension.*

PROOF. First, assume that X and Y are isomorphic under the isomorphism T. If $B = (x_s)_{s \in S}$ is a Hamel basis of X, then showing that $T \circ B = (T(x_s))_{s \in S}$ is a Hamel basis of Y will prove that X and Y have the same dimension. For any finite subset S' of S,

$$\sum_{s \in S'} \alpha^s T(x_s) = 0 \quad \Rightarrow \quad T\left(\sum_{s \in S'} \alpha^s x_s\right) = 0$$

$$\Rightarrow \sum_{s \in S'} \alpha^s x_s = 0 \quad \Rightarrow \quad \alpha^s = 0 \quad \text{for all } s \in S'.$$

Therefore, $T \circ B$ is a linearly independent family of points in Y. To show that the range of $T \circ B$ spans Y, take any point $y \in Y$. Since T maps X onto Y there is a point $x \in X$ such that $T(x) = y$. The point x has a representation of the form

$$x = \sum_{s \in S'} \alpha^s x_s \quad \text{for some finite subset } S' \text{ of } S.$$

Then

$$y = T(x) = T\left(\sum_{s \in S'} \alpha^s x_s\right) = \sum_{s \in S'} \alpha^s T(x_s).$$

This shows that $T \circ B$ is a Hamel basis of Y and, therefore, X and Y have the same dimension.

Now assume that X and Y have the same dimension and let $B = (x_s)_{s \in S}$ and $C = (y_s)_{s \in S}$ be Hamel bases of X and Y, respectively. Define a one-to-one function T from $\{x_s : s \in S\}$ onto $\{y_s : s \in S\}$ by letting $T(x_s) = y_s$. Now extend T to a function from X into Y as follows. For each $x \in X$, if $x = \sum_{s \in S'} \alpha^s x_s$ where S' is a finite subset of S, let

$$T(x) = \sum_{s \in S'} \alpha^s T(x_s) = \sum_{s \in S'} \alpha^s y_s.$$

It is easy to verify that T is an isomorphism from X onto Y and we leave this to the reader. ▌

As a result of this theorem we can state that in a sense $V_n(\mathbf{R})$ is the only n-dimensional linear space over $\mathbf{R}$.

In the set of all linear transformations from a linear space X into a linear space Y, introduce the operations of addition and scalar multiplication defined in the usual way:

$$[T + S](x) = T(x) + S(x) \quad \text{and} \quad [\alpha T](x) = \alpha T(x), \qquad x \in X, \quad \alpha \in \mathbf{F}.$$

With these operations, the set of all linear transformations from X into Y is a linear space over $\mathbf{F}$. This space will be denoted by $L(X, Y)$.

Suppose X and Y are finite-dimensional linear spaces over the field $\mathbf{F}$ with $\dim X = n$ and $\dim Y = m$ and suppose that T is a linear transformation from X into Y. Let $(x_1, \cdots, x_n)$ and $(y_1, \cdots, y_m)$ be bases of X and Y, respectively. If $T(x_i) = \sum_{j=1}^m \tau_i{}^j y_j$, define the function φ from $L(X, Y)$ into $\mathbf{F}_n{}^m$, the space of all $m \times n$ matrices with entries in $\mathbf{F}$, as follows: $\varphi(T) = [\tau_i{}^j]$. It is easy to verify that φ is an isomorphism of $L(X, Y)$ onto $\mathbf{F}_n{}^m$. Thus $L(X, Y)$ is isomorphic to $\mathbf{F}_n{}^m$ and $\dim L(X, Y) = mn$.

In the case where $Y = \mathbf{F}$, the space $L(X, \mathbf{F})$ is called the **algebraic dual** of X. The elements of $L(X, \mathbf{F})$ are called **linear functionals** on X. Note that if X is of finite dimension, then $L(X, \mathbf{F})$ has the same dimension since $\dim \mathbf{F} = 1$.

PROBLEMS

1. Define the k-th projection function P_k from $V_n(\mathbf{R})$ into $\mathbf{R}$ by $P_k(x) = \alpha^k$ where $x = (\alpha^1, \cdots, \alpha^n)$. Show that P_k is a linear transformation from $V_n(\mathbf{R})$ into $\mathbf{R}$. Also show that (P_k) is a Hamel basis for $L(V_n(\mathbf{R}), \mathbf{R})$.

2. If the function T from $C[a, b]$ into $C[a, b]$ is defined by

$$T(f)(x) = \int_a^x f,$$

show that T is linear.

3. If T is a linear transformation show that

$$T\left(\sum_{i=1}^n \alpha^i x_i\right) = \sum_{i=1}^n \alpha^i T(x_i).$$

4. If T is a linear transformation from X into Y, let

$$\mathcal{N} = \{x \in X : T(x) = 0\}.$$

Show that $\mathcal{N}$ is a subspace of X. This subspace is called the **null space** or **kernel** of T.

5. If X is the linear space of polynomials of degree less than n with real coefficients (Problem 4, p. 44), show that X is isomorphic to $V_n(\mathbf{R})$.

6. Complete the proof of Theorem 4 · 5 by verifying that T is an isomorphism of X onto Y.

7. Verify that $L(X, Y)$ is a linear space over $\mathbf{F}$ if X and Y are linear spaces over $\mathbf{F}$.

8. If X and Y are finite-dimensional linear spaces, say $\dim X = n$ and $\dim Y = m$, verify that $L(X, Y)$ is isomorphic to $\mathbf{F}_n{}^m$, the space of $m \times n$ matrices.

9. Let T be the transformation from $V_2(\mathbf{R})$ into $V_2(\mathbf{R})$ defined by

$$T(x, y) = (x, -y).$$

a. Show that T is linear.
b. Determine the matrix which corresponds to T with respect to the standard basis of $V_2(\mathbf{R}) : ((1, 0), (0, 1))$.

10. Determine and describe the linear transformation from $V_2(\mathbf{R})$ into $V_2(\mathbf{R})$ which has the corresponding matrix

$$\frac{1}{\sqrt{2}}\begin{bmatrix} 1 & -1 \\ 1 & 1 \end{bmatrix}$$

with respect to the standard basis of $V_2(\mathbf{R})$ (Problem 9b).

11. Show that $\dim L(X, \mathbf{F}) \geq \dim X$. If X is finite-dimensional show that $L(X, \mathbf{F})$ is isomorphic to X.

5 · ALGEBRAS

In many vector spaces in addition to the vector space operations there occurs another operation usually called multiplication. For example, the real and complex number systems **R** and **C** have a multiplication operation as well as the vector space operations. A vector space with a multiplication properly related to the vector space operations is called an algebra.

5 · 1 Definition. *An **algebra** is a linear space X with a function, called multiplication, from $X \times X$ into X which has the following properties for all $x, y, z \in X$ and $\alpha \in \mathbf{F}$:*

(1) $x(yz) = (xy)z$;
(2) $x(y + z) = xy + xz$ *and* $(x + y)z = xz + yz$;
(3) $\alpha(xy) = (\alpha x)y = x(\alpha y)$.

Thus, an algebra has both the structure of a linear space and the structure of a ring with these structures related by (3). If the multiplication is commutative then the algebra is called a **commutative algebra**. An algebra X is called an **algebra with unit (identity)** if there exists a nonzero element in X, denoted by e and called the multiplicative **unit** or **identity**, such that for all $x \in X$

$$ex = x = xe.$$

If for an element x in an algebra X with unit there exists an element $y \in X$ such that

$$xy = e = yx$$

then x is said to be **invertible** and y is called the **inverse** of x and is denoted by x^{-1}. Clearly **R** and **C** are examples of commutative algebras with unit. In these algebras every nonzero element has an inverse.

If X is the linear space of all functions from a set T into a field **F**, then X is a commutative algebra with unit when we define multiplication pointwise:

$$[fg](t) = f(t) \cdot g(t).$$

In the linear space $L(Y, Y)$ of linear transformations from a linear space Y into itself if we define multiplication to be composition of functions

$$ST = S \circ T$$

then $L(Y, Y)$ is an algebra with unit. The unit in this algebra is the linear transformation I such that $I(y) = y$ for all $y \in Y$. In general this is not a commutative algebra.

A **subalgebra** of an algebra X is a subspace of X which is closed under multiplication. For example, $C[a, b]$ is a subalgebra of the algebra of all real-valued functions defined on the closed interval $[a, b]$.

Two algebras X and X' with the same scalar field are said to be **isomorphic** if there is a one-to-one mapping f from X onto X' such that for all $x, y \in X$ and $\alpha \in \mathbf{F}$, $f(x + y) = f(x) + f(y), f(xy) = f(x)f(y)$, and $f(\alpha x) = \alpha f(x)$. We have already seen that as vector spaces $L(V_n(\mathbf{F}), V_n(\mathbf{F}))$ and $\mathbf{F}_n{}^n$ are isomorphic. In fact, if $(e_1, \cdots, e_n)$ is the standard basis for $V_n(\mathbf{F})$; that is, $e_k = (\delta_k{}^1, \cdots, \delta_k{}^n)$, and

$$T(e_k) = \sum_{j=1}^{n} \tau_k{}^j e_j,$$

then the function f defined by

$$f(T) = [\tau_k{}^j]$$

is an isomorphism of the linear space $L(V_n(\mathbf{F}), V_n(\mathbf{F}))$ onto $\mathbf{F}_n{}^n$. Suppose now that we define multplication in $\mathbf{F}_n{}^n$ by

$$[\sigma_j{}^i][\tau_k{}^j] = [\rho_k{}^i] \quad \text{where} \quad \rho_k{}^i = \sum_{j=1}^{n} \sigma_j{}^i \, \tau_k{}^j.$$

Then $\mathbf{F}_n^n$ is an algebra. Also, if for $S \in L(V_n(\mathbf{F}), V_n(\mathbf{F}))$

$$S(e_j) = \sum_{i=1}^{n} \sigma_j{}^i e_i,$$

then

$$(ST)(e_k) = S\left(\sum_{j=1}^{n} \tau_k{}^j e_j\right) = \sum_{j=1}^{n} \tau_k{}^j S(e_j) = \sum_{j=1}^{n} \tau_k{}^j \sum_{i=1}^{n} \sigma_j{}^i e_i$$
$$= \sum_{i=1}^{n} \left(\sum_{j=1}^{n} \sigma_j{}^i \tau_k{}^j\right) e_i .$$

Thus, if $\rho_k{}^i = \sum_{j=1}^{n} \sigma_j{}^i \tau_k{}^j$, then

$$f(ST) = [\rho_k{}^i] = [\sigma_j{}^i][\tau_k{}^j] = f(S)f(T)$$

and, hence, f is an isomorphism of the algebra $L(V_n(\mathbf{F}), V_n(\mathbf{F}))$ onto the algebra $\mathbf{F}_n^n$.

PROBLEMS

1. Verify that $L(Y, Y)$ is an algebra with unit.

2. If x and y are invertible elements in an algebra with unit, show that xy is invertible and

$$(xy)^{-1} = y^{-1}x^{-1}.$$

3. If T is a one-to-one linear transformation of the linear space Y onto itself, show that its inverse transformation T^{-1} is the inverse of T in the algebra $L(Y, Y)$.

4. Verify that $\mathbf{F}_n^n$ is an algebra with unit.

5. If X and X' are algebras with unit which are isomorphic under the isomorphism f, show that:

a. the units correspond: $f(e) = e'$

b. if x has an inverse x^{-1} in X, then $f(x^{-1}) = (f(x))^{-1}$.

6. Let $T \in L(V_n(\mathbf{F}), V_n(\mathbf{F}))$, $(e_1, \cdots, e_n)$ be the standard basis of $V_n(\mathbf{F})$, and $[\tau_k{}^j]$ be the matrix which corresponds to T with respect to this basis.

a. If $x = (\alpha^1, \cdots, \alpha^n) \in V_n(\mathbf{F})$, determine $T(x)$.

b. If $x = (\alpha^1, \cdots, \alpha^n) \in V_n(\mathbf{F})$ is identified with the $n \times 1$ matrix $[\alpha^k]$ in the isomorphic space $\mathbf{F}_1^n$, show that

$$T(x) = [\tau_k{}^j][\alpha^k].$$

4 | *Metric Spaces*

1 · INTRODUCTION

If in the linear space $V_m(\mathbf{R})$ we define the distance between points $x = (\alpha^1, \cdots, \alpha^m)$ and $y = (\beta^1, \cdots, \beta^m)$ by

$$|x - y| = \left[\sum_{k=1}^{m} (\alpha^k - \beta^k)^2\right]^{1/2},$$

then we obtain the m-dimensional Euclidean space denoted by $\mathbf{R}^m$. Having a distance defined, we can then define convergence of a sequence (x_n) of points in $\mathbf{R}^m$ as follows: $\lim x_n = x$ if for each $\varepsilon > 0$ there exists a positive integer n_0 such that

$$|x_n - x| < \varepsilon \quad \text{whenever } n \geq n_0.$$

Continuity of a function from one Euclidean space into another is also defined in terms of distance between points in these spaces: if f is a function from $\mathbf{R}^n$ into $\mathbf{R}^m$, then f is continuous at the point $x \in \mathbf{R}^n$ if for each $\varepsilon > 0$ there exists a $\delta > 0$ such that

$$|f(x) - f(y)| < \varepsilon \quad \text{whenever } |x - y| < \delta.$$

This suggests that convergence of sequences and continuity of functions can be defined whenever we have the notion of distance between elements of a set. A set in which distance between elements is defined is called a metric space.

In this chapter we study the properties of metric spaces and define continuity of functions and convergence of sequences in these spaces. Considering these topics in the general setting of a metric space has two advantages. It highlights the essential properties of these concepts and shows how broad is their range of applicability. The theory can then be expanded by introducing additional properties into the space. In Chapter 7 we study metric spaces which have the algebraic structure of a linear space.

2 · METRIC SPACES

A distance or metric function on an arbitrary set X will be defined as a function having certain properties which are possessed by the distance function in a Euclidean space. The following properties of Euclidean distance are taken to be basic:

(1) $|x - y| \geq 0$; $|x - y| = 0$ if and only if $x = y$;
(2) $|x - y| = |y - x|$;
(3) $|x - y| \leq |x - z| + |z - y|$.

Properties (1) and (2) follow directly from the definition of Euclidean distance. We now prove property (3). For any $\lambda \in \mathbf{R}$,

$$0 \leq \sum_{k=1}^{m} (\alpha^k - \lambda\beta^k)^2 = \sum_{k=1}^{m} (\alpha^k)^2 - 2\lambda \sum_{k=1}^{m} \alpha^k\beta^k + \lambda^2 \sum_{k=1}^{m} (\beta^k)^2.$$

If $\sum_{k=1}^{m}(\beta^k)^2 \neq 0$, then letting $\lambda = (\sum_{k=1}^{m} \alpha^k\beta^k)(\sum_{k=1}^{m}(\beta^k)^2)^{-1}$, we obtain

$$\left(\sum_{k=1}^{m} \alpha^k\beta^k\right)^2 \leq \sum_{k=1}^{m} (\alpha^k)^2 \sum_{k=1}^{m} (\beta^k)^2.$$

If $\sum_{k=1}^{m}(\beta^k)^2 = 0$, then the last inequality holds trivially $(0 = 0)$. This is the **Cauchy-Schwarz Inequality.** Setting $x - z = (\alpha^1, \cdots, \alpha^m)$ and $z - y = (\beta^1, \cdots, \beta^m)$ and using the Cauchy-Schwarz Inequality, we have

$$\begin{aligned}
|x - y|^2 &= \sum_{k=1}^{m} (\alpha^k + \beta^k)^2 = \sum_{k=1}^{m} (\alpha^k)^2 + 2\sum_{k=1}^{m} \alpha^k\beta^k + \sum_{k=1}^{m} (\beta^k)^2 \\
&\leq \sum_{k=1}^{m} (\alpha^k)^2 + 2\left[\sum_{k=1}^{m} (\alpha^k)^2\right]^{1/2}\left[\sum_{k=1}^{m} (\beta^k)^2\right]^{1/2} + \sum_{k=1}^{m} (\beta^k)^2 \\
&= |x - z|^2 + 2|x - z|\,|z - y| + |z - y|^2 \\
&= (|x - z| + |z - y|)^2.
\end{aligned}$$

Therefore,

$$|x - y| \leq |x - z| + |z - y|.$$

2 · 1 Definition. *A **metric** or **distance function** on a set X is a real-valued function d defined on $X \times X$ which has the following properties: for all x, y, and z in X,*

D_1. $d(x, y) \geq 0$; $d(x, y) = 0$ *if and only if* $x = y$;
D_2. $d(x, y) = d(y, x)$;
D_3. $d(x, y) \leq d(x, z) + d(z, y)$.

Properties D_1 and D_2 are called, respectively, the positive definite and symmetric properties. Thus, a metric on X is a symmetric, positive definite, real-valued function on $X \times X$ which satisfies the triangle inequality D_3.

2 · 2 Definition. *A **metric space** (X, d) is a nonempty set X and a metric d defined on X.*

The elements of X are called the points of (X, d). When no confusion will arise as a result, the metric space (X, d) will be denoted simply by X.

Since the definition of a metric space was patterned on certain properties of Euclidean spaces, it is clear that these spaces are metric spaces. The complex number system $\mathbf{C}$ is also a metric space. As a metric space, $\mathbf{C}$ is the same as $\mathbf{R}^2$. Some other examples of metric spaces follow.

(1) Let X be any nonempty set and let d be defined by

$$d(x, y) = \begin{cases} 0 & \text{if } x = y \\ 1 & \text{if } x \neq y. \end{cases}$$

Then d is a metric on X, called the **discrete metric,** and the metric space (X, d) is called a **discrete space.**

(2) If X is the set of all m-tuples of real numbers and, if for $x = (\alpha^1, \cdots, \alpha^m)$ and $y = (\beta^1, \cdots, \beta^m)$,

$$d(x, y) = \max \{|\alpha^k - \beta^k| : k = 1, \cdots, m\},$$

then (X, d) is a metric space.

(3) Let X be the set of all real-valued functions which are defined and continuous on the closed interval $[a, b]$ in $\mathbf{R}$ and let

$$d(f, g) = \max\{|f(t) - g(t)| : t \in [a, b]\}.$$

Then (X, d) is a metric space which is usually denoted by $C[a, b]$.

Notice that in example (2) we have defined a metric on the set of m-tuples of real numbers which is different from the Euclidean metric. The resulting metric space differs from the Euclidean space $\mathbf{R}^m$. On the other hand, there are metric spaces which for all practical purposes are the same even though they may have different sets of points.

2 · 3 Definition. *Let (X_1, d_1) and (X_2, d_2) be two metric spaces. A function f from X_1 into X_2 is called an **isometry** if*

$$d_2(f(x), f(y)) = d_1(x, y) \qquad \textit{for all } x, y \in X_1.$$

*Two metric spaces (X_1, d_1) and (X_2, d_2) are said to be **isometric** if there exists an isometry of X_1* onto *X_2.*

Since an isometry preserves distance between points, it is clearly a one-to-one mapping. Considered as metric spaces, two isometric spaces are essentially the same and may be identified if we so choose.

This discussion of isometry suggests a method for defining a metric for certain sets. Let (X_2, d_2) be a metric space, X_1 be a set, and f be a one-to-one function from X_1 into X_2. If we define

$$d_1(x, y) = d_2(f(x), f(y)) \qquad \text{for all } x, y \in X_1,$$

then (X_1, d_1) is a metric space. This is easily verified.

Consider the extended real line. This is the real line with two points adjoined. These points are called (plus) infinity and minus infinity and are denoted by $(+)\infty$ and $-\infty$. The order relation in the real number system is extended by defining $-\infty < x < \infty$ for all $x \in \mathbf{R}$. The usefulness of the extended real line is illustrated by the fact that we can say that any set of real numbers has a least upper bound in the extended real number system. In

Chapter 2 we proved that a nonempty set A of real numbers which is bounded above has a least upper bound. If the set A is empty, we take the least upper bound to be $-\infty$, and, if A is not bounded above, then we take the least upper bound to be ∞.

In order to introduce a metric in the extended real line, we define

$$f(-\infty) = -1$$

$$f(x) = \frac{x}{1 + |x|}, \quad x \in \mathbf{R}$$

$$f(\infty) = 1.$$

Then f is a one-to-one mapping of the extended real line into $\mathbf{R}$ and, letting

$$d(x, y) = |f(x) - f(y)| \qquad \text{for all } x, y \in \mathbf{R} \cup \{-\infty, \infty\},$$

we obtain a metric space which is denoted by $\overline{\mathbf{R}}$. Notice that the extended real line $\overline{\mathbf{R}}$ is isometric to the metric space consisting of the set $[-1, 1]$ with the distance between points given by the Euclidean metric. This space, denoted simply by $[-1, 1]$, is called a subspace of $\mathbf{R}$.

In general, given a metric space (X, d), we can obtain new metric spaces as follows. Let E be a nonempty subset of X and let d_1 be the restriction of d to $E \times E$. Then (E, d_1) is a metric space and is said to be a **subspace** of (X, d).

In a metric space (X, d) we define the distance between a point and a set and the distance between two sets as follows. If $x \in X$ and B is a nonempty subset of X, then the distance between x and B, denoted by $d(x, B)$, is

$$d(x, B) = \inf\{d(x, y) : y \in B\}.$$

And, if A and B are nonempty subsets of X, the distance between A and B, denoted by $d(A, B)$, is

$$\begin{aligned} d(A, B) &= \inf\{d(x, B) : x \in A\} \\ &= \inf\{d(x, y) : x \in A, y \in B\}. \end{aligned}$$

PROBLEMS

1. Show that the functions d defined in examples (1), (2), and (3), satisfy the properties of a metric.

2. Let w, x, y, z be four points in a metric space. Establish the **quadrilateral inequality.**

$$|d(w, x) - d(x, y)| \le d(w, z) + d(z, y).$$

In $\mathbf{R}^2$ this inequality may be interpreted as stating that the sum of the lengths of two sides of a quadrilateral is greater than or equal to the absolute value of the difference in length of the other two sides.

3. Let X be the set of all bounded sequences of real numbers. If $x = (\alpha^k)$ and $y = (\beta^k)$ are elements of X, show that the function d defined by

$$d(x, y) = \sup\{|\alpha^k - \beta^k| : k \in \mathbf{N}\}$$

is a metric on X.

4. Let A be any set and let X be the set of all bounded real-valued functions defined on A. Show that

$$d(f, g) = \sup\{|f(t) - g(t)| : t \in A\}$$

defines a metric on X.

5. Let $B(x; r) = \{y : d(x, y) < r\}$, $r > 0$.

a. If (X, d) is a discrete space, describe $B(x; 1)$ and $B(x; 2)$.

b. Describe $B(x; r)$ in $\mathbf{R}^n$.

c. Describe $B(x; r)$ in the space of example (2).

6. In $\overline{\mathbf{R}}$ determine $d(0, 1)$, $d(5, 6)$, $d(-\infty, 1)$ and $d(-\infty, \infty)$.

7. If d is a real-valued function on $X \times X$ which for all x, y, and z in X satisfies

$$d(x, y) = 0 \text{ if and only if } x = y$$
$$d(x, y) + d(x, z) \geq d(y, z),$$

show that d is a metric on X.

8. Give an example of sets A and B in a metric space such that $A \cap B = \varnothing$ but $d(A, B) = 0$.

9. Extend the complex plane $\mathbf{C}$ by adjoining a point called infinity and denoted ∞.

a. If

$$f(z) = \frac{1}{1 + |z|^2}(x, y, |z|^2), \quad z \in \mathbf{C}$$

$$f(\infty) = (0, 0, 1),$$

show that f is a one-to-one mapping of the extended complex plane onto the sphere $S = \{(x, y, u) : x^2 + y^2 + u^2 = u\}$ in $\mathbf{R}^3$.

b. The metric d defined by $d(z_1, z_2) = |f(z_1) - f(z_2)|$ is called the **chordal metric** and the extended complex plane with this metric is denoted by $\overline{\mathbf{C}}$. Show that

$$d(z_1, z_2) = \frac{|z_1 - z_2|}{(1 + |z_1|^2)^{1/2}(1 + |z_2|^2)^{1/2}}, \quad z_1, z_2 \in \mathbf{C}$$

$$d(z, \infty) = \frac{1}{(1 + |z|^2)^{1/2}}, \quad z \in \mathbf{C}.$$

3 · OPEN AND CLOSED SETS

For any point x in a metric space (X, d) and any positive real number r, the **open ball** with center x and radius r is defined to be the set:

$$B(x; r) = \{y : d(x, y) < r\}.$$

If E is any set in X, we define:

x is an **interior point** of E if some open ball with center x is contained in E;

x is a **boundary point** of E if every open ball with center x contains at least one point of E and at least one point of $\sim E$, the complement of E;

x is an **exterior point** of E if some open ball with center x is contained in $\sim E$.

The sets of all interior, boundary, and exterior points of E are called, respectively, the **interior, boundary,** and **exterior** of E and are denoted by $\text{Int}(E)$, $\text{Bdy}(E)$, and $\text{Ext}(E)$.

Clearly, Int(E), Bdy(E), and Ext(E), constitute a partition of X; that is, they are pairwise disjoint and

$$X = \text{Int}(E) \cup \text{Bdy}(E) \cup \text{Ext}(E).$$

Also, $\text{Int}(E) \subset E$ and $\text{Ext}(E) \subset \sim E$; in fact, $\text{Ext}(E)$ is the interior of $\sim E$. A boundary point of E may be in either E or $\sim E$.

If E is a set such that all points of E are interior points of E, then E is said to be **open;** that is, E is open if $E = \text{Int}(E)$. Since any point of E is either an interior point or a boundary point of E, we may describe an open set as one that does not contain any of its boundary points.

A set E is said to be **closed** if its complement is open; that is, E is closed if $\sim E = \text{Ext}(E)$ or, equivalently, $E = \text{Int}(E) \cup \text{Bdy}(E)$. Thus, a set is closed if and only if it contains all of its boundary points.

If a set contains some but not all of its boundary points, it is neither open nor closed.

3 · 1 Proposition. *An open ball $B(a; r)$ is an open set.*

Proof. If $x \in B(a; r)$, then $d(a, x) < r$. Let $d(a, x) = s$. Then $B(x; r - s) \subset B(a; r)$. For, if $y \in B(x; r - s)$, then

$$d(a, y) \le d(a, x) + d(x, y) < s + (r - s) = r$$

and, hence, $y \in B(a; r)$. This shows that $B(a; r)$ is open. ∎

3 · 2 Proposition. If *E is any set in (X, d), then $\text{Int}(E)$ is open.*

Proof. For any $x \in \text{Int}(E)$ there exists an open ball $B(x; r) \subset E$. Then

$$B(x; r) = \text{Int}(B(x; r)) \subset \text{Int}(E)$$

and, hence, x is an interior point of $\text{Int}(E)$. ∎

Since $\text{Ext}(E)$ is the interior of $\sim E$, $\text{Ext}(E)$ is an open set. Thus, its complement, $\text{Int}(E) \cup \text{Bdy}(E)$, is a closed set, called the **closure** of E and denoted by $\bar{E}$. Since a set E is closed if and only if $E = \text{Int}(E) \cup \text{Bdy}(E)$, we may now state that E is closed if and only if $E = \bar{E}$.

3 · 3 Proposition. *If E is a set in the metric space (X, d), then $\bar{E} = \{x : d(x, E) = 0\}$.*

PROOF. Since $\bar{E} = \text{Int}(E) \cup \text{Bdy}(E)$, it is clear that $x \in \bar{E}$ if and only if every open ball with center x intersects E. Thus, if $x \in \bar{E}$, then for every $r > 0$ there exists a $y \in B(x; r) \cap E$ and, therefore, $d(x, E) = 0$. If $d(x, E) = 0$ then for every $r > 0$ there is a $y \in E$ such that $d(x, y) < r$; that is, $B(x; r) \cap E \neq \varnothing$. Thus, $x \in \bar{E}$. ∎

In any metric space (X, d) the empty set is open since each of its points (there are none) is an interior point. Clearly the whole space X is also open. The other basic properties of open sets are given in the following two theorems.

3 · 4 Theorem. *The union of an arbitrary collection of open sets, $\{E_s : s \in S\}$, in a metric space is open.*

PROOF. If $x \in \bigcup_{s \in S} E_s$, then $x \in E_s$ for some $s \in S$. Since E_s is open, there exists an open ball $B(x; r)$ such that

$$B(x; r) \subset E_s \subset \bigcup_{s \in S} E_s. \blacksquare$$

3 · 5 Theorem. *The intersection of a finite collection $\{E_i : i = 1, \cdots n\}$ of open sets in a metric space is open.*

PROOF. If $x \in \bigcap_{i=1}^{n} E_i$, then $x \in E_i$ for each $i = 1, \cdots, n$. Since E_i is open, there exists an open ball $B(x; r_i) \subset E_i$. If $r = \min\{r_i : i = 1, \cdots, n\}$, then

$$B(x; r) \subset \bigcap_{i=1}^{n} E_i. \blacksquare$$

In general, the intersection of an infinite collection of open sets need not be open. For example, the intersection of the open intervals $(-1/n, 1/n)$ in $\mathbf{R}$, where n is any positive integer, is the set $\{0\}$ which is not open.

Since $\varnothing = \sim X$ and $X = \sim\varnothing$ we see that the empty set and the whole space X are closed as well as open sets. Using DeMorgan's laws we obtain from Theorems 3 · 4 and 3 · 5 corresponding statements for closed sets.

3 · 6 Theorem. *The intersection of an arbitrary collection of closed sets is closed.*

3 · 7 Theorem. *The union of a finite collection of closed sets is closed.*

3 · 8 Propositon. *In a discrete space X every set is open and, therefore, every set is also closed.*

PROOF. We show this by proving that every singleton $\{x\} \subset X$ is open and then applying Theorem 3 · 4. Since $B(x; 1) \subset \{x\}$, the set $\{x\}$ is open. Each set $E \subset X$ is the union of singletons and is therefore open. Also, for any set $E \subset X$ the set $\sim E$ is open and, hence, E is closed. ▮

If A is a nonempty subset of a metric space X, an open set which contains A is called an **open neighborhood** of A. Any set containing an open neighborhood of A is called a **neighborhood** of A. If A is the singleton $\{x\}$, we speak of neighborhoods of the point x. Clearly, an open ball with center x is an open neighborhood of x. Also, if N is any neighborhood of x, then there exists an open ball $B(x; r)$ such that $B(x; r) \subset N$.

3 · 9 Proposition. *Any open set G in a metric space X is the union of a collection of open balls.*

PROOF. For any point $x \in G$ there exists an open ball $B(x; r_x)$ such that $B(x; r_x) \subset G$. Then

$$G = \bigcup_{x \in G} B(x; r_x). \quad ▮$$

A collection $\mathscr{B} = \{G_s : s \in S\}$ of nonempty open sets is called an **open base** for the collection of open sets in a metric space if each open set is the union of a subset of $\mathscr{B}$. Proposition 3 · 9 states that the collection of open balls is an open base.

A set E in a metric space (X, d) is said to be **dense** in X if $\bar{E} = X$. A metric space with a countable dense subset is said to be **separable.** For example, the set of m-tuples of rational numbers is a countable dense set in $\mathbf{R}^m$ and, hence, $\mathbf{R}^m$ is a separable space (Problem 9).

3 · 10 Proposition. *If X is a separable metric space, then X has a countable open base consisting of open balls.*

PROOF. Let E be a countable dense set in X. Then the set of open balls with a point of E as center and rational radius is countable. Let G be an open set in X and take $x \in G$. There exists a rational number $r(x)$ such that $B(x; r(x)) \subset G$. Since E is dense in X there is a point $y(x) \in E$ such that $d(x, y(x)) < \frac{1}{2}r(x)$. Then

$$x \in B(y(x); \tfrac{1}{2}r(x)) \subset B(x; r(x)) \subset G.$$

Thus $G = \bigcup_{x \in G} B(y(x); \frac{1}{2}r(x))$. ▮

PROBLEMS

1. In the metric space **R** show that:

a. Any open interval of the form (a, b), (a, ∞), or $(-\infty, b)$ is an open set.

b. A closed interval $[a, b]$ is a closed set.

c. Any interval of the form $[a, \infty)$ is a closed set.

2. If A is a closed set in a metric space (X, d) and $x \notin A$, show that $d(x, A) > 0$.

3. In a metric space (X, d) a **closed ball** with center x and radius r is defined to be the set

$$B[x; r] = \{y : d(x, y) \leq r\}.$$

a. Show that $B[x; r]$ is a closed set.

b. Give an example where $B[x; r] \neq \overline{B(x; r)}$.

4. Let E be a set in a metric space.

a. Show that $\text{Int}(E)$ is the largest open set contained in E; that is, that $\text{Int}(E)$ is an open subset of E and if F is an open subset of E then $F \subset \text{Int}(E)$.

b. Show that $\bar{E}$ is the smallest closed set containing E.

5. If A and B are sets in a metric space, show that:

a. $A \subset B$ implies $\text{Int}(A) \subset \text{Int}(B)$ and $\bar{A} \subset \bar{B}$.

b. $\text{Int}(A) \cup \text{Int}(B) \subset \text{Int}(A \cup B)$ and $\text{Int}(A) \cap \text{Int}(B) = \text{Int}(A \cap B)$.

c. $\overline{(A \cup B)} = \bar{A} \cup \bar{B}$ and $\overline{(A \cap B)} \subset \bar{A} \cap \bar{B}$.

d. Give an example of two sets A and B in a metric space such that $\text{Int}(A) \cup \text{Int}(B) \neq \text{Int}(A \cup B)$.

6. If A is a nonempty set in a metric space X and if $r > 0$, show that

$$N(A; r) = \{x \in X : d(x, A) < r\}$$

is an open neighborhood of A.

7. If A is open and $A \cap \bar{B} \neq \varnothing$, show that $A \cap B \neq \varnothing$.

8. Show that $\text{Bdy}(\bar{A}) \subset \text{Bdy}(A)$ and $\text{Bdy}(\text{Int}(A)) \subset \text{Bdy}(A)$. Give an example on the real line where these three sets are distinct.

9. Show that $\mathbf{R}^m$ is a separable metric space.

10. Show that a collection $\{G_s : s \in S\}$ of nonempty open sets in a metric space X is an open base if and only if for each open set G and for each $x \in G$ there exists an $s \in S$ such that $x \in G_s \subset G$.

11. If X is a metric space with a countable open base, show that X is separable.

4 · CONTINUITY

4 · 1 Definition. *Let (X, d_1) and (Y, d_2) be two metric spaces. A function f from X into Y is said to be* ***continuous at the point*** *$a \in X$ if for each $\varepsilon > 0$ there* *exists a* $\delta > 0$ *such that*

$$d_2(f(x), f(a)) < \varepsilon \quad \textit{whenever} \quad d_1(x, a) < \delta.$$

This definition can also be given in the following equivalent forms. The equivalence can be easily checked by the reader.

(1) A function f from X into Y is continuous at $a \in X$ if for each $\varepsilon > 0$ there exists a $\delta > 0$ such that

$$f(B(a; \delta)) \subset B(f(a); \varepsilon).$$

(2) A function f from X into Y is continuous at $a \in X$ if for each neighborhood N of $f(a)$ in Y there exists a neighborhood M of a in X such that

$$f(M) \subset N.$$

(3) A function f from X into Y is continuous at $a \in X$ if, for each neighborhood N of $f(a)$ in Y, $f^{-1}(N)$ is a neighborhood of a in X.

We say that a function is continuous if it is continuous at each point of its domain. Also a function f is continuous on a subset E of its domain if the restricted function $f \mid E$ is continuous.

If f is a continuous mapping of the metric space (X, d_1) into (Y, d_2) and if E is a subset of X, then it is easy to see that $f \mid E$ is a continuous mapping of (E, d_1') into (Y, d_2) where d_1' is the restriction of d_1 to $E \times E$.

We now give a characterization of continuous mappings from one metric space into another in terms of open sets. This characterization is taken as a point of departure for further generalization of the concept of continuity to mappings from one topological space to another.

4·2 Theorem. *Let f be a function from a metric space X into a metric space Y. Then f is continuous if and only if $f^{-1}(G)$ is open in X whenever G is open in Y.*

PROOF. First assume that f is continuous and let G be an open set in Y. If $f^{-1}(G) = \varnothing$ then $f^{-1}(G)$ is open. Assuming $f^{-1}(G) \neq \varnothing$, take $x \in f^{-1}(G)$. Then $f(x) \in G$ and, since G is open, there exists an open neighborhood $B(f(x); \varepsilon) \subset G$. The continuity of f at x implies that there exists an open neighborhood $B(x; \delta)$ such that

$$f(B(x; \delta)) \subset B(f(x); \varepsilon);$$

that is,

$$B(x; \delta) \subset f^{-1}(B(f(x); \varepsilon)) \subset f^{-1}(G).$$

Thus $f^{-1}(G)$ is open.

Now assume $f^{-1}(G)$ is open in X whenever G is open in Y and let x be any point in X. The ball $B(f(x); \varepsilon)$ is open in Y and hence $f^{-1}(B(f(x); \varepsilon))$ is open in X. Thus, there exists an open ball $B(x; \delta)$ such that $B(x; \delta) \subset f^{-1}(B(f(x); \varepsilon))$. This shows that f is continuous. ∎

We will use this characterization of continuity to prove an important property of metric spaces. First we prove a lemma.

4 · 3 Lemma. *If A is a nonempty set in the metric space (X, d), then the function f defined by $f(x) = d(x, A)$ is a continuous function from X into* **R**.

PROOF. Take $x \in X$. We will show that

$$|d(x, A) - d(y, A)| \le d(x, y) \qquad \text{for each } y \in X.$$

For each $z \in A$, $d(x, z) \le d(x, y) + d(y, z)$. Hence,

$$\begin{aligned} d(x, A) &= \inf_{z \in A} d(x, z) \le \inf_{z \in A} [d(x, y) + d(y, z)] \\ &= d(x, y) + \inf_{z \in A} d(y, z) = d(x, y) + d(y, A). \end{aligned}$$

Similarly, $d(y, A) \le d(x, y) + d(x, A)$ and, therefore,

$$|d(x, A) - d(y, A)| \le d(x, y).$$

Hence, for any $\varepsilon > 0$

$$|f(x) - f(y)| < \varepsilon \quad \text{whenever} \quad d(x, y) < \varepsilon. \blacksquare$$

4 · 4 Theorem. *In a metric space (X, d) any two nonempty disjoint closed sets have disjoint open neighborhoods.*

PROOF. Let A and B be nonempty disjoint closed sets and let $U = \{x \in X : d(x, A) < d(x, B)\}$ and $V = \{x \in X : d(x, B) < d(x, A)\}$. Clearly U and V are disjoint. To show that $A \subset U$, take $x \in A$. Then $x \notin B$ and, since B is closed, $d(x, B) > 0 = d(x, A)$. Thus, $x \in U$ and $A \subset U$. Similarly, $B \subset V$. It remains to show that U and V are open. Let

$$g(x) = d(x, A) - d(x, B).$$

Then g is a continuous function from X into **R** and, since $U = g^{-1}((-\infty, 0))$, U is open. Similarly, V is open. ▮

Continuity of a function can be characterized in terms of closed sets as well as open sets.

4 · 5 Theorem. *The function f from the metric space X into the metric space Y is continuous if and only if $f^{-1}(F)$ is closed in X whenever F is closed in Y.*

The proof is left to the reader (Problem 4).

By Theorems 4 · 2 and 4 · 5, under a continuous mapping the inverse image of an open set is open and the inverse image of a closed set is closed. It is not true in general that a continuous function maps open sets onto open sets and closed sets onto closed sets. For example, if $f(x) = x^2$ then f maps $(-1, 1)$ onto $[0, 1)$ and if $g(x) = 1/x$ then g maps $[1, \infty)$ onto $(0, 1]$.

A function which maps open sets onto open sets is called an **open mapping**.

4 · 6 Definition. *A function f from a metric space X into a metric space Y is said to be a* ***homeomorphism*** *if it is one-to-one, continuous, and open. Two metric spaces X and Y are said to be* ***homeomorphic*** *if there exists a homeomorphism from X* onto *Y.*

Since a homeomorphism f is one-to-one, it has an inverse f^{-1}. In such a case the statement that f is open is equivalent to the statement that f^{-1} is continuous. Thus, a homeomorphism f may be described as a one-to-one bicontinuous function where bicontinuous means both f and f^{-1} are continuous. It is easy to show that a one-to-one continuous mapping is a homeomorphism if and only if it maps closed sets onto closed sets.

An isometry f of X onto Y is a one-to-one continuous mapping of X onto Y. Since f^{-1} is also an isometry, f^{-1} is continuous. Thus, an isometry is a homeomorphism. However, a homeomorphism need not be an isometry (Problem 6).

The notion of uniform continuity can also be defined for functions from a metric space into a metric space.

4 · 7 Definition. *A function f from a metric space* (X, d_1) *into a metric space* (Y, d_2) *is said to be* ***uniformly continuous on a set*** $E \subset X$ *if for each* $\varepsilon > 0$ *there exists a* $\delta > 0$ *such that*

$$d_2(f(x), f(y)) < \varepsilon \quad \textit{whenever} \quad x, y \in E \quad \textit{and} \quad d_1(x, y) < \delta.$$

Clearly if f is unformly continuous on E then f is continuous on E, but the converse is not true in general.

PROBLEMS

1. Show the equivalence of the different forms of the definition of continuity given on p. 65.

2. Show that the following functions from $\mathbf{R}^2$ to $\mathbf{R}$ are continuous:

a. $f(x, y) = x + y$

b. $f(x, y) = xy$.

3. Show that any function from a discrete metric space X into a metric space Y is continuous.

4. Prove Theorem 4 · 5.

5. Prove that a function f from the metric space X into the metric space Y is continuous if and only if for each $A \subset X$, $f(\bar{A}) \subset \overline{f(A)}$.

6. Let X be the set of all m-tuples of real numbers and let

$$d(x, y) = \max\{|\alpha^k - \beta^k| : k = 1, \cdots, m\},$$

where $x = (\alpha^1, \cdots, \alpha^m)$ and $y = (\beta^1, \cdots, \beta^m)$. Show that (X, d) and the Euclidean space $\mathbf{R}^m$ are homeomorphic but not isometric under the identity mapping I if $m > 1$.

7. Show that **R** is homeomorphic to its subspace (0, 1).

8. Let f be a continuous real-valued function on a metric space X, $c \in \mathbf{R}$, and $E = \{x \in X : f(x) = c\}$. Show that E is a closed set.

9. If f is a one-to-one mapping from the metric space X into the metric space Y, show that f is a homeomorphism if and only if, for each $A \subset X$, $f(\bar{A}) = \overline{f(A)}$.

10. If X is separable and Y is homeomorphic to X, show that Y is separable.

11. If x_0 is a point in the metric space (X, d), show that the function $f: X \to \mathbf{R}$ defined by $f(x) = d(x, x_0)$ is uniformly continuous on X.

12. Show that the function $f: \mathbf{R} \to \mathbf{R}$ defined by $f(x) = x^2$ is uniformly continuous on any interval $[a, b]$ but is not uniformly continuous on **R**.

13. If $f: (X, d_1) \to (Y, d_2)$ and $g: (Y, d_2) \to (Z, d_3)$ are uniformly continuous on X and Y, respectively, show that $g \circ f$ is uniformly continuous on X.

5 · TOPOLOGICAL SPACES

In Theorem 4 · 2 we gave a characterization of continuity in terms of open sets. This suggests that in defining continuity we do not really need a metric but that a notion of open set would suffice. A set with a distinguished collection of subsets having the basic properties of the collection of open sets in a metric space is called a topological space.

5 · 1 Definition. *Let X be any nonempty set and let $\mathscr{T}$ be a collection of subsets of X satisfying the following conditions:*

(1) *the empty set and X itself belong to $\mathscr{T}$;*
(2) *the intersection of any finite collection of sets in $\mathscr{T}$ is a set in $\mathscr{T}$;*
(3) *the union of any collection of sets in $\mathscr{T}$ is a set in $\mathscr{T}$.*

The collection $\mathscr{T}$ is called a ***topology*** *for X and the pair $(X, \mathscr{T})$ is called a* ***topological space****.*

The elements of X are called the **points** of the topological space $(X, \mathscr{T})$ and the members of the collection $\mathscr{T}$ are called the **open sets** of the space. Frequently a topological space $(X, \mathscr{T})$ is denoted simply as X.

Since we patterned the properties of the collection $\mathscr{T}$ on those of the collection of open sets in a metric space, it is clear that a metric space (X, d) becomes a topological space $(X, \mathscr{T})$ when we take as the topology the collection of open sets with respect to the metric. Speaking somewhat imprecisely, we may say that a metric space is a topological space. The converse of this is not true in general. A topological space $(X, \mathscr{T})$ is said to be **metrizable** if a metric can be defined on X in such a way that the open sets defined in terms of this metric coincide with the open sets in $\mathscr{T}$. Below we give an example of a topological space which is not metrizable.

Let X be any set. The collection $\mathscr{T} = \{\varnothing, X\}$ is a topology for X, called the **indiscrete** (or **trivial**) **topology**. The space $(X, \mathscr{T})$ is then called an **indiscrete topological space**. At the other extreme, the collection $\mathscr{T}$ of all subsets of X is a topology on X. This is called the **discrete topology** for X and the space $(X, \mathscr{T})$ is called a **discrete topological space.** Clearly a discrete topological space is metrizable, using the discrete metric, and corresponds to a discrete metric space. However, an indiscrete topological space may fail to be metrizable. Let $X = \{a, b\}$, where $a \neq b$, and suppose d is a metric on X. Then $d(a, b) > 0$. If $0 < r < d(a, b)$, then $B(a; r) = \{a\}$ and, hence, $\{a\}$ is an open set. This shows that the indiscrete topological space X is not metrizable if X contains two distinct points.

If $\mathscr{T}_1$ and $\mathscr{T}_2$ are two topologies on a set X and $\mathscr{T}_1 \subset \mathscr{T}_2$, then $\mathscr{T}_1$ is said to be **weaker** than $\mathscr{T}_2$ and $\mathscr{T}_2$ is said to be **stronger** than $\mathscr{T}_1$. Since every topology on X contains $\varnothing$ and X, the indiscrete topology is the weakest topology for X. The discrete topology is the strongest topology for X.

Let $(X, \mathscr{T})$ be a topological space and let E be a nonempty subset of X. The subset E can be topologized in a natural way. Let $\mathscr{T}'$ be the collection of all intersections with E of sets of $\mathscr{T}$. Then $\mathscr{T}'$ is a topology for E and is called the **subspace topology** for E. The topological space $(E, \mathscr{T}')$ is called a **subspace** of $(X, \mathscr{T})$.

5 · 2 Definition. *A function f from a topological space X into a topological space Y is* ***continuous*** *if, for each open set B in Y, $f^{-1}(B)$ is an open set in X.*

If E is a nonempty set in the topological space X, then f is said to be continuous on E if $f \mid E$, the restriction of f to E, is continuous in the subspace topology on E.

Let f be a continuous function from a topological space X into a topological space Y and let E be a nonempty subset of X. Then f is continuous on the subspace E. For, if B is an open set in Y, then $(f \mid E)^{-1}(B) = f^{-1}(B) \cap E$ is an open set in the subspace topology for E.

Defining an **open mapping** to be a function which maps open sets onto open sets, we can define, a homeomorphism as follows.

5 · 3 Definition. *A function f mapping a topological space X into a topological space Y is said to be a* ***homeomorphism*** *if it is a one-to-one continuous open mapping. Two topological spaces X and Y are said to be* ***homeomorphic*** *if there exists a homemorphism of X* onto *Y.*

If the topological spaces X and Y are homeomorphic, then there is a one-to-one correspondence between the points and the open sets of one space and those of the other. Thus, the two spaces are not essentially different as topological spaces and may be identified if we so choose. In topological spaces homeomorphisms play the rôle played by isometries in the case of metric spaces. We call a property a **topological property** if whenever it holds in a given space then it holds in every space which is homeomorphic to the given space. For example, neighborhood, interior, boundary, exterior, closure, and denseness are topological properties.

For metric spaces we know that an isometry is a homeomorphism but a homeomorphism is not necessarily an isometry. Let (X, d_1) and (X, d_2) be two metric spaces with the same set of points. If the identity mapping of X onto itself is a homeomorphism of (X, d_1) onto (X, d_2), then d_1 and d_2 are said to be **(topologically) equivalent** metrics on X. In such a case the two metrics determine the same family $\mathscr{T}$ of open sets and hence the same topological space $(X, \mathscr{T})$. For example, $\mathbf{R}^m$ and the space (X, d), where X is the set of all m-tuples of real numbers and, for $x = (\alpha^1, \cdots, \alpha^m)$ and $y = (\beta^1, \cdots, \beta^m)$, $d(x, y) = \max\{|\alpha^k - \beta^k| : k = 1, \cdots, m\}$, are homeomorphic under the identity mapping of X onto itself (Problem 6, p. 67). Hence, as topological spaces, these two spaces are the same. In particular, this means that if X is the set of all ordered pairs of real numbers then we get the same topological space whether we use the interior of circles or the interior of squares as the open balls.

Another type of equivalence which is important in certain considerations is uniform equivalence. The metrics d_1 and d_2 on X are said to be **uniformly equivalent** if the identity mapping of (X, d_1) onto (X, d_2) and its inverse are uniformly continuous.

If X_1 and X_2 are topological spaces there is a standard way of topologizing $X_1 \times X_2$ and the resulting topological space is called the product space, also denoted by $X_1 \times X_2$. We will not consider the general problem of topologizing $X_1 \times X_2$ but will limit ourselves to the case where X_1 and X_2 are metric spaces. Suppose (X_1, d_1) and (X_2, d_2) are metric spaces. If $x = (x_1, x_2)$ and $y = (y_1, y_2)$ are points in $X_1 \times X_2$, define:

$$(5 \cdot 4) \qquad d(x, y) = d_1(x_1, y_1) + d_2(x_2, y_2)$$

$$(5 \cdot 5) \qquad d'(x, y) = \max\{d_1(x_1, y_1), d_2(x_2, y_2)\}.$$

It is easy to check that d and d' are metrics on $X_1 \times X_2$. Also, since

$$d'(x, y) \leq d(x, y) \leq 2d'(x, y),$$

these two metrics are topologically equivalent (Problem 6). The topological space corresponding to either $(X_1 \times X_2, d)$ or $(X_1 \times X_2, d')$ is the **product space** $X_1 \times X_2$.

We now generalize to a topological space $(X, \mathscr{T})$ some of the concepts we have discussed in the context of metric spaces. A **closed set** is the complement of an open set. A **neighborhood** of a set E is a set which contains an open set containing E. An **interior point** of a set E is a point such that E is a neighborhood of the point. An **exterior point** of E is a point such that $\sim E$ is a neighborhood of the point. A **boundary point** of E is a point such that neither E nor $\sim E$ is a neighborhood of the point.

For our purposes metric spaces are sufficiently general and we will not develop further the theory of topological spaces. However, even though we will be working in the context of metric spaces, in many instances it turns out that a topological approach is easier than a metric approach. For example, the proof that the composition of continuous functions is continuous is neater in topological form.

5 · 6 Proposition. *Let X, Y, and Z be topological spaces, f be a continuous function from X into Y, and g be a continuous function from Y into Z. Then $g \circ f$ is a continuous function from X into Z.*

PROOF. If B is open in Z, then $g^{-1}(B)$ is open in Y and $[g \circ f]^{-1}(B) = f^{-1}(g^{-1}(B))$ is open in X. ∎

PROBLEMS

1. Let X be the set of all positive integers and let $\mathscr{T}$ be the collection consisting of $\varnothing$ and all sets which contain all but a finite number of elements of X. Show that $(X, \mathscr{T})$ is a topological space. Is this space metrizable?

2. If f is a function from a topological space X into an indiscrete topological space Y, show that f is continuous.

3. If $\mathscr{T}_1$ and $\mathscr{T}_2$ are two topologies for X, show that $\mathscr{T}_1 \cap \mathscr{T}_2$ is a topology for X.

4. If E is a set in a topological space X, the **closure** of E is defined to be the set of points $x \in X$ such that every neighborhood of x contains a point of E. Show that:

- **a.** $\bar{E} = E \cup \operatorname{Bdy}(E)$
- **b.** E is closed if and only if $E = \bar{E}$.

5. Let X_1 and X_2 be metric spaces and let $X_1 \times X_2$ be the product space with metric defined by 5 · 4 or 5 · 5.

- **a.** If $E_1 \subset X_1$ and $E_2 \subset X_2$, show that $\overline{E_1 \times E_2} = \bar{E}_1 \times \bar{E}_2$.
- **b.** Show that $X_1 \times X_2$ is separable if and only if X_1 and X_2 are separable.

6. Show that the metric spaces (X, d_1) and (X, d_2) are topologically equivalent, in fact, uniformly equivalent if there exist positive numbers a and b such that for all x and y in X

$$ad_2(x, y) \leq d_1(x, y) \leq bd_2(x, y).$$

7. If (X_1, d_1) and (X_2, d_2) are metric spaces and d is the metric on $X_1 \times X_2$ defined by 5 · 4, define the projections

$$P_1 : (X_1 \times X_2, d) \to (X_1, d_1) \quad \text{and} \quad P_2 : (X_1 \times X_2, d) \to (X_2, d_2)$$

as follows:

$$P_1(x_1, x_2) = x_1 \quad \text{and} \quad P_2(x_1, x_2) = x_2.$$

- **a.** Show that P_1 and P_2 are uniformly continuous on $X_1 \times X_2$.
- **b.** If f is a function from $(Y, \tilde{d})$ into $(X_1 \times X_2, d)$ show that f is continuous if and only if $P_1 \circ f$ and $P_2 \circ f$ are continuous.

8. On **C** show that the Euclidean metric $d(z_1, z_2) = |z_1 - z_2|$ and the chordal metric (Problem 9, p. 60)

$$d'(z_1, z_2) = \frac{|z_1 - z_2|}{(1 + |z_1|^2)^{1/2}(1 + |z_2|^2)^{1/2}}$$

are equivalent.

Suggestion. If I is the identity from $(\mathbf{C}, d)$ to $(\mathbf{C}, d')$ it is easy to show that I is continuous. To show that I^{-1} is continuous at a point z_0 first show that z is bounded if $d'(z, z_0)$ is sufficiently small; e.g.

$$d'(z, z_0) < \frac{1}{\sqrt{2}(1 + |z_0|^2)^{1/2}} \quad \Rightarrow \quad |z| < 2|z_0| + (1 + 2|z_0|^2)^{1/2}.$$

9. Given a set X, let $\mathscr{S}$ be a collection of subsets of X. Let $\mathscr{B}$ be the collection of all sets which are the intersection of a finite subcollection of $\mathscr{S}$. Let $\mathscr{T}$ be the collection of all sets which are the union of a subcollection of $\mathscr{B}$. Show that $\mathscr{T}$ is a topology for X. This is called the topology generated by $\mathscr{S}$.

10. Let f be a function from the set X into the topological space $(Y, \mathscr{T}')$. Let $\mathscr{S} = \{f^{-1}(G) : G \in \mathscr{T}'\}$. If $\mathscr{T}$ is the topology generated by $\mathscr{S}$ (Problem 9), show that f is a continuous function from $(X, \mathscr{T})$ into $(Y, \mathscr{T}')$. Also show that if f is a continuous function from $(X, \mathscr{T}_1)$ into $(Y, \mathscr{T}')$, then $\mathscr{T}_1 \supset \mathscr{T}$.

6 · CONVERGENCE AND COMPLETENESS

Returning now to metric spaces we consider the convergence of sequences for such spaces. A sequence of points in a metric space (X, d) is a function from the set of positive integers into X.

6 · 1 Definition. *A point b in X is the* **limit** *of the sequence (x_n) in the metric space (X, d), written* $\lim x_n = b$, *if for each $\varepsilon > 0$ there exists a positive integer n_0 such that $d(x_n, b) < \varepsilon$ whenever $n \geq n_0$.*

If a sequence has a limit it is said to be **convergent** and if it is not convergent it is said to be **divergent.**

From the definition it is clear that $\lim x_n = b$ if and only if $\lim d(x_n, b) = 0$. It is easy to show that the limit of a convergent sequence in a metric space is unique (Problem 1).

The definition of the limit of a sequence can be cast in topological terms as follows: the point $b \in X$ is the limit of the sequence (x_n) in (X, d) if for each neighborhood N of b there exists a positive integer n_0 such that $x_n \in N$ whenever $n \geq n_0$. This topological definition can be taken as the definition of limit of a sequence in a topological space. However, in general topological spaces, many of the familiar properties of limits do not hold. For example, a sequence may converge to more than one point (Problem 6). For metric spaces the usual properties are satisfied.

It follows readily from the definition that the continuous image of a convergent sequence is a convergent sequence.

6 · 2 Proposition. *If f is a continuous function from the metric space X into the metric space Y and (x_n) is a sequence of points in X which converges to x, then the sequence $(f(x_n))$ converges to $f(x)$ in Y.*

PROOF. Take $\varepsilon > 0$. The continuity of f implies that there is a $\delta > 0$ such that $f(B(x; \delta)) \subset B(f(x); \varepsilon)$. Since $\lim x_n = x$, there is a positive integer n_0 such that $x_n \in B(x; \delta)$ whenever $n \geq n_0$. Then, if $n \geq n_0$, $f(x_n) \in B(f(x); \varepsilon)$. ▌

We define a Cauchy sequence in a metric space (X, d) in the expected way.

6 · 3 Definition. *A sequence (x_n) of points in a metric space (X, d) is called a **Cauchy sequence** if for each $\varepsilon > 0$ there exists a positive integer n_0 such that*

$$d(x_n, x_m) < \varepsilon \quad \textit{whenever} \quad n, m \geq n_0.$$

6 · 4 Proposition. *A convergent sequence in a metric space is a Cauchy sequence.*

PROOF. If $\lim x_n = b$, then for each $\varepsilon > 0$ there exists a positive integer n_0 such that $d(x_n, b) < \varepsilon/2$ whenever $n \geq n_0$. Thus, if $n, m \geq n_0$

$$d(x_n, x_m) \leq d(x_n, b) + d(b, x_m) < \varepsilon. \ ▌$$

The converse of this theorem does not hold in every metric space.

6 · 5 Definition. *A metric space X is **complete** if every Cauchy sequence in X converges to a point in X.*

We have shown that the real line **R** is a complete metric space. However, the subspace (0, 1) of **R** is not complete; the sequence $(1/n)$ is a Cauchy sequence in (0, 1) which does not converge to a point in this space. Since **R** and (0, 1) are homeomorphic, completeness is not a topological property.

We now prove the completeness of some important spaces.

6 · 6 Proposition. *The Euclidean space $\mathbf{R}^m$ is a complete metric space.*

PROOF. Let (x_n) be a Cauchy sequence in $\mathbf{R}^m$ where $x_n = (\alpha_n^1, \cdots, \alpha_n^m)$. Then for each $\varepsilon > 0$ there exists a positive integer n_0 such that

$$d(x_r, x_s) = \left[\sum_{i=1}^{m} (\alpha_r^i - \alpha_s^i)^2\right]^{1/2} < \varepsilon \quad \text{whenever } r, s \geq n_0.$$

Thus, for each $i = 1, \cdots, m$,

$$|\alpha_r^i - \alpha_s^i| < \varepsilon \quad \text{whenever } r, s \geq n_0.$$

That is, each component sequence $(\alpha_n{}^i)$ is a Cauchy sequence in $\mathbf{R}$. Since $\mathbf{R}$ is complete there exists a real number α^i such that $\alpha^i = \lim_{n\to\infty} \alpha_n{}^i$. We now show that the point $x = (\alpha^1, \cdots, \alpha^m)$ is the limit of the sequence (x_n). Take $\varepsilon > 0$. Then, for each $i = 1, \cdots, m$, there exists a positive integer n_i such that

$$|\alpha_n{}^i - \alpha^i| < \varepsilon/\sqrt{m} \quad \text{whenever} \quad n \geq n_i.$$

Hence, if $n \geq \max\{n_i : i = 1, \cdots, m\}$, then

$$d(x_n, x) = \left[\sum_{i=1}^{m} (\alpha_n{}^i - \alpha^i)^2\right]^{1/2} < \left[\sum_{i=1}^{m} \frac{\varepsilon^2}{m}\right]^{1/2} = \varepsilon. \blacksquare$$

Consider now convergence in the metric space $C[a, b]$. This is the space of continuous real-valued functions defined on $[a, b]$ with metric

$$d(f, g) = \max\{|f(t) - g(t)| : t \in [a, b]\}.$$

In this space $\lim f_n = f$ means: for each $\varepsilon > 0$ there exists a positive integer n_0 such that

$$\max\{|f_n(t) - f(t)| : t \in [a, b]\} < \varepsilon \quad \text{whenever} \quad n \geq n_0;$$

that is, for all $t \in [a, b]$

$$|f_n(t) - f(t)| < \varepsilon \quad \text{whenever} \quad n \geq n_0.$$

Thus, convergence of (f_n) to f in $C[a, b]$ is equivalent to uniform convergence of (f_n) to f on $[a, b]$.

6 · 7 Proposition. *The metric space $C[a, b]$ is complete.*

PROOF. Let (f_n) be a Cauchy sequence in $C[a, b]$. Then for $\varepsilon > 0$ there exists an n_0 such that for all $t \in [a, b]$

$$(6 \cdot 8) \qquad |f_n(t) - f_m(t)| < \varepsilon \quad \text{whenever } n, m \geq n_0.$$

Thus, for each $t \in [a, b]$, $(f_n(t))$ is a Cauchy sequence in $\mathbf{R}$ and, hence, converges to a real number $f(t)$. Taking the limit with respect to m in 6 · 8 we obtain, for all $t \in [a, b]$,

$$|f_n(t) - f(t)| \leq \varepsilon \quad \text{whenever} \quad n \geq n_0.$$

Thus (f_n) converges uniformly to f on $[a, b]$. Since each of the functions f_n is continuous on $[a, b]$, f is continuous on $[a, b]$; that is, $f \in C[a, b]$ (Problem 13). Thus, we have shown that (f_n) converges to f in $C[a, b]$. $\blacksquare$

A set E in a metric space X is said to be complete if the subspace E of X is complete. We will establish a connection between a set being complete and being closed. Before doing that we give a characterization of closed sets in terms of the convergence of sequences.

6 · 9 Proposition. *A set E in a metric space X is closed if and only if every sequence of points in E which is convergent in X converges to a point in E.*

PROOF. Suppose E is closed and let (x_n) be a sequence of points in E such that $\lim x_n = b$. Then every neighborhood of b intersects E and, hence, $b \in \bar{E} = E$.

Suppose now that E has the property that $b \in E$ whenever (x_n) is a sequence of points in E which converges to b. Take any point $x \in \bar{E}$. Then, for each positive integer n, $B(x; 1/n) \cap E \neq \varnothing$. Choose a point $x_n \in B(x; 1/n) \cap E$. The sequence (x_n) so chosen converges to x and, hence, $x \in E$. Thus, $\bar{E} \subset E$ and E is closed. ▮

6 · 10 Proposition. *If E is a complete subspace of a metric space X, then E is closed in X.*

PROOF. Take $x \in \bar{E}$. Select a point $x_n \in B(x; 1/n) \cap E$. Then (x_n) converges to x in X. Since (x_n) is a Cauchy sequence in E, it must converge to a point $y \in E$. But the limit of a convergent sequence is unique and, hence, $y = x$. Thus, $x \in E$ and E is closed. ▮

6 · 11 Proposition. *A closed set in a complete metric space is complete.*

PROOF. Let E be a closed subset of X. Any Cauchy sequence of points in E is convergent in X and, hence, converges to a point in E (Proposition 6 · 9). Thus, E is complete. ▮

Propositions 6 · 10 and 6 · 11 characterize closed sets in a complete metric space: A set E in a complete metric space X is closed if and only if E is complete.

PROBLEMS

1. Show that a convergent sequence in a metric space has a unique limit.

2. If $\lim x_n = x$ and $\lim y_n = y$, show that $\lim d(x_n, y_n) = d(x, y)$.

3. If $\lim x_n = b$ and (x_{n_k}) is a subsequence of (x_n), show that $\lim x_{n_k} = b$.

4. Define a point x in a metric space X to be a **limit point** of a sequence (x_n) if there exists some subsequence (x_{n_k}) of (x_n) which converges to x.

- **a.** Show that x is a limit point of a sequence (x_n) if and only if every neighborhood of x contains infinitely many terms of (x_n).
- **b.** If (x_n) converges, show that (x_n) has only one limit point. Give an example of a sequence with only one limit point which does not converge.
- **c.** Show that if x is a limit point of a Cauchy sequence (x_n), then (x_n) converges to x.

5. If (x_n) and (y_n) are convergent sequences in **R**, show that:

a. $\lim(x_n + y_n) = \lim x_n + \lim y_n$

b. $\lim x_n y_n = (\lim x_n)(\lim y_n)$.

Suggestion. Use Problem 2, p. 67.

6. Show that in an indiscrete topological space X every sequence (x_n) converges to every point $x \in X$.

7. Show that a discrete metric space is complete.

8. Let X be the set of all m-tuples of real numbers and let d_∞ and d_1 be the metrics defined by

$$d_\infty(x, y) = \max\{|\alpha^i - \beta^i| : i = 1, \cdots, m\}$$

and

$$d_1(x, y) = \sum_{i=1}^{m} |\alpha^i - \beta^i|$$

where $x = (\alpha^1, \cdots, \alpha^m)$ and $y = (\beta^1, \cdots, \beta^m)$. Show that (X, d_∞) and (X, d_1) are complete.

9. Let X be the set of all continuous functions from $[a, b]$ into $\mathbf{R}^n$ and let d be defined by

$$d(f, g) = \max\{|f(t) - g(t)| : t \in [a, b]\}.$$

Show that (X, d) is a complete metric space.

10. Let X be the set of all bounded sequences $x = (\alpha^i)$ of real numbers and, for $x = (\alpha^i)$ and $y = (\beta^i)$ let

$$d(x, y) = \sup\{|\alpha^i - \beta^i| : i \in \mathbf{N}\}.$$

Show that (X, d) is a complete metric space.

11. Let f be a function from the metric space X into the metric space Y. Suppose whenever a sequence (x_n) converges to x then $(f(x_n))$ converges to $f(x)$. Show that f is continuous at x.

12. A sequence (B_n) of balls is said to be **nested** if $B_n \supset B_{n+1}$ for all n. Show that a metric space X is complete if and only if every nested sequence of closed balls in X with radii tending to zero has nonempty intersection.

13. If (f_n) is a sequence of continuous real-valued functions on $[a, b]$ which converges to f uniformly on $[a, b]$, show that f is continuous on $[a, b]$.

14. If $f: (X, d_1) \to (Y, d_2)$ is uniformly continuous and (x_n) is a Cauchy sequence in (X, d_1), show that $(f(x_n))$ is a Cauchy sequence in (Y, d_2).

15. If f is a homeomorphism of (X, d_1) onto (Y, d_2) which is uniformly continuous and (Y, d_2) is complete, show that (X, d_1) is complete. Thus if d_1 and d_2 are uniformly equivalent metrics on X, then (X, d_1) and (X, d_2) are both complete or are both incomplete.

16. If X and Y are metric spaces, show that $X \times Y$ is complete if and only if X and Y are complete (c.f. Problem 7, p. 71). Use this result to obtain another proof that $\mathbf{R}^n$ is complete.

7 · COMPLETION OF A METRIC SPACE

By the completion of a metric space (X, d) we mean the construction of a complete metric space $(\tilde{X}, \tilde{d})$ which contains an isometric image of (X, d). Since isometric metric spaces are essentially the same, we may consider (X, d) to be a subspace of $(\tilde{X}, \tilde{d})$ and say that (X, d) is embedded in $(\tilde{X}, \tilde{d})$. The construction of the real numbers from the rational numbers given in Chapter 2 is an example of such an embedding and provides the method for the completion of an arbitrary metric space.

Let (X, d) be a metric space. If $\mathscr{C}$ is the set of all Cauchy sequences of elements of (X, d), define a relation R on $\mathscr{C}$ as follows:

$$(7 \cdot 1) \qquad (x_n)R(y_n) \quad \text{if} \quad \lim d(x_n, y_n) = 0.$$

It is easy to verify that this is an equivalence relation in $\mathscr{C}$ (Problem 1). Thus, $\mathscr{C}$ can be partitioned into equivalence classes. Let $[(x_n)]$ denote the equivalence class containing (x_n) and let $\tilde{X}$ be the quotient set $\mathscr{C}/R$.

Before defining a metric on $\tilde{X}$ we observe that if (x_n) and (y_n) are in $\mathscr{C}$, then $(d(x_n, y_n))$ is a Cauchy sequence of real numbers since by the quadrilateral inequality (Problem 2, p. 59),

$$|d(x_n, y_n) - d(x_m, y_m)| \leq d(x_n, x_m) + d(y_m, y_n).$$

Then, the completeness of the real field implies that the sequence $(d(x_n, y_n))$ converges to some real number.

We define a metric $\tilde{d}$ on $\tilde{X}$ as follows: if $\tilde{x} = [(x_n)]$ and $\tilde{y} = [(y_n)]$ are elements of $\tilde{X}$ then

$$(7 \cdot 2) \qquad \tilde{d}(\tilde{x}, \tilde{y}) = \lim d(x_n, y_n).$$

We must verify that this definition does not depend on the particular representatives of $\tilde{x}$ and $\tilde{y}$ used. If (x_n') and (y_n') are elements of $\mathscr{C}$ such that $(x_n')\ R\ (x_n)$ and $(y_n')\ R\ (y_n)$, then $\lim d(x_n, x_n') = 0$ and $\lim d(y_n, y_n') = 0$. Using the quadrilateral inequality

$$|d(x_n', y_n') - d(x_n, y_n)| \leq d(x_n', x_n) + d(y_n, y_n'),$$

we obtain

$$\lim d(x_n', y_n') = \lim d(x_n, y_n).$$

The verification that $\tilde{d}$ is a metric on $\tilde{X}$ is left to the reader (Problem 2).

Let f be the function from (X, d) into $(\tilde{X}, \tilde{d})$ such that, for each $x \in X$, $f(x) = [(\bar{x})]$ where $(\bar{x})$ denotes the Cauchy sequence all of whose terms are x. Then, for each $x, y \in X$,

$$\tilde{d}(f(x), f(y)) = d(x, y)$$

and, hence, f is an isometry. Thus, (X, d) is embedded in $(\tilde{X}, \tilde{d})$.

It remains to be shown that $(\tilde{X}, \tilde{d})$ is complete. First we prove a lemma. In this lemma f denotes the function defined above.

7 · 3 Lemma. *If $\tilde{x} = [(x_n)]$, then* $\lim f(x_n) = \tilde{x}$.

PROOF. Take $\varepsilon > 0$. Since (x_n) is a Cauchy sequence, there exists a positive integer n_0 such that

$$d(x_n, x_m) < \varepsilon \quad \text{whenever} \quad m, n \geq n_0.$$

Then, for any $n \geq n_0$,

$$\tilde{d}(f(x_n), \tilde{x}) = \lim_{m\to\infty} d(x_n, x_m) \leq \varepsilon. \blacksquare$$

Recall that a set A is dense in a metric space (Y, d) if $\bar{A} = Y$ where $\bar{A}$ is the closure of A. Thus, the set A in (Y, d) is dense in (Y, d) if and only if for each $y \in Y$ and each $\varepsilon > 0$ there exists a point $a \in A$ such that $d(y, a) < \varepsilon$. Lemma 7 · 3 implies that $f(X)$ is dense in $(\tilde{X}, \tilde{d})$.

7 · 4 Proposition. *The metric space* $(\tilde{X}, \tilde{d})$ *is complete.*

PROOF. Let $(\tilde{x}_n)$ be a Cauchy sequence in $(\tilde{X}, \tilde{d})$. Since $f(X)$ is dense in $(\tilde{X}, \tilde{d})$, for each positive integer n there exists a point $x_n \in X$ such that $\tilde{d}(f(x_n), \tilde{x}_n) < 1/n$. We now show that (x_n) is a Cauchy sequence in (X, d). Take $\varepsilon > 0$. There exists a positive integer $n_0 > 3/\varepsilon$ such that

$$\tilde{d}(\tilde{x}_n, \tilde{x}_m) < \varepsilon/3 \quad \text{whenever} \quad m, n \geq n_0.$$

Then, if $m, n \geq n_0$

$$d(x_n, x_m) = \tilde{d}(f(x_n), f(x_m)) \leq \tilde{d}(f(x_n), \tilde{x}_n) + \tilde{d}(\tilde{x}_n, \tilde{x}_m) + \tilde{d}(\tilde{x}_m, f(x_m))$$
$$< \frac{1}{n} + \frac{\varepsilon}{3} + \frac{1}{m} < \varepsilon.$$

Thus, (x_n) is a Cauchy sequence in (X, d). Letting $\tilde{x} = [(x_n)]$, we will show that $(\tilde{x}_n)$ converges to $\tilde{x}$. By Lemma 7 · 3, $\lim f(x_n) = \tilde{x}$. Then, since

$$\tilde{d}(\tilde{x}, \tilde{x}_n) \leq \tilde{d}(\tilde{x}, f(x_n)) + \tilde{d}(f(x_n), \tilde{x}_n) < \tilde{d}(\tilde{x}, f(x_n)) + \frac{1}{n},$$

$\lim \tilde{d}(\tilde{x}, \tilde{x}_n) = 0$; that is, $\lim \tilde{x}_n = \tilde{x}$. $\blacksquare$

We have now shown that any metric space (X, d) can be embedded in a complete metric space $(\tilde{X}, \tilde{d})$ in such a way that X is dense in $(\tilde{X}, \tilde{d})$.

PROBLEMS

1. Show that the relation 7 · 1 is an equivalence relation.

2. Show that the function $\tilde{d}$ defined in 7 · 2 is a metric on $\tilde{X}$.

3. Determine the completion of the space $(0, 1)$.

4. Let A be a dense set in the metric space X. If f and g are continuous functions on X such that $f(x) = g(x)$ for all $x \in A$ show that $f = g$.

5. Show that the completion of a metric space is unique; that is, if X is dense in the complete space (X', d') then (X', d') and $(\tilde{X}, \tilde{d})$ are isometric.

6. Let

$$d(z_1, z_2) = \frac{|z_1 - z_2|}{(1 + |z_1|^2)^{1/2}(1 + |z_2|^2)^{1/2}}$$

be the chordal metric on **C** (Problem 9, p. 60).

a. If $z_n = n$, show that (z_n) is a Cauchy sequence in $(\mathbf{C}, d)$ which does not converge and, hence, $(\mathbf{C}, d)$ is not complete (c.f. Problem 8, p. 72).

b. Show that the completion of $(\mathbf{C}, d)$ is obtained by adding a single point (usually denoted by ∞).

Suggestion. If (z_n) is a Cauchy sequence such that, for some M, $|z_n| \leq M$ for all n, then (z_n) converges. If (z_n) is a Cauchy sequence such that for each M there exists an n such that $|z_n| > M$, then $\lim d(z_n, n) = 0$.

8 · COMPACTNESS

An important property of Euclidean spaces is given by the Heine-Borel Theorem: any open cover $\{U_s : s \in S\}$ of a closed and bounded set E in $\mathbf{R}^n$ contains a finite subcover of E. Although this theorem may be familiar to the reader, a proof will be given in §10. After some preliminary definitions we give an example to show that the Heine-Borel Theorem does not hold in an arbitrary metric space.

The **diameter** of a set E in a metric space X, denoted by $d(E)$ is defined as follows:

$$d(E) = \sup\{d(x, y) : x, y \in E\}.$$

The diameter of a set may be finite or infinite. If $d(E)$ is finite, then E is said to be **bounded.** A collection $\{U_s : s \in S\}$ of open sets in X is an **open cover** of a set E in X if $E \subset \bigcup_{s \in S} U_s$. A finite subcollection $\{U_s : s \in F\}$ of sets in the open cover $\{U_s : s \in S\}$ is called a **finite subcover** of E if $E \subset \bigcup_{s \in F} U_s$.

Consider the metric space l_∞ consisting of all bounded sequences of real numbers with metric

$$d(x, y) = \sup\{|\alpha^k - \beta^k| : k = 1, 2, \cdots\} \quad \text{where } x = (\alpha^k) \text{ and } y = (\beta^k).$$

Let $E = \{e_j : e_j = (\delta_j^{\,k}), j = 1, 2, \cdots\}$ where $\delta_j^{\,k} = 0, j \neq k$ and $\delta_j^{\,j} = 1$. The set E is closed and bounded in l_∞. But the collection $\{B(e_j\,; 1/2) : j = 1, 2, \cdots\}$ is an open cover of E which has no finite subcover of E since $d(e_j, e_k) = 1$ for $j \neq k$. Thus, the Heine-Borel Theorem does not hold in the space l_∞.

Many of the results concerning closed and bounded sets in a Euclidean space actually depend on the fact that these sets have the Heine-Borel property: any open cover has a finite subcover. Thus, these results will hold in a metric space with the Heine-Borel property.

8 · 1 Definition. *A subset E of a metric space X is said to be* **compact** *if every open cover of E contains a finite subcover. A metric space X is said to be compact if the set X is compact.*

The preceding example shows that a closed and bounded set in a metric space need not be compact. We now show that a compact set is closed and bounded.

8 · 2 Proposition. *A compact set in a metric space is closed and bounded.*

PROOF. Let E be a compact set in the metric space X. To show that E is closed, take $x \in \sim E$. For each $y \in E$ there exist open neighborhoods $N(y)$ and $N_y(x)$ of y and x, respectively, such that $N(y) \cap N_y(x) = \varnothing$. The collection $\{N(y) : y \in E\}$ is an open cover of E. Since E is compact, this open cover has a finite subcover $\{N(y) : y \in F\}$. Then, the set $\bigcap_{y \in F} N_y(x)$ is an open neighborhood of x which does not intersect E. Thus, $\sim E$ is open and consequently E is closed.

To show that E is bounded, consider the open cover $\{B(y; 1) : y \in E\}$ of E. This has a finite subcover $\{B(y; 1) : y \in F_1\}$. Thus,

$$d(E) \le d(F_1) + 2$$

which shows that E is bounded since $d(F_1)$ is finite. ▌

8 · 3 Proposition. *A closed set in a compact metric space is compact.*

PROOF. Let E be a closed set in the compact metric space X. Let $\{U_s : s \in S\}$ be an open cover of E. Adjoining the open set $\sim E$ to the collection $\{U_s : s \in S\}$, we obtain an open cover of X. Since X is compact, this open cover contains a finite subcover of X. Thus, the removal of $\sim E$ (if it is present) from this finite subcover yields a finite subcover of the cover $\{U_s : s \in S\}$ of E. ▌

Since a nonempty set in a metric space is itself a metric space, we can state results in terms of metric spaces rather than sets in a metric space. Compactness is a topological property; that is, the homeomorphic image of a compact space is compact. In fact, we have the following result.

8 · 4 Theorem. *The continuous image of a compact space is compact.*

PROOF. Let X be a compact space and let f be a continuous function from X into a metric space Y. Let $\{U_s : s \in S\}$ be an open cover of $f(X)$. Then, $\{f^{-1}(U_s) : s \in S\}$ is an open cover of X. Since X is compact, there is a finite subcover $\{f^{-1}(U_s) : s \in F\}$ of X. Then $\{U_s : s \in F\}$ is a finite subcover of $f(X)$. Thus, $f(X)$ is compact. ▌

8 · 5 Corollary. *A one-to-one continuous mapping of a compact metric space is a homeomorphism.*

PROOF. Let f be a one-to-one continuous mapping of the compact metric space X into the metric space Y. To show that f is a homeomorphism all we need to show is that f maps closed sets onto closed sets. If E is a closed set in X, then E is compact. Then, $f(E)$ is compact, and, hence, closed in Y. ∎

If f is a continuous real-valued function on the compact space X, then f is bounded on X and, furthermore, f has a maximum and a minimum value on X (Problem 8). Let X be a compact metric space and let $C(X, \mathbf{R})$ be the set of continuous real-valued functions on X with

$$d(f, g) = \max\{|f(x) - g(x)| : x \in X\}.$$

Then $C(X, \mathbf{R})$ is a complete metric space (Problem 9). It is a generalization of $C[a, b]$.

PROBLEMS

1. Determine the diameter of the following sets:

a. the open ball $B(a; r)$ in a metric space X

b. $E = \{(x, y) \in \mathbf{R}^2 : 0 < x < 1; 0 < y < 1\}$

c. the set of irrational numbers in $[0, 1]$.

2. Show that a set E in the metric space X is bounded if and only if, for some $a \in X$, there exists an open ball $B(a; r)$ such that $E \subset B(a; r)$.

3. If E is a bounded set in the metric space X, show that $\bar{E}$ is bounded.

4. Show that $E = \{1/n : n \text{ a positive integer}\}$ is not compact in $\mathbf{R}$ but $E \cup \{0\}$ is compact.

5. Show that the finite union of compact sets in a metric space X is compact.

6. Show that a discrete metric space X is not compact unless X is finite. Show that an indiscrete metric space is always compact.

7. Prove that every finite subset of a metric space is compact.

8. Show that if f is a continuous real-valued function on the compact space X, then there exist points $x_1, x_2 \in X$ such that $f(x_1) = \inf\{f(x) : x \in X\}$ and $f(x_2) = \sup\{f(x) : x \in X\}$.

* **9.** If X is compact prove that $C(X, \mathbf{R})$ is a complete metric space.

10. Is $C[0, 1]$ compact?

11. A collection $\mathscr{F}$ of sets is said to have the **finite intersection property** if the intersection of any finite subcollection of $\mathscr{F}$ is nonempty. Prove that a metric space X is compact if and only if the intersection of any collection of closed sets of X with the finite intersection property is nonempty.

12. If X is a separable metric space, show that every open cover of X has a countable subcover.

9 · SEQUENTIAL COMPACTNESS

A set E in $\mathbf{R}$ is closed and bounded if and only if every sequence of points in E has a subsequence which converges to a point in E. This leads to another notion of compactness which we call sequential compactness. However, we will show that *in a metric space* this type of compactness is actually equivalent to that defined in terms of the Heine-Borel property.

9 · 1 Definition. *A metric space X is* ***sequentially compact*** *if every sequence in X has a convergent subsequence.*

This definition can also be stated in the following way: a metric space X is sequentially compact if every sequence in X has a limit point. A point x is said to be a **limit point** of the sequence (x_n) if there exists some subsequence of (x_n) which converges to x. It is not difficult to show that a point x is a limit point of a sequence (x_n) if and only if every neighborhood of x contains infinitely many terms of (x_n) (Problem 4, p. 75).

9 · 2 Theorem. *A compact metric space is sequentially compact.*

PROOF. Let X be a compact metric space and let (x_n) be a sequence in X. Suppose (x_n) has no convergent subsequence; that is, (x_n) has no limit point. Then for each $x \in X$ there is an open neighborhood $N(x)$ which contains only a finite number of terms of (x_n). The collection $\{N(x) : x \in X\}$ is an open cover of X and hence has a finite subcover. This implies that (x_n) has only a finite number of terms. Thus, the supposition that (x_n) has no convergent subsequence cannot hold. ∎

The converse of this theorem is more difficult and we do some preliminary work before proving it.

A metric space X is said to be **totally bounded** if for each $\varepsilon > 0$ there is a finite collection $\{B(x_i ; \varepsilon) : i = 1, \cdots, k\}$ of open balls which covers X.

9 · 3 Proposition. *A sequentially compact metric space X is totally bounded.*

PROOF. Suppose X is not totally bounded. Then, there is an $\varepsilon > 0$ such that for each finite collection $\{x_i : i = 1, \cdots, k\}$ there is a point $x \in X$ such that $d(x, x_i) \geq \varepsilon$ for each $i = 1, \cdots, k$. Thus, we can choose a sequence (y_n) such that $d(y_n, y_m) \geq \varepsilon$ for $m \neq n$. Clearly, this sequence can have no convergent subsequence and, hence, X is not sequentially compact. ∎

9 · 4 Proposition. *If $\{U_s : s \in S\}$ is an open cover of a sequentially compact metric space X, then there exists a positive integer n such that for each $x \in X$ the open ball $B(x; 1/n)$ is contained in some set U_s of the open cover.*

PROOF. Suppose for each positive integer n there exists a point $x_n \in X$ such that $B(x_n ; 1/n)$ is not contained in any set U_s of the given open cover of X.

Since X is sequentially compact, the sequence (x_n) has a limit point x. The point $x \in U_s$ for some $s \in S$ and, since U_s is open, there exists an open ball $B(x; \delta)$ such that $B(x; \delta) \subset U_s$. Since x is a limit point of (x_n), there exists a positive integer $n > 2/\delta$ such that $x_n \in B(x; \delta/2)$. Then

$$B(x_n ; 1/n) \subset B(x; \delta) \subset U_s.$$

This contradicts the property on which the selection of x_n was based. ▮

If we define a **Lebesgue number** of an open cover $\{U_s : s \in S\}$ of X to be a positive number δ such that any set in X with diameter less than δ lies in some one of the sets U_s, then we can restate Proposition 9 · 4 as follows: Any open cover of a sequentially compact metric space has a Lebesgue number.

We now prove that a sequentially compact space is compact.

9 · 5 Theorem. *A sequentially compact metric space is compact.*

PROOF. Let X be a sequentially compact metric space and let $\{U_s : s \in S\}$ be an open cover of X with a Lebesgue number δ. Since X is totally bounded there exists a finite collection $\{B(x_i ; \delta/2) : i = 1, \cdots, k\}$ of open balls which covers X. Since each $B(x_i ; \delta/2)$ is contained in some set, say U_{s_i}, of the given open cover, then $\{U_{s_i} : i = 1, \cdots, k\}$ is a finite subcover of X. This shows that X is compact. ▮

Thus, we have shown that for metric spaces sequential compactness and compactness are equivalent.

The fact that any open cover of a compact metric space has a Lebesgue number can be used to prove a generalization of an important theorem in the theory of real-valued functions of a real variable.

9 · 6 Theorem. *If f is a continuous function from a compact metric space (X, d_1) into a metric space (Y, d_2), then f is uniformly continuous on X.*

PROOF. Take $\varepsilon > 0$. For each $x \in X$ there exists a δ_x such that $f(B(x; \delta_x)) \subset B(f(x); \varepsilon/2)$. The collection $\{B(x; \delta_x) : x \in X\}$ is an open cover of the compact space X and, hence, has a Lebesgue number δ. Then, if $d_1(x, y) < \delta$, there exists some $z \in X$ such that $\{x, y\} \subset B(z; \delta_z)$ and, thus, $\{f(x), f(y)\} \subset B(f(z); \varepsilon/2)$. Therefore

$$d_2(f(x), f(y)) < \varepsilon \quad \text{whenever } d_1(x, y) < \delta. \ ▮$$

We now give another characterization of compactness for metric spaces.

9 · 7 Theorem. *A metric space is compact if and only if it is complete and totally bounded.*

PROOF. First assume that the metric space X is compact. Then, it is sequentially compact and, hence, totally bounded. Also, if (x_n) is a Cauchy sequence in X then (x_n) has a limit point x and, therefore by Problem 4c, p. 75, (x_n) converges to x.

Now assume that X is complete and totally bounded. Let (x_n) be a sequence in X. Since X is complete, to show that (x_n) has a convergent subsequence it suffices to show that (x_n) has a Cauchy subsequence. The total boundedness of X implies that there is a finite collection $\{B(a_j^1; 1) : j = 1, \cdots, k_1\}$ of open balls of radius 1 which covers X. At least one of these balls, call it B_1, must contain a subsequence (x_n^1) of (x_n). A finite collection $\{B(a_j^2; 1/2) : j = 1, \cdots, k_2\}$ also covers X. At least one of these balls, call it B_2, contains a subsequence (x_n^2) of (x_n^1). Continuing in this way we obtain a subsequence (x_n^k) of (x_n^{k-1}) contained in an open ball B_k of radius $1/k$ for $k = 2, 3, \cdots$. From the sequences $(x_n^1), (x_n^2), \cdots$ take the diagonal sequence (x_n^n). This is a Cauchy subsequence of (x_n) since for $n > n_0$, x_n^n is in the open ball B_{n_0} of radius $1/n_0$. ∎

PROBLEMS

1. If $a < b$, show that the open interval (a, b) in $\mathbf{R}$ is not sequentially compact.

2. Show that a totally bounded set in a metric space is bounded. Give an example of a bounded set which is not totally bounded.

3. The intervals $[0, 1/10)$, $(1/2, 1]$ and $(1/(n+1), 1/(n-1))$ for $n = 2, 3, \cdots$ constitute an open cover of the metric space $[0, 1]$. Determine a Lebesgue number of this cover.

4. Prove that a compact metric space is separable.

5. If (X_1, d_1) and (X_2, d_2) are compact metric spaces, show that the product metric space with the distance given by $d(x, y) = d_1(x_1, y_1) + d_2(x_2, y_2)$, where $x = (x_1, x_2)$ and $y = (y_1, y_2)$, is compact.

6. If (F_n) is a sequence of nonempty closed sets in a compact metric space X such that $F_n \supset F_{n+1}$, show that $\bigcap_{n=1}^{\infty} F_n$ is nonempty.

7. A metric space X is said to be **Fréchet compact** or to have the **Bolzano-Weierstrass property** if every infinite subset of X has a point of accumulation in X where x is a point of accumulation of a set $S \subset X$ if every neighborhood of x contains points of S distinct from x. Show that a metric space X is sequentially compact if and only if it has the Bolzano-Weierstrass property.

8. If the sequence (x_n) in a compact metric space has a unique limit point x, show that (x_n) converges to x.

9. Let (X, d_1) and (Y, d_2) be compact metric spaces and let $C(X, Y)$ be the space of continuous functions from X into Y with metric

$$d(f, g) = \sup_{x \in X} d_2(f(x), g(x)).$$

Show that $C(X, Y)$ is a complete metric space.

10 · HEINE-BOREL AND ARZELÀ-ASCOLI THEOREMS

We now consider two theorems which give useful characterizations for compactness in special spaces. The first is the Heine-Borel Theorem which applies to Euclidean spaces.

10 · 1 Theorem (Heine-Borel). *If E is a closed and bounded set in* $\mathbf{R}^m$, *then E is compact.*

PROOF. We will show that E is complete and totally bounded. Then Theorem 9 · 7 implies that E is compact. Since $\mathbf{R}^m$ is complete (Proposition 6 · 6, p. 73) and E is a closed set in $\mathbf{R}^m$, the subspace E is complete (Proposition 6 · 11, p. 75). The fact that E is bounded implies that there is an open ball $B(0; r)$ which contains E. To show that E is totally bounded we will show that for each $\varepsilon > 0$ the open ball $B(0; r)$ can be covered by a finite number of open balls of radius ε. If q is an integer greater than $\sqrt{m}/\varepsilon$, we claim that $B(0; r)$ is covered by the finite family

$$\{B((p_1/q, p_2/q, \cdots, p_m/q); \varepsilon) : -rq \le p_i \le rq, p_i \text{ an integer}\}.$$

For if $x = (\alpha^1, \alpha^2, \cdots. \alpha^m) \in B(0; r)$, then for each α^i there is a rational number of the form p_i/q whose distance from α^i is less than $1/q < \varepsilon/\sqrt{m}$. Thus,

$$\left| x - \left(\frac{p_1}{q}, \frac{p_2}{q}, \cdots, \frac{p_m}{q}\right)\right| = \left[\sum_{i=1}^{m} \left(\alpha^i - \frac{p_i}{q}\right)^2\right]^{1/2} < \varepsilon. \blacksquare$$

Since the complex space $\mathbf{C}$ can be identified with $\mathbf{R}^2$ as a metric space, the Heine-Borel Theorem applies to $\mathbf{C}$ also. We can now show that the complex field is algebraically closed; that is, any nonconstant polynomial over $\mathbf{C}$ can be factored into a product of linear factors. If

$$P(z) = \sum_{k=0}^{n} a_k z^k$$

where $a_k \in \mathbf{C}$, $n \ge 1$, and $a_n \ne 0$, then we will prove that

$$P(z_0) = 0 \quad \text{for some } z_0 \in \mathbf{C}.$$

This result is known as the **Fundamental Theorem of Algebra**. It follows that $P(z) = (z - z_0)Q(z)$ where Q is a polynomial of degree $n - 1$. If $n - 1 \ge 1$ then $Q(z)$ has a zero and thus has a linear factor. In this way we show that any nonconstant polynomial over $\mathbf{C}$ can be written as a product of linear factors.

10 · 2 Theorem (Euler-Gauss). *If $P(z)$ is a nonconstant polynomial over* $\mathbf{C}$, *then $P(z_0) = 0$ for some $z_0 \in$* $\mathbf{C}$.

PROOF. Let $P(z) = \sum_{k=0}^{n} a_k z^k$ where $a_n \ne 0$ and $n \ge 1$ and let

$$m = \inf\{|P(z)| : z \in \mathbf{C}\}.$$

Since

$$|P(re^{it})| \geq r^n(|a_n| - r^{-1}|a_{n-1}| - \cdots - r^{-n}|a_0|),$$

$\lim_{r\to\infty} |P(re^{it})| = \infty$. Therefore, there exists a real number R such that

$$|P(re^{it})| > m + 1 \quad \text{whenever } r > R.$$

If $E = \{re^{it} : r \leq R\}$, then E is compact in the metric space $\mathbf{C}$ and $m = \inf\{|P(z)| : z \in E\}$. Since $|P|$ is a continuous real-valued function on E, $|P|$ has a minimum value on E (Problem 8, p. 81); that is, there exists a $z_0 \in E$ such that

$$|P(z_0)| = m.$$

If $m = 0$, then the theorem is proved.

Suppose $m \neq 0$ and let

$$Q(z) = \frac{P(z + z_0)}{P(z_0)}, \quad z \in \mathbf{C}.$$

Then Q is a polynomial of degree n and

(10 · 3) $$|Q(z)| \geq 1 \quad \text{for all } z \in \mathbf{C}.$$

Since $Q(0) = 1$ we can write $Q(z)$ in the form:

$$Q(z) = 1 + b_k z^k + \cdots + b_n z^n$$

where k is the smallest positive integer $\leq n$ such that $b_k \neq 0$. Since the absolute value of $-|b_k|/b_k$ is 1, there exists a $t_0 \in [0, 2\pi/k)$ such that

$$e^{ikt_0} = -\frac{|b_k|}{b_k}.$$

Then

$$\begin{aligned} Q(re^{it_0}) &= 1 + b_k r^k e^{ikt_0} + b_{k+1} r^{k+1} e^{i(k+1)t_0} + \cdots + b_n r^n e^{int_0} \\ &= 1 - r^k |b_k| + b_{k+1} r^{k+1} e^{i(k+1)t_0} + \cdots + b_n r^n e^{int_0} \end{aligned}$$

and, hence, if $r^k|b_k| < 1$

$$|Q(re^{it_0})| \leq 1 - r^k(|b_k| - r|b_{k+1}| - \cdots - r^{n-k}|b_n|).$$

Thus, for r sufficiently small $|Q(re^{it_0})| < 1$ which contradicts 10 · 3. Therefore, the assumption $m \neq 0$ cannot hold and $P(z_0) = 0$. ∎

The Arzelà-Ascoli Theorem applies to a space $C(X, \mathbf{R})$ of continuous real-valued functions on a compact metric space X. This theorem will be proved after some preliminary discussion.

Let E be a set of continuous real-valued functions defined on a compact metric space X. Since each function f in E is uniformly continuous on X

(Theorem 9 · 6), for each $\varepsilon > 0$ there exists a $\delta > 0$ such that $|f(x_1) - f(x_2)| < \varepsilon$ whenever $x_1, x_2 \in X$ and $d(x_1, x_2) < \delta$. In general, the number δ depends not only on ε but also on the particular function $f \in E$. If, for each given $\varepsilon > 0$, there is a $\delta > 0$ which will work for all the functions in E, then we say that E is equicontinuous.

10 · 4 Definition. *A set E in $C(X, \mathbf{R})$ is said to be* ***equicontinuous*** *if for each $\varepsilon > 0$ there exists a $\delta > 0$ such that*

$$|f(x_1) - f(x_2)| < \varepsilon \qquad \text{whenever } x_1, x_2 \in X, \quad d(x_1, x_2) < \delta, \text{ and } f \in E:$$

10 · 5 Theorem (Arzelà-Ascoli). *A set E in $C(X, \mathbf{R})$ is compact if and only if it is closed, bounded, and equicontinuous.*

PROOF. First assume that E is compact. Then E is closed and bounded. Also, E is totally bounded. Thus, for each $\varepsilon > 0$ there is a finite collection $\{B(f_i; \varepsilon) : i = 1, \cdots, k\}$ which covers E. Since each function f_i is uniformly continuous on X, there exists a number $\delta_i > 0$ such that

$$|f_i(x_1) - f_i(x_2)| < \varepsilon \quad \text{whenever } x_1, x_2 \in X \text{ and } d(x_1, x_2) < \delta_i.$$

Let $\delta = \min\{\delta_i : i = 1, \cdots, k\}$ and take any $f \in E$. Then $f \in B(f_i; \varepsilon)$ for some $i = 1, \cdots, k$ and

$$|f(x_1) - f(x_2)| \le |f(x_1) - f_i(x_1)| + |f_i(x_1) - f_i(x_2)| + |f_i(x_2) - f(x_2)| < 3\varepsilon$$

whenever $x_1, x_2 \in X$ and $d(x_1, x_2) < \delta$. Thus, E is equicontinuous.

Now assume that E is closed, bounded, and equicontinuous. Since E is a closed subspace of the complete metric space $C(X, \mathbf{R})$, E is complete (Proposition 6 · 11). Then, E is compact if it is totally bounded. The fact that E is bounded in $C(X, \mathbf{R})$ implies that there exists a number M such that

$$|f(x)| \le M \quad \text{for all } f \in E \text{ and all } x \in X.$$

Since E is equicontinuous, corresponding to each $\varepsilon > 0$ there exists a $\delta > 0$ such that

$$|f(x_1) - f(x_2)| < \varepsilon/4 \qquad \text{whenever } x_1, x_2 \in X, \quad d(x_1, x_2) < \delta, \quad \text{and} \quad f \in E.$$

Since X is compact, it is totally bounded and, hence, there is a finite collection $\{B(x_i; \delta) : i = 1, \cdots, p\}$ which covers X. Let $P = \{y_0, y_1, \cdots, y_q\}$ be a partition of $[-M, M]$ into subintervals of length less than $\varepsilon/4$. The set $\mathscr{P}$ of all p-tuples $(y_{k_1}, \cdots, y_{k_p})$ of points in P is a finite set. Corresponding to each function $g \in E$ there exists a p-tuple in $\mathscr{P}$ such that

$$(10 \cdot 6) \qquad |g(x_i) - y_{k_i}| < \varepsilon/4 \quad \text{for } i = 1, \cdots, p.$$

Let $\mathscr{P}'$ be the subset of $\mathscr{P}$ consisting of all of those p-tuples for which there exists at least one $g \in E$ satisfying 10 · 6. Now corresponding to each p-tuple in $\mathscr{P}'$ select a single $g \in E$ satisfying 10 · 6 and let F be the set of g's so selected.

Then F is also a finite set. For each function $f \in E$ there is a p-tuple $(y_{k_1}, \cdots, y_{k_p}) \in \mathscr{P}'$ such that

$$|f(x_i) - y_{k_i}| < \varepsilon/4 \quad \text{for } i = 1, \cdots, p.$$

Let g be the element of F which corresponds to $(y_{k_1}, \cdots, y_{k_p})$. Then, for each $x \in X$ there exists an x_i such that $x \in B(x_i; \delta)$ and, hence,

$$|f(x) - g(x)| \leq |f(x) - f(x_i)| + |f(x_i) - y_{k_i}| + |y_{k_i} - g(x_i)| + |g(x_i) - g(x)| < \varepsilon.$$

Thus, the finite set $\{B(g; \varepsilon) : g \in F\}$ covers E and E is totally bounded. ▮

PROBLEMS

1. Let (X, d) be the metric space consisting of m-tuples of real numbers with metric

$$d(x, y) = \max\{|\alpha^k - \beta^k| : k = 1, \cdots, m\}$$

where $x = (\alpha^1, \cdots, \alpha^m)$ and $y = (\beta^1, \cdots, \beta^m)$. In this space is every closed and bounded set compact?

2. A metric space X is said to be **locally compact** if each point $x \in X$ has a compact neighborhood. Show that $\mathbf{R}^n$ is locally compact.

3. Show that any nonconstant polynomial over the real field can be factored into a product of linear and quadratic factors.

4. Let E be a set of differentiable functions in $C[a, b]$ with uniformly bounded derivatives; i.e., there exists a number M, independent of f in E, such that $|f'(x)| \leq M$ for all $x \in [a, b]$ and all $f \in E$. Show that E is equicontinuous.

5. If E is equicontinuous in $C(X, \mathbf{R})$, show that $\bar{E}$ is equicontinuous.

6. Let (X, d) be the metric space consisting of all continuous functions from a compact metric space X into $\mathbf{R}^m$ with metric

$$d(f, g) = \max\{|f(x) - g(x)| : x \in X\}.$$

Prove the Arzelà-Ascoli Theorem for this space.

7. Let X and Y be compact metric spaces. Show that a set E in $C(X, Y)$ (Problem 9, p. 84) is compact if and only if it is closed and equicontinuous.

11 · CONNECTEDNESS

The intuitive meaning of a metric space X being connected is that it is in one piece; that is, it is not possible to represent it as the union of two separated sets A and B. By separated sets A and B we mean $\bar{A} \cap B = \varnothing$ and $A \cap \bar{B} = \varnothing$. This is a stronger condition on A and B than disjointness but not as strong as requiring that the distance between them should be positive. For example, in $\mathbf{R}$ the sets $[0, 1]$ and $(1, 2]$ are disjoint but not separated. On the other hand the sets $[0, 1)$ and $(1, 2]$ are separated but the distance between them is zero.

Thus, a space X is not connected if there are two nonempty sets A and B such that $X = A \cup B$ and $\bar{A} \cap B = \varnothing$ and $A \cap \bar{B} = \varnothing$. If such is the case then $\bar{A} \subset \sim B = A$ and, hence, A is closed. Similarly B must be a closed set. If A and B are closed sets then the statements $\bar{A} \cap B = \varnothing$ and $A \cap \bar{B} = \varnothing$ are both equivalent to the single statement $A \cap B = \varnothing$. Thus, we can say that a space X is connected if there exists no pair of closed sets A and B in X such that $X = A \cup B$ and $A \cap B = \varnothing$. Since the sets A and B in such a pair are complementary we can equally well say that they are both open and the definition is usually given in these terms.

11 · 1 Definition. *The space X is* ***connected*** *if there exists no pair of nonempty open sets A and B in X such that $X = A \cup B$ and $A \cap B = \varnothing$.*

From this definition and the discussion which precedes it, it is clear that a space X is connected if and only if $\varnothing$ and X are the only sets in X which are both open and closed.

A set E in a metric space X is said to be connected if the subspace E is connected. Thus, the set E is connected if there do not exist open sets A and B in X such that $E \subset A \cup B$, $A \cap E \neq \varnothing$, $B \cap E \neq \varnothing$, and $A \cap B \cap E = \varnothing$. If a set E is not connected, it is said to be **disconnected**.

Connectedness is a topological property. In fact, connectedness is preserved under continuous mappings; that is, the continuous image of a connected set is connected.

11 · 2 Theorem. *If f is a continuous function from the connected metric space X onto the metric space Y, then Y is connected.*

Proof. If Y is not connected, then there exist nonempty open sets A and B in Y such that $Y = A \cup B$ and $A \cap B = \varnothing$. Then

$$X = f^{-1}(Y) = f^{-1}(A \cup B) = f^{-1}(A) \cup f^{-1}(B)$$

and

$$f^{-1}(A) \cap f^{-1}(B) = f^{-1}(A \cap B) = f^{-1}(\varnothing) = \varnothing.$$

Also, since f is continuous and onto, $f^{-1}(A)$ and $f^{-1}(B)$ are open and nonempty. Thus, if Y is not connected, then X is not connected. ▮

The connected sets in **R** are simple to describe; they are the intervals. Recall that a set I in **R** is called an interval if, for any two points a and b in I with $a < b$, $\{x : a < x < b\} \subset I$.

11 · 3 Proposition. *A nonempty set E in* **R** *is connected if and only if it is an interval.*

Proof. First assume that E is not an interval. Then for some points a and b in E with $a < b$ there exists an $x \in (a, b)$ such that $x \notin E$. Then the sets

$A = E \cap (-\infty, x)$ and $B = E \cap (x, \infty)$ are nonempty open sets in the subspace E such that $E = A \cup B$ and $A \cap B = \emptyset$. Thus, if E is not an interval, then it is not connected.

Now assume that E is not connected. Then there exist nonempty open sets A and B in the subspace E such that $E = A \cup B$ and $A \cap B = \emptyset$. Take $a \in A$ and $b \in B$. We may assume that $a < b$. We will show that $(a, b) \not\subset E$ and therefore E is not an interval. Suppose $(a, b) \subset E$ and, hence, $[a, b] \subset E$. Let $c = \sup(A \cap [a, b])$. Then $c \in [a, b] \subset E$. And, since A is closed in the subspace E, $c \in A$. Using the fact that $b \in B$ we can state that $c \in A \cap [a, b)$. Since A is open in the subspace E there exists an $\varepsilon > 0$ such that $c + \varepsilon \in A \cap [a, b)$ which contradicts the fact that $c = \sup(A \cap [a, b])$. ∎

Using Theorem 11 · 2 and Proposition 11·3 we can obtain some interesting results. The first is an extension of an important theorem of Weierstrass in the theory of real-valued functions of a real variable.

11 · 4 Proposition. (The Intermediate Value Theorem). *Let f be a continuous function from a metric space X into* **R**. *If a and b belong to a connected set E in X and $f(a) < f(b)$, then for any $t \in (f(a), f(b))$ there exists a point $c \in E$ such that $f(c) = t$.*

PROOF. Since $f(E)$ is a connected set in **R** it is an interval. Hence, if $f(a)$ and $f(b)$ belong to $f(E)$ then $(f(a), f(b)) \subset f(E)$. ∎

If we define an **arc** in a metric space to be the homeomorphic image of a closed interval in **R**, then we have the following result.

11 · 5 Proposition. *If E is an arc in a metric space, then E is connected.*

PROOF. Since E is the continuous image of a connected set in **R**, it is connected. ∎

This leads to the consideration of another type of connectedness.

11 · 6 Definition. *A set E in a metric space is said to be* ***arcwise connected*** *if for any two points x and y in E there is an arc C in E which contains x and y.*

We will show that if a set is arcwise connected, then it is connected. Since arcwise connectedness is relatively easy to verify in many instances, this provides a useful method for showing connectedness. The definition of connectedness itself, since it involves the nonexistence of something, is usually difficult to apply to specific cases.

11 · 7 Theorem. *If E is an arcwise connected set in a metric space, then E is connected.*

PROOF. Assume that E is not connected. Then there exist nonempty open sets A and B in the subspace E such that $E = A \cup B$ and $A \cap B = \varnothing$. Let a and b be points in A and B, respectively, and let C be any set in E which contains a and b. Then $C \cap A$ and $C \cap B$ are nonempty open sets in the subspace C such that

$$(C \cap A) \cup (C \cap B) = C \cap (A \cup B) = C \cap E = C$$

and

$$(C \cap A) \cap (C \cap B) = C \cap (A \cap B) = \varnothing.$$

Thus, C is not connected and hence by Proposition 11 · 5, C cannot be an arc. Therefore, there is no arc in E containing a and b. That is, if E is not connected then E is not arcwise connected. ∎

The converse of this theorem does not hold: a set may be connected but not be arcwise connected (Problem 8).

We now use Theorem 11 · 7 to show that the Euclidean space $\mathbf{R}^n$ is connected. Any two points x and y in $\mathbf{R}^n$ can be joined by a line segment $\{x + t(y - x) : t \in [0, 1]\}$. Clearly this line segment is an arc in $\mathbf{R}^n$. Hence, $\mathbf{R}^n$ is arcwise connected and consequently connected. A set E in $\mathbf{R}^n$ is said to be **convex** if for each pair of points in E the line segment joining them is in E. Thus, any convex set in $\mathbf{R}^n$ is connected.

If a connected set is a set which consists of one piece, then the opposite extreme is a set in which no single piece contains more than one point. We call a set **totally disconnected** if every subset containing more than one point is not connected. A discrete space is totally disconnected. Also, in $\mathbf{R}$ the set of rational numbers and the set of irrational numbers are totally disconnected. For example, between any two distinct rational numbers there is an irrational number. Thus, any set of rational numbers containing more than one point is not an interval and, hence, is not connected.

PROBLEMS

1. If E_i, $i = 1, \cdots, n$, are connected sets in a metric space X and $\bigcap_{i=1}^n E_i \neq \varnothing$, show that $\bigcup_{i=1}^n E_i$ is connected.

2. If E is connected in X and if $E \subset F \subset \bar{E}$, show that F is connected. Thus, in particular if E is connected then $\bar{E}$ is connected.

3. In a metric space X define a relation R as follows: $xRy \Leftrightarrow$ there exists a connected set in X which contains x and y.

- **a.** Show that R is an equivalence relation in X.
- **b.** Call the sets in the associated partition of X the **(connected) components** of X. If $C(x)$ denotes the component containing x, show that $C(x)$ is the union of all connected sets containing x.
- **c.** Show that $C(x)$ is a maximal connected set; that is, $C(x)$ is connected and if $C(x) \subset E$ where E is connected then $C(x) = E$.

d. Show that $C(x)$ is closed.
e. If E is connected and $E \cap C(x) \neq \varnothing$, show that $E \subset C(x)$.
f. X is connected $\Leftrightarrow$ it has only one component.

4. Show that an open ball $B(a; r)$ in $\mathbf{R}^n$ is convex and hence connected.

5. If E is an open set in $\mathbf{R}^n$, show that the components of the subspace E are open and connected in $\mathbf{R}^n$.

6. Show that any open set E in $\mathbf{R}$ is the countable union of disjoint open intervals.

7. If X and Y are metric spaces, show that $X \times Y$ is connected $\Leftrightarrow X$ and Y are connected.

Suggestion. Use the projection mappings to show that $X \times Y$ connected implies that X and Y are connected. If X and Y are connected, take any two points (x_1, y_1) and (x_2, y_2) in $X \times Y$. Show that if $I_{x_1}(y) = (x_1, y)$ and $I_{y_2}(x) = (x, y_2)$, then I_{x_1} and I_{y_2} are continuous mappings of Y onto $\{x_1\} \times Y$ and X onto $X \times \{y_2\}$, respectively. Conclude that $(\{x_1\} \times Y) \cup (X \times \{y_2\})$ is connected and that $X \times Y$ has only one component.

8. Show that the set

$$E = \left\{(x, y) \in \mathbf{R}^2 : y = \sin \frac{1}{x}, x > 0\right\} \cup \{(0, y) : -1 \leq y \leq 1\}$$

is connected but is not arcwise connected in $\mathbf{R}^2$.

9. If E is an open connected set in $\mathbf{R}^n$, show that E is arcwise connected.

10. If X is an arcwise connected metric space, Y is a metric space, and f is a continuous mapping of X onto Y, show that Y is arcwise connected. Hence, arcwise connectedness is a topological property.

11. If X is a connected metric space which contains at least two points, then X contains an uncountable number of points.

Suggestion. Consider the function f from X into $\mathbf{R}$ defined by $f(x) = d(x, x_0)$ where x_0 is a fixed-point of X.

5 | *A Fixed-Point Theorem; Applications*

1 · INTRODUCTION

Many mathematical problems can be formulated as follows: solve the equation $T(x) = x$. If T is a transformation of a space X into itself, then a point x in X such that $T(x) = x$ is called a **fixed-point** of T. In this chapter we prove a theorem which ensures, under appropriate conditions, the existence and uniqueness of a fixed-point. This theorem also provides a method for obtaining an approximation to the fixed-point.

The fixed-point theorem will be applied to the problem of the solution of a system of simultaneous linear algebraic equations by methods of successive approximation. The fixed-point theorem will also be applied to prove an existence and uniqueness theorem for the solution of a differential equation. In this method of proof the differential equation is converted to an integral equation and this leads to a brief discussion of integral equations.

2 · A FIXED-POINT THEOREM

There are a number of important fixed-point theorems in mathematics. The one considered here concerns contractions on a complete metric space and is called the Banach Fixed-Point Theorem after the Polish mathematician Stefan Banach (1892–1945).

2 · 1 Definition. *Let* (X, d_1) *and* (Y, d_2) *be metric spaces. A transformation* T *from* X *into* Y *is called a* ***contraction*** *if there exists a number* $\lambda \in (0, 1)$ *such that*

$$d_2(T(x), T(y)) \leq \lambda d_1(x, y) \quad \textit{for all } x, y \in X.$$

A contraction T transforms each pair of points (x, y) into a pair $(T(x), T(y))$ whose members are closer together. It is clear from the definition that a contraction is continuous.

If f is a real-valued function of a real variable which is differentiable on an interval $[a, b]$ and if

$$|f'(x)| \leq \lambda < 1 \quad \text{for all } x \in [a, b],$$

then f is a contraction on $[a, b]$. This follows from the Mean Value Theorem: for each $x, y \in [a, b]$ there is a $z \in [a, b]$ such that

$$|f(y) - f(x)| = |f'(z)(y - x)| \leq \lambda |y - x|.$$

$d_2(y,x)$ $\leq \lambda\, d_1(y,x)$

2 · 2 Theorem (Banach Fixed-Point Theorem). *Let T be a contraction of a complete metric space X into itself:*

$$d(T(x), T(y)) \leq \lambda d(x, y), \quad \lambda \in (0, 1).$$

Then T has a unique fixed-point $\bar{x}$. Moreover, if x_0 is any point in X and the sequence (x_n) is defined recursively by the formula $x_n = T(x_{n-1})$, $n = 1, 2, \cdots$, then $\lim x_n = \bar{x}$ *and*

$$d(\bar{x}, x_n) \leq \frac{\lambda}{1-\lambda} d(x_{n-1}, x_n) \leq \frac{\lambda^n}{1-\lambda} d(x_0, x_1).$$

PROOF. Since T is a contraction it cannot have more than one fixed point. For, if $\bar{x}_1$ and $\bar{x}_2$ are fixed-points of T

$$d(\bar{x}_1, \bar{x}_2) = d(T(\bar{x}_1), T(\bar{x}_2)) \leq \lambda d(\bar{x}_1, \bar{x}_2) \quad \text{where } \lambda \in (0, 1).$$

Thus, $\bar{x}_1 = \bar{x}_2$. We now prove that T has a fixed-point. Take any point $x_0 \in X$ and let $x_n = T(x_{n-1})$, $n = 1\ 2, \cdots$. Then, for $k > 0$,

$$d(x_k, x_{k+1}) = d(T(x_{k-1}), T(x_k)) \leq \lambda d(x_{k-1}, x_k).$$

Using this formula repeatedly we obtain

$$d(x_k, x_{k+1}) \leq \lambda^i d(x_{k-i}, x_{k-i+1}), \qquad i = 1, \cdots, k.$$

In general, if $m > n \geq 1$, then

$$(2 \cdot 3) \qquad d(x_n, x_m) \leq \sum_{j=0}^{m-n-1} d(x_{n+j}, x_{n+j+1}) \leq \sum_{j=0}^{m-n-1} \lambda^{j+1} d(x_{n-1}, x_n)$$

$$\leq \frac{\lambda}{1-\lambda} d(x_{n-1}, x_n) \leq \frac{\lambda^n}{1-\lambda} d(x_0, x_1).$$

Since $\lambda \in (0, 1)$, for each $\varepsilon > 0$ there exists a positive integer n_0 such that

$$\frac{\lambda^n}{1-\lambda} d(x_0, x_1) < \varepsilon \quad \text{whenever } n \geq n_0.$$

Thus, (x_n) is a Cauchy sequence in the complete space X and hence converges to a point $\bar{x} \in X$. Since T is continuous,

$$\bar{x} = \lim x_n = \lim T(x_{n-1}) = T(\bar{x}).$$

Therefore $\bar{x}$ is a fixed-point of T in X.

Using 2 · 3 and the fact that for fixed $x_n \in X$ the function g defined by $g(x) = d(x, x_n)$ is continuous, we have

$$d(\bar{x}, x_n) = \lim_{m \to \infty} d(x_m, x_n) \leq \frac{\lambda}{1-\lambda} d(x_{n-1}, x_n) \leq \frac{\lambda^n}{1-\lambda} d(x_0, x_1). \quad \blacksquare$$

If we use the notation $T^1 = T$ and $T^n = T \circ T^{n-1}$ when $n > 1$ for the iterates of T, then the sequence (x_n) above can be written $(T^n(x_0))$. The Banach Fixed-Point Theorem not only states that under appropriate conditions the equation $T(x) = x$ has a unique solution but it also provides a method for obtaining an approximation to this solution and a formula for measuring the accuracy of the approximation. The method of approximating a solution by forming the sequence $(T^n(x_0))$ is called the method of successive approximations.

For example, consider the problem of determining the solutions of the equation $f(x) = 0$ where f is a real-valued function of a real variable. Suppose that for some numbers a and b with $a < b$ we know that $f(a)$ and $f(b)$ have different signs, say $f(a) < 0 < f(b)$. Then, if f is continuous on $[a, b]$, the Intermediate Value Theorem implies that f has a zero at some point in the interval (a, b). Suppose now that we have the additional condition that f is differentiable on $[a, b]$ and $0 < m_1 \le f'(x) \le m_2$ for all $x \in [a, b]$. Then f has a unique zero in (a, b) and we can apply the Banach Fixed-Point Theorem to approximate this zero.

Convert the given problem to a fixed-point problem by letting

$$g(x) = x - \frac{1}{m} f(x).$$

Clearly, a point is a zero of f if and only if it is a fixed-point of g. We will show that, for an appropriate choice of m, g is a contraction of $[a, b]$ into itself. Since

$$g'(x) = 1 - \frac{1}{m} f'(x),$$

if we take $m \ge m_2$, then

$$0 \le 1 - \frac{m_2}{m} \le g'(x) \le 1 - \frac{m_1}{m} < 1 \quad \text{for all } x \in [a, b],$$

while if $0 < m_2 < 2m_1$ and we take $m \in (\frac{1}{2}m_2, m_1]$, then

$$-1 < 1 - \frac{m_2}{m} \le g'(x) \le 1 - \frac{m_1}{m} \le 0.$$

In either case for such a choice of m, g is a contraction of $[a, b]$ into itself. Thus, by the Banach Fixed-Point Theorem the unique fixed-point $\bar{x}$ of g in $[a, b]$ can be approximated by choosing any point x_0 in $[a, b]$ and then forming the sequence (x_n) according to the rule $x_n = g(x_{n-1})$. The fixed-point $\bar{x}$ is the limit of this sequence. If we take x_n as an approximation to $\bar{x}$, the formula

$$d(\bar{x}, x_n) \le \frac{\lambda}{1 - \lambda} d(x_{n-1}, x_n),$$

where $\lambda = 1 - m_1/m$ if $m \geq m_2$ and $\lambda = (m_2/m) - 1$ if $m \in (\frac{1}{2}m_2, m_1]$, gives a measure of the accuracy of the approximation. In Figures 1 and 2 we illustrate the successive approximations in the cases where $0 \leq g'(x) \leq 1 - m_1/m < 1$ and $-1 < 1 - m_2/m \leq g'(x) \leq 0$, respectively.

PROBLEMS

1. Show that a contraction is continuous.

2. Show that the equation

$$\cos x = x$$

has a unique solution, that it occurs in $[0, \pi/2]$, and indicate how an approximation to the solution can be obtained by successive approximation.

3. Determine an approximate solution of

$$x^4 + 2x^3 + 2x^2 + 3x - 4 = 0$$

in $(0, 1)$ by successive approximation so that the error is less than .01.

4. Newton's Method. Consider the equation

$$f(x) = 0$$

where f is a real-valued function of a real variable. Let x_0 be any initial approximation of the solution and let

$$x_{n+1} = x_n - \frac{f(x_n)}{f'(x_n)}.$$

Show that if there is a positive number a such that for all $x \in [x_0 - a, x_0 + a]$

$$\left| \frac{f(x)f''(x)}{(f'(x))^2} \right| \leq \lambda < 1 \quad \text{and} \quad \left| \frac{f(x_0)}{f'(x_0)} \right| \leq (1 - \lambda)a$$

then the sequence (x_n) converges to a solution of $f(x) = 0$.

5. Use Newton's Method to approximate $\sqrt{2}$ with an error less than 10^{-4}.

6. Prove the following generalization of the Banach Fixed-Point Theorem: If T is a transformation of a complete metric space X into itself such that the nth iterate, T^n, is a contraction for some positive integer n, then T has a unique fixed-point.

3 · LINEAR ALGEBRAIC EQUATIONS

In this section we consider an application of the Banach Fixed-Point Theorem to the solution of systems of linear algebraic equations. Let $X = V_m(\mathbf{R})$, the vector space of m-tuples of real numbers. There are a number of ways of defining a metric on X. As we have seen in Problem 8, p. 76, the metric spaccs (X, d_∞) and (X, d_1), where

$$d_\infty(x, y) = \max \{|x^i - y^i| : i = 1, \cdots, m\}$$

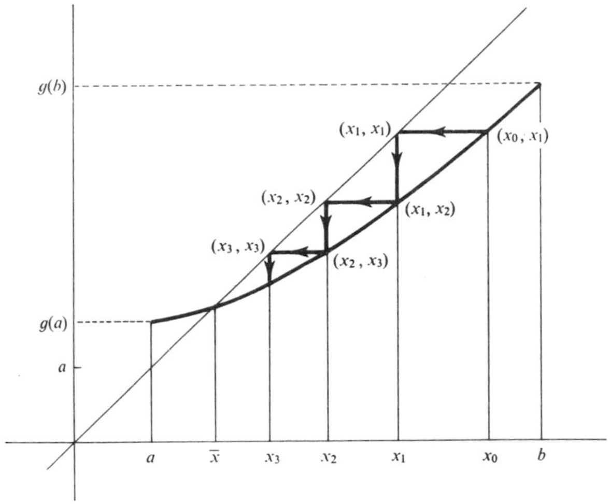

FIGURE 1

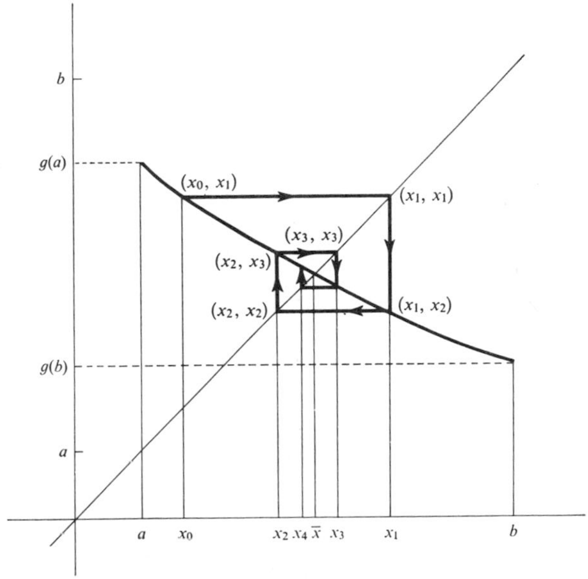

FIGURE 2

and

$$d_1(x, y) = \sum_{i=1}^{m} |x^i - y^i|,$$

are complete metric spaces. We shall use these two metrics here.

Consider the system of equations

$$(3 \cdot 1) \qquad \sum_{j=1}^{m} a_j{}^i x^j = y^i, \qquad i = 1, \cdots, m$$

which in matrix notation is

$$(3 \cdot 2) \qquad Ax = y,$$

where A is the $m \times m$ matrix $[a_j{}^i]$ and x and y are the $m \times 1$ matrices $[x^j]$ and $[y^i]$, respectively. The problem of solving Equation 3 · 2 can be replaced by the equivalent problem of finding the fixed-points of the transformation S_y from X into itself defined by

$$(3 \cdot 3) \qquad S_y(x) = (I - A)x + y = Bx + y$$

where $I = [\delta_j{}^i]$ is the $m \times m$ identity matrix and $B = I - A = [b_j{}^i]$. A point $x \in X$ is a solution of Equation 3 · 2 if and only if it is a fixed-point of the transformation S_y. If S_y is a contraction on the complete metric space (X, d_∞) [or on (X, d_1)], then by the Banach Fixed-Point Theorem, S_y has a unique fixed-point and Equation 3 · 2 has a unique solution.

Let $x_1 = (x_1{}^1, x_1{}^2, \cdots, x_1{}^m)$, $x_2 = (x_2{}^1, x_2{}^2, \cdots, x_2{}^m)$ and let $y_1 = S_y(x_1)$, $y_2 = S_y(x_2)$. Then

$$\begin{aligned} d_\infty(y_1, y_2) &= \max_i |y_1{}^i - y_2{}^i| = \max_i \left| \sum_{j=1}^{m} b_j{}^i (x_1{}^j - x_2{}^j) \right| \\ &\le \max_i \sum_{j=1}^{m} |b_j{}^i| \, |x_1{}^j - x_2{}^j| \\ &\le \left(\max_i \sum_{j=1}^{m} |b_j{}^i| \right) \left(\max_j |x_1{}^j - x_2{}^j| \right) \\ &= \lambda_\infty \, d_\infty(x_1, x_2) \end{aligned}$$

where $\lambda_\infty = \max_i \sum_{j=1}^{m} |b_j{}^i|$. If $\lambda_\infty < 1$, then S_y is a contraction of (X, d_∞) into itself and so has a unique fixed-point $\bar{x}$.

If x_0 is any point in X and the sequence (x_n) is defined recursively by the formula $x_n = S_y(x_{n-1})$, $n = 1, 2, \cdots$, then the sequence converges to $\bar{x}$ and

$$d_\infty(\bar{x}, x_n) \le \frac{\lambda_\infty{}^n}{1 - \lambda_\infty} \, d_\infty(x_1, x_0).$$

Moreover, since $|\bar{x}^i - x_n{}^i| \le d_\infty(\bar{x}, x_n)$ for each $i = 1, \cdots, m$, we conclude that the error in each component when $\bar{x}$ is approximated by x_n does not exceed $(\lambda_\infty{}^n/(1 - \lambda_\infty)) \, d_\infty(x_1, x_0)$.

In the space (X, d_1), we have

$$\begin{aligned} d_1(y_1, y_2) &= \sum_{i=1}^{m} |y_1{}^i - y_2{}^i| = \sum_{i=1}^{m} \left| \sum_{j=1}^{m} b_j{}^i(x_1{}^j - x_2{}^j) \right| \\ &\leq \sum_{i=1}^{m} \sum_{j=1}^{m} |b_j{}^i| \, |x_1{}^j - x_2{}^j| \\ &\leq \left(\max_j \sum_{i=1}^{m} |b_j{}^i| \right) \left(\sum_{j=1}^{m} |x_1{}^j - x_2{}^j| \right) \\ &= \lambda_1 \, d_1(x_1, x_2) \end{aligned}$$

where $\lambda_1 = \max_j \sum_{i=1}^{m} |b_j{}^i|$. If $\lambda_1 < 1$, then S_y is a contraction of (X, d_1) into itself and so has a unique fixed-point $\bar{x}$. Again, the sequence (x_n) defined by $x_n = S_y(x_{n-1})$ converges to $\bar{x}$ for any choice of x_0 and

$$d_1(\bar{x}, x_n) \leq \frac{\lambda_1{}^n}{1 - \lambda_1} \, d_1(x_1, x_0).$$

Since $|\bar{x}^i - x_n{}^i| \leq d_1(\bar{x}, x_n)$ for each $i = 1, \cdots, m$, the error in each component when $\bar{x}$ is approximated by x_n does not exceed $(\lambda_1{}^n/(1 - \lambda_1)) \, d_1(x_1, x_0)$.

If A is a symmetric matrix, i.e., if $a_j{}^i = a_i{}^j$ for all $i, j = 1, \cdots, m$, then $\lambda_1 = \lambda_\infty$, but if A is not symmetric then λ_1 and λ_∞ need not be equal. If either of them is less than 1, then S_y has a unique fixed-point and Equation $3 \cdot 2$ has a unique solution. Theoretically this solution can be found to any desired degree of accuracy by successive approximation. Practically in using successive approximation to solve problems on a computer the accuracy is limited by the number of digits one can conveniently use in the computation and by roundoff errors. We have the theoretical bound $(\lambda^n/(1 - \lambda)) \, d(x_1, x_0)$, on the error of the nth iterate where λ is the smaller of λ_1 and λ_∞ and d is the corresponding distance function.

In some important applications, the matrix A of Equation $3 \cdot 2$ is diagonally dominant or even strictly diagonally dominant [40]. A matrix A is said to be **diagonally dominant** if

$$|a_i{}^i| \geq \sum_{\substack{j=1 \\ j \neq i}}^{m} |a_j{}^i|$$

for each $i = 1, \cdots, m$ with strict inequality for at least one i. A is said to be **strictly diagonally dominant** if the strict inequality holds for each value of i. Suppose that A is strictly diagonally dominant. Without loss of generality we may assume that all diagonal elements of A are 1's since otherwise we can divide the ith equation in $3 \cdot 1$ by $a_i{}^i$ without changing the solution. Then the diagonal elements of $B = I - A$ are all 0's and

$$\sum_{j=1}^{m} |b_j{}^i| = \sum_{\substack{j=1 \\ j \neq i}}^{m} |a_j{}^i| < 1, \qquad i = 1, \cdots, m.$$

Hence $\lambda_\infty = \max_i \sum_{j=1}^m |b_j^i| < 1$ and Equation 3 · 2 has a unique solution which can be approximated by the successive iterates $x_n = S_y(x_{n-1})$. Any convenient choice can be made for x_0. Often one takes $x_0 = y$. This corresponds to taking $x_{-1} = 0$.

The iterative scheme determined by 3 · 3 is closely related to the iteration proposed by the German mathematician Karl G. J. Jacobi (1804–1851). Let $A = L + D + U$ where L is the lower triangular matrix with elements $l_j^i = a_j^i$ for $j < i$ and $l_j^i = 0$ for $j \geq i$, D is the diagonal matrix with elements $d_j^i = a_j^i$ for $i = j$ and $d_j^i = 0$ for $i \neq j$, and U is the upper triangular matrix with elements $u_j^i = a_j^i$ for $j > i$ and $u_j^i = 0$ for $j \leq i$. If all diagonal elements a_i^i are nonzero, then the matrix D is nonsingular and the Jacobi iteration is defined by

$$Dx_n = y - (L + U)x_{n-1}$$

or

$$(3 \cdot 4) \qquad x_n = D^{-1}y - D^{-1}(L + U)x_{n-1}.$$

In this notation the iteration determined by 3 · 3 becomes

$$(3 \cdot 5) \qquad x_n = y - (L + U + D - I)x_{n-1}.$$

If $D = I$, and this reduction can be made if $a_i^i \neq 0$ for all $i = 1, \cdots, m$, then the iterations 3 · 4 and 3 · 5 are the same.

4 · INTEGRAL EQUATIONS

Consider a system of n first-order differential equations in n unknowns with given initial values:

$$\dot{x}^i(t) = f^i(t, x^1(t), \cdots, x^n(t)); \qquad x^i(t_0) = v_0^{\,i};$$

$i = 1, \cdots, n$. In vector form this initial value problem can be written

$$(4 \cdot 1) \qquad \dot{x}(t) = f(t, x(t)), \qquad x(t_0) = v_0$$

where f is a given function from $\mathbf{R}^{n+1}$ into $\mathbf{R}^n$ and x is an unknown function from $\mathbf{R}$ into $\mathbf{R}^n$. In order to obtain a theorem on the uniqueness of a solution to this problem we assume that f satisfies a Lipschitz condition in the second variable. If there exists a number L such that for all x

$$|f(x, y_1) - f(x, y_2)| \leq L|y_1 - y_2|,$$

then we say that f satisfies a **Lipschitz condition** in the second variable.

It is easy to see that the differential equation 4 · 1 with initial condition is equivalent to the integral equation

$$(4 \cdot 2) \qquad x(t) = \int_{t_0}^{t} f(s, x(s))\, ds + v_0\,;$$

that is, x is a solution of the differential equation $4 \cdot 1$ satisfying the initial condition if and only if it is a solution of the integral equation $4 \cdot 2$. We will now discuss solutions of integral equations. The results to be obtained can be applied to the differential equation $4 \cdot 1$ and we will do that in the next section.

The integral equation $4 \cdot 2$ is of the Volterra type, named after the Italian mathematician Vito Volterra (1860–1940) who instituted a systematic study of such equations. It is a special case of the general **nonlinear Volterra integral equation of the second kind**:

$$x(t) = \lambda \int_{t_0}^{t} f(t, s, x(s))\, ds + g(t). \tag{4 · 3}$$

4 · 4 Theorem. *Let the function* $g : \mathbf{R} \to \mathbf{R}^n$ *satisfy a Lipschitz condition*

$$|g(t_1) - g(t_2)| \le L_1 |t_1 - t_2|$$

in a neighborhood N_1 *of* t_0 *and let the function* $f : \mathbf{R}^{n+2} \to \mathbf{R}^n$ *be continuous and satisfy a Lipschitz condition*

$$|f(t, s, v_1) - f(t, s, v_2)| \le L_2 |v_1 - v_2|$$

in a neighborhood N_2 *of* $(t_0, t_0, g(t_0))$. *Then, for each* $\lambda \in \mathbf{R}$ *and each* $a > 0$ *such that*

$$I = [t_0 - a, t_0 + a] \subset N_1$$

and

$$S = I \times I \times B[g(t_0); b] \subset N_2 \quad \textit{for some } b > 0,$$

the integral equation $4 \cdot 3$ *has a unique solution on* $I_1 = [t_0 - a_1, t_0 + a_1]$ *where* $a_1 = \min\{a, b/[|\lambda| M + L_1]\}$ *and* M *is a bound for* $|f|$ *on* S.

PROOF. Let X be the space of all continuous functions from the interval I_1 into the closed ball $B[g(t_0); b]$ with distance defined by†

$$d(x, y) = \max_{t \in I_1}(|x(t) - y(t)| e^{-\alpha|t - t_0|})$$

where $\alpha > 0$. It is easily shown that X is a complete metric space (Problem 1). If we let T be the transformation defined on X by the rule

$$T(x)(t) = \lambda \int_{t_0}^{t} f(t, s, x(s))\, ds + g(t),$$

then the solutions of the integral equation are the fixed-points of T.

We now show that T maps X into itself. Since $|f(t, s, v)| \le M$ for all $(t, s, v) \in S$, for each $x \in X$, $y = T(x)$ satisfies

$$|y(t) - g(t_0)| = \left| \lambda \int_{t_0}^{t} f(t, s, x(s))\, ds + g(t) - g(t_0) \right|$$

$$\le (|\lambda|\, M + L_1)|t - t_0| \le (|\lambda|\, M + L_1)\, a_1 \le b$$

for all $t \in I_1$. Thus $T(X) \subset X$.

†The choice of this metric was suggested to us by Professor Karl Nickle.

It remains to be shown that T is a contraction on X for sufficiently large α. Take $x_1, x_2 \in X$ and let $y_1 = T(x_1)$ and $y_2 = T(x_2)$. Then, for all $t \in I_1$

$$\begin{aligned}
|y_1(t) - y_2(t)| &= \left|\lambda \int_{t_0}^{t} [f(t, s, x_1(s)) - f(t, s, x_2(s))]\, ds\right| \\
&\leq |\lambda| \left|\int_{t_0}^{t} |f(t, s, x_1(s)) - f(t, s, x_2(s))|\, ds\right| \\
&\leq |\lambda|\, L_2 \left|\int_{t_0}^{t} |x_1(s) - x_2(s)|\, ds\right| \\
&\leq |\lambda|\, L_2\, d(x_1, x_2) \left|\int_{t_0}^{t} e^{\alpha|s-t_0|}\, ds\right| \\
&= \frac{1}{\alpha} |\lambda|\, L_2\, d(x_1, x_2)(e^{\alpha|t-t_0|} - 1) \\
&\leq \frac{1}{\alpha} |\lambda|\, L_2\, d(x_1, x_2) e^{\alpha|t-t_0|}.
\end{aligned}$$

Therefore

$$d(y_1, y_2) = d(T(x_1), T(x_2)) \leq \frac{1}{\alpha} |\lambda|\, L_2\, d(x_1, x_2).$$

Hence, T is a contraction on X if $\alpha > |\lambda|\, L_2$. Thus, if we choose $\alpha > |\lambda|\, L_2$, then T satisfies the conditions of the Banach Fixed-Point Theorem and, hence, T has a unique fixed-point which is the unique solution on I_1 of the integral equation. ∎

Another type of integral equation which has been studied extensively is:

$$(4 \cdot 5) \qquad x(t) = \lambda \int_a^b K(t, s) x(s)\, ds + g(t).$$

This is called a **Fredholm linear integral equation of the second kind** after the Swedish mathematician Erik Ivar Fredholm (1866–1927) who first studied such equations. If $g = 0$, then the equation is said to be **homogeneous**. The function K is called the **kernel**.

4·6 Theorem. *Let g be a continuous function from the closed interval $[a, b]$ in $\mathbf{R}$ into $\mathbf{R}^n$ and K be a continuous function from the square $S = \{(x, y) : x, y \in [a, b]\}$ in $\mathbf{R}^2$ into $\mathbf{R}^n$. Then the integral equation*

$$x(t) = \lambda \int_a^b K(t, s)x(s)\,ds + g(t)$$

has a unique continuous solution x on $[a, b]$ for each real λ such that $|\lambda| < 1/(M(b-a))$ where M is a bound for K on S.

PROOF. Since K is continuous on the compact set S, it is bounded on S; i.e., there exists a number M such that $|K(t, s)| \leq M$ for all $(t, s) \in S$. Let X be the complete metric space of all continuous functions from $[a, b]$ into $\mathbf{R}^n$ with distance defined by

$$d(x, y) = \max\{|x(t) - y(t)| : t \in [a, b]\}.$$

If T is the mapping of X into itself defined by

$$T(x)(t) = \lambda \int_a^b K(t, s)x(s)\,ds + g(t),$$

then the solutions of the integral equation are the fixed-points of T. Since

$$\begin{aligned} d(T(x_1), T(x_2)) &= \max_{t \in [a, b]} |T(x_1)(t) - T(x_2)(t)| \\ &\leq |\lambda|\, M(b-a)d(x_1, x_2), \end{aligned}$$

T is a contraction if $|\lambda| < 1/(M(b-a))$. Applying the Banach Fixed-Point Theorem, we conclude that this integral equation has a unique solution for each real λ such that $|\lambda| < 1/(M(b-a))$. ∎

Notice that for the Volterra equation we were able to show that there exists a unique solution for each real λ. On the other hand for the Fredholm equation we were able to show the existence of a unique solution only for $|\lambda|$ sufficiently small. In fact, it is not true in general that the Fredholm equation has a unique solution for all real λ.

PROBLEMS

1. Let X be the set of all continuous functions from $I_1 = [t_0 - a_1, t_0 + a_1]$ into the closed ball $B[g(t_0); b] \subset \mathbf{R}^n$. Show that for each $\alpha > 0$ the rule

$$d(x, y) = \max_{t \in I_1}(|x(t) - y(t)|\, e^{-\alpha|t - t_0|})$$

defines a metric on X and that the metric space (X, d) is complete.

2. Let the function $g : \mathbf{R} \to \mathbf{R}^n$ satisfy a Lipschitz condition in a neighborhood N_1 of t_0 and let the function $K : \mathbf{R}^2 \to \mathbf{R}$ be continuous in a neighborhood N_2 of (t_0, t_0). Show that for some $a > 0$ the linear Volterra integral equation

$$x(t) = \lambda \int_{t_0}^{t} K(t, s)x(s)\, ds + g(t)$$

has a unique solution on $I = [x_0 - a, x_0 + a]$ for each $\lambda \in \mathbf{R}$.

3. Consider the nonlinear Fredholm equation

$$x(t) = \lambda \int_{a}^{b} f(t, s, x(s))\, ds + g(t)$$

where $g : \mathbf{R} \to \mathbf{R}^n$ is continuous on $[a, b]$ and $f : \mathbf{R}^{n+2} \to \mathbf{R}^n$ is continuous and satisfies a Lipschitz condition:

$$|f(t, s, v_1) - f(t, s, v_2)| \leq L|v_1 - v_2|$$

on the set $S = \{(t, s, v) : t, s \in [a, b], v \in \mathbf{R}^n\}$. Show that the integral equation has a unique solution on $[a, b]$ if $|\lambda| < 1/(L(b - a))$.

5 · DIFFERENTIAL EQUATIONS

Returning to the system 4 · 1 of n first-order differential equations with initial conditions:

$$(5 \cdot 1) \qquad \dot{x}(t) = f(t, x(t)), \qquad x(t_0) = v_0$$

we can now obtain an existence and uniqueness theorem for this system. Since 5 · 1 is equivalent to the Volterra integral equation

$$(5 \cdot 2) \qquad x(t) = \int_{t_0}^{t} f(s, x(s))\, ds + v_0,$$

we can apply Theorem 4 · 4 to obtain:

5 · 3 Theorem. *Let the function $f : \mathbf{R}^{n+1} \to \mathbf{R}^n$ be continuous and satisfy a Lipschitz condition:*

$$|f(t, v_1) - f(t, v_2)| \leq L|v_1 - v_2|$$

in a neighborhood N of the point (t_0, v_0). Then, for each $a > 0$ such that

$$S = \{(t, v) : |t - t_0| \leq a,\ |v - v_0| \leq b\} \subset N \quad \textit{for some } b > 0,$$

there exists a unique solution of 5 · 1 on $I_1 = [t_0 - a_1, t_0 + a_1]$ where $a_1 =$ *min $\{a, b/M\}$ and M is a bound for $|f|$ on S.*

We now use the Arzelà–Ascoli Theorem to prove another existence theorem for differential equations due to the Italian logician and mathematician Giuseppe Peano (1858–1932). We will not assume that f satisfies a Lipschitz condition but only that f is continuous. However, in this case, we cannot conclude that the solution is unique.

5 · 4 Theorem (Peano). *If f is a continuous function from a neighborhood N of (t_0, v_0) in $\mathbf{R}^{n+1}$ into $\mathbf{R}^n$, then for some $a > 0$ there is a solution on $[t_0 - a, t_0 + a]$ of the differential equation $\dot{x}(t) = f(t, x(t))$ satisfying $x(t_0) = v_0$.*

PROOF. Let a_1 and b be positive numbers such that the set

$$S = \{(t, v) : |t - t_0| \leq a_1, \ |v - v_0| \leq b\} \subset N.$$

Since f is continuous on the compact set S in $\mathbf{R}^{n+1}$, f is bounded on S; that is, there exists a number M such that $|f(t, v)| \leq M$ for all $(t, v) \in S$. Let $a = \min\{a_1, b/M\}$. We will show that the differential equation has a solution on $I = [t_0, t_0 + a]$ by consideration of the equivalent integral equation:

$$x(t) = \int_{t_0}^{t} f(s, x(s))\, ds + v_0, \quad t \in I. \tag{5 · 5}$$

A similar argument shows the existence of a solution on $[t_0 - a, t_0]$.

Define a sequence (x_k) of functions on I as follows:

$$x_k(t) = \begin{cases} v_0 & \text{for } t \in [t_0, t_0 + a/k] \\ \displaystyle\int_{t_0}^{t - a/k} f(s, x_k(s))\, ds + v_0 & \text{for } t \in (t_0 + a/k, t_0 + a]. \end{cases}$$

Observe that these formulas determine the function x_k on I, since its values on $(t_0 + ja/k, t_0 + (j + 1)a/k]$ are determined by its values on $[t_0, t_0 + ja/k]$, $j = 1, \cdots, k - 1$, and its values on $[t_0, t_0 + a/k]$ are given. The set $E = \{x_k\}$ is bounded in $C(I)$, since for all $x_k \in E$

$$|x_k(t)| \leq \int_{t_0}^{t} |f(s, x_k(s))|\, ds + |v_0| \leq aM + |v_0| \quad \text{for all } t \in I.$$

Take $\varepsilon > 0$ and $t_1, t_2 \in I$. Then, for all $x_k \in E$, if $\tau_i = \max\{t_i - a/k, t_0\}$

$$|x_k(t_1) - x_k(t_2)| = \left| \int_{\tau_2}^{\tau_1} f(s, x_k(s))\, ds \right| \leq M\, |t_1 - t_2| < \varepsilon$$

whenever $|t_1 - t_2| < \varepsilon/M$. Thus, E is equicontinuous.

Then the set $\bar{E}$ is closed, bounded, and equicontinuous. Thus, by the Arzelà–Ascoli Theorem $\bar{E}$ is compact. This implies that the sequence (x_k) has a subsequence (x_{k_i}) which converges in $\bar{E}$, say to $\bar{x}$. We now show that $\bar{x}$ satisfies the integral equation 5 · 5. Clearly, $\bar{x}(t_0) = v_0$ since $x_k(t_0) = v_0$ for all k. For $t \in (t_0 + a/k_i, t_0 + a]$

$$x_{k_i}(t) = v_0 + \int_{t_0}^{t} f(s, x_{k_i}(s))\, ds - \int_{t - a/k_i}^{t} f(s, x_{k_i}(s)) ds. \tag{5 · 6}$$

Since

$$\left| \int_{t - a/k_i}^{t} f(s, x_{k_i}(s))\, ds \right| \leq \frac{Ma}{k_i},$$

the limit as $i \to \infty$ of the last term in $5 \cdot 6$ is zero. Moreover, since f is uniformly continuous on S and (x_{k_i}) converges uniformly to $\bar{x}$ on I, the sequence $(f(s, x_{k_i}(s)))$ converges uniformly to $f(s, \bar{x}(s))$ on I. Therefore

$$\lim_{i \to \infty} \int_{t_0}^{t} f(s, x_{k_i}(s))\, ds = \int_{t_0}^{t} f(s, \bar{x}(s))\, ds.$$

Thus, taking the limit as $i \to \infty$ in $5 \cdot 6$ we obtain

$$\bar{x}(t) = v_0 + \int_{t_0}^{t} f(s, \bar{x}(s))\, ds. \blacksquare$$

6 | The Lebesgue Integral

1 · INTRODUCTION

In analysis it is convenient to have a generalization of the integral considered in elementary analysis which was introduced in 1854 by the German mathematician Bernhard Riemann (1826–1866). The Riemann integral lacks some desirable properties which are possessed by its generalization introduced in 1904 by the French mathematician Henri Lebesgue (1875–1941). Consider the following example. Let $\{r_1, r_2, \cdots\}$ be an enumeration of the rational numbers in $[0, 1]$ and let

$$f_n(x) = \begin{cases} 1 & \text{for } x \in \{r_1, \ldots, r_n\} \\ 0 & \text{for } x \in [0, 1] \sim \{r_1, \ldots, r_n\}. \end{cases}$$

For all $x \in [0, 1]$, $\lim f_n(x) = f(x)$ where

$$f(x) = \begin{cases} 1 & \text{for } x \text{ rational} \\ 0 & \text{for } x \text{ irrational.} \end{cases}$$

Each of the functions f_n is Riemann integrable on $[0, 1]$ but f is not Riemann integrable. Thus, the nondecreasing sequence $(f_n(x))$ converges to $f(x)$ for each $x \in [0, 1]$, but $(\int_0^1 f_n)$ does not converge to $\int_0^1 f$, this integral does not even exist. However, the Lebesgue integral $\int_0^1 f$ does exist and $\lim \int_0^1 f_n = \int_0^1 f$ in the Lebesgue theory. In fact, we shall see that if (f_n) is a monotonic sequence of Lebesgue integrable functions which converges to a function f and if $(\int f_n)$ is bounded, then f is also Lebesgue integrable and $\lim \int f_n = \int \lim f_n$; i.e., the order of integration and limit may be interchanged.

A metric can be defined on the space $\mathbf{R}[0, 1]$ of functions which are Riemann integrable on $[0, 1]$ in the following way. We first alter the usual definition of equality of functions so that $f = g$ in $\mathbf{R}[0, 1]$ means that $\int_0^1 |f - g| = 0$. Then the metric is defined by $d(f, g) = \int_0^1 |f - g|$ for $f, g \in \mathbf{R}[0, 1]$.

Unfortunately the resulting metric space is not complete. For example, if $f_n(x) = 0$ for $x \in [0, 1/n)$ and $f_n(x) = 1/\sqrt{x}$ for $x \in [1/n, 1]$, then (f_n) is a Cauchy sequence which does not converge in $\mathbf{R}[0,1]$. However, if the Riemann integral is replaced by the Lebesgue integral, then the corresponding space is complete and we have a more satisfactory situation.

In this chapter we will first recall some facts about the Riemann integral of a function from $\mathbf{R}^n$ into $\mathbf{R}$. We will then define the Lebesgue integral for such functions and develop the theory for these integrals. In the course of this development we will show that if a function is Riemann integrable on a set then it is Lebesgue integrable on the set and the two integrals are equal.

Once the Lebesgue theory has been developed for functions from $\mathbf{R}^n$ to $\mathbf{R}$ it is an easy step to extend it to functions from $\mathbf{R}^n$ to $\mathbf{C}$.

2 · THE RIEMANN INTEGRAL

We recall briefly some facts about the Riemann integral of a function from $\mathbf{R}^n$ into $\mathbf{R}$. If $a = (a^1, \cdots, a^n)$ and $b = (b^1, \cdots, b^n)$ are points in $\mathbf{R}^n$ such that $a^i \leq b^i$ for $i = 1, \cdots, n$, then the closed interval $[a, b]$ is the set $\{x : a^i \leq x^i \leq b^i\}$ and the open interval (a, b) is the set $\{x : a^i < x^i < b^i\}$. If $I = (a, b)$ or $I = [a, b]$, the **volume** of I is defined as follows:

$$v(I) = \prod_{i=1}^{n} (b^i - a^i).$$

Observe that the empty set $\varnothing$ may be considered to be an open interval and $v(\varnothing) = 0$.

A set of open intervals $I_1, \cdots, I_r$ will be called a **partition** of $[a, b]$ if they are pairwise disjoint and $[a, b] = \bigcup_{i=1}^{r} \bar{I}_i$, where the bar denotes closure. Note that this use of the term partition differs somewhat from that in Chapter 1 since the points on the faces of the subintervals are in none of these open subintervals. A partition $P' = \{I_j' : j = 1, \cdots, s\}$ of $[a, b]$ is said to be a **refinement** of the partition P, denoted by $P \subset P'$, if for each j there exists an i such that $I_j' \subset I_i$. If P and P' are any two partitions of $[a, b]$, then the partition P'' consisting of the intervals of the form $I_i \cap I_j'$ is a common refinement of P and P'.

Let f be a real-valued function which is bounded on an interval $[a, b]$ in $\mathbf{R}^n$ and let $m = \inf\{f(x) : x \in [a, b]\}$ and $M = \sup\{f(x) : x \in [a, b]\}$. For a partition $P = \{I_1, \cdots, I_r\}$ of $[a, b]$ define

$$m_i = \inf\{f(x) : x \in \bar{I}_i\} \quad \text{and} \quad M_i = \sup\{f(x) : x \in \bar{I}_i\};$$

$$L(f, P) = \sum_{i=1}^{r} m_i v(I_i) \quad \text{and} \quad U(f, P) = \sum_{i=1}^{r} M_i v(I_i).$$

The sums $L(f, P)$ and $U(f, P)$ are called **lower** and **upper sums**, respectively, for f corresponding to the partition P. If P' is a refinement of P then

$$L(f, P) \leq L(f, P') \quad \text{and} \quad U(f, P') \leq U(f, P).$$

Thus, for any two partitions P_1 and P_2 of $[a, b]$, using a common refinement of P_1 and P_2, we can show that

$$mv([a, b]) \leq L(f, P_1) \leq U(f, P_2) \leq Mv([a, b]).$$

Let $\mathscr{P}$ denote the set of all partitions of $[a, b]$ and define the **lower** and **upper integrals** of f as follows:

$$\underline{\int_a^b} f = \sup\{L(f, P) : P \in \mathscr{P}\} \quad \text{and} \quad \overline{\int_a^b} f = \inf\{U(f, P) : P \in \mathscr{P}\}.$$

Then, for each $P \in \mathscr{P}$, we have

$$L(f, P) \le \underline{\int_a^b} f \le \overline{\int_a^b} f \le U(f, P).$$

The function f is said to be **Riemann integrable** on $[a, b]$ if the lower and upper integrals of f are equal and in this case the **integral** of f is defined to be the common value; that is,

$$\int_a^b f = \underline{\int_a^b} f = \overline{\int_a^b} f.$$

From this definition one can show that if f and g are integrable on $[a, b]$ and α and β are any real numbers, then $\alpha f + \beta g$ is integrable on $[a, b]$ and

$$\int_a^b (\alpha f + \beta g) = \alpha \int_a^b f + \beta \int_a^b g;$$

that is, the set of Riemann integrable functions on $[a, b]$ is a real linear space and the integral is a linear functional on this space.

The **mesh** of a partition P, denoted by $|P|$, is defined to be the largest diameter of any of the intervals in the partition; that is, if $P = \{I_1, \cdots, I_r\}$ then

$$|P| = \max\{d(I_i) : i = 1, \cdots, r\}.$$

2 · 1 Proposition. *If f is bounded on $[a, b]$, then there exists a sequence (P_k) of partitions of $[a, b]$ such that $P_k \subset P_{k+1}$, $\lim |P_k| = 0$, and*

$$\lim L(f, P_k) = \underline{\int_a^t} f \quad \textit{and} \quad \lim U(f, P_k) = \overline{\int_a^b} f.$$

PROOF. There exist partitions P_1' and P_1'' such that

$$\underline{\int_a^b} f - 1 < L(f, P_1') \le \underline{\int_a^b} f \quad \text{and} \quad \overline{\int_a^b} f \le U(f, P_1'') < \overline{\int_a^b} f + 1.$$

Let P_1 be a common refinement of P_1' and P_1'' for which $|P_1| < 1$. Then

$$\underline{\int_a^b} f - 1 < L(f, P_1) \le \underline{\int_a^b} f \quad \text{and} \quad \overline{\int_a^b} f \le U(f, P_1) < \overline{\int_a^b} f + 1.$$

Similarly there exist partitions P_2' and P_2'' such that

$$\underline{\int_a^b} f - \tfrac{1}{2} < L(f, P_2') \le \underline{\int_a^b} f \quad \text{and} \quad \overline{\int_a^b} f \le U(f, P_2'') < \overline{\int_a^b} f + \tfrac{1}{2}.$$

Let P_2 be a common refinement of P_1, P_2', and P_2'' for which $|P_2| < 1/2$. Then

$$\underline{\int_a^b} f - \tfrac{1}{2} < L(f, P_2) \le \underline{\int_a^b} f \quad \text{and} \quad \overline{\int_a^b} f \le U(f, P_2) < \overline{\int_a^b} f + \tfrac{1}{2}.$$

Continuing in this way we obtain a sequence of partitions having the desired properties. ∎

We will have more to say concerning the Riemann integral but we interrupt our discussion to introduce some concepts which are useful in that discussion as well as in the development of the Lebesgue integral.

PROBLEMS

1. If $\chi_{(c,d)}$ is the function from **R** into **R** such that

$$\chi_{(c,d)}(x) = \begin{cases} 1 & \text{for } x \in (c, d) \\ 0 & \text{otherwise} \end{cases}$$

and $(c, d) \subset [a, b]$, show that

$$\int_a^b \chi_{(c,d)} = d - c.$$

2. If f is the function from **R** into **R** such that

$$f(x) = \begin{cases} 1 & \text{for } x \text{ rational} \\ 0 & \text{for } x \text{ irrational} \end{cases}$$

show that f is not Riemann integrable on $[0, 1]$.

3. If χ_c is the function from $\mathbf{R}^n$ into **R** such that

$$\chi_c(x) = \begin{cases} 1 & \text{if } x = c \\ 0 & \text{if } x \neq c \end{cases}$$

and $c \in [a, b]$, show that

$$\int_a^b \chi_c = 0.$$

4. If f and g are functions from $\mathbf{R}^n$ into **R** which are Riemann integrable on $[a, b]$ and which differ at only a finite number of points in $[a, b]$, show that

$$\int_a^b f = \int_a^b g.$$

5. If $\{I_1, I_2\}$ is a partition of the interval $[a, b]$ in $\mathbf{R}^n$, verify that

$$v(I_1) + v(I_2) = v([a, b]).$$

6. Let f be a function from $\mathbf{R}^n$ to **R** which is bounded on $[a, b]$. Show that f is Riemann integrable on $[a, b]$ if and only if for each $\varepsilon > 0$ there is a partition P such that

$$U(f, P) - L(f, P) < \varepsilon.$$

7. If f is a function from **R** to **R** which is increasing on $[a, b]$, show that f is Riemann integrable on $[a, b]$.

3 · STEP FUNCTIONS

3 · 1 Definition. *A **step function** g on $\mathbf{R}^n$ is a real-valued function such that there exist an interval $[a, b]$, a partition of $[a, b]$ into open intervals $I_1, \cdots, I_r$, and real numbers $c_1, \cdots, c_r$ such that*

$$g(x) = c_i \quad \textit{for } x \in I_i \quad \textit{and} \quad g(x) = 0 \textit{ for } x \notin [a, b].$$

Clearly, a step function g does not uniquely determine the interval $[a, b]$ and partition $I_1, \cdots, I_r$ appearing in the above definition. A particular choice of these is called a representation of the step function.

We now define a real-valued function on the set of step functions. This function is called an integral and is denoted by the usual integral sign. If g is a step function with the representation given in Definition 3 · 1, then we define

$$\int g = \sum_{i=1}^{r} c_i v(I_i). \tag{3·2}$$

Actually this integral corresponds to the Riemann integral of g over $[a, b]$ but we do not need to use this fact and we will not prove it at this time. We must however, show that the integral of g is independent of the particular representation of g used.

Suppose that in addition to the above representation the step function g has the representation: there exist an interval $[a', b']$, a partition of $[a', b']$ into open intervals $I'_1, \cdots, I'_s$, and real numbers $c'_1, \cdots, c'_s$ such that

$$g(x) = c'_j \quad \text{for } x \in I'_j \quad \text{and} \quad g(x) = 0 \quad \text{for } x \notin [a', b'].$$

Letting $a''^k = \max\{a^k, a'^k\}$ and $b''^k = \min\{b^k, b'^k\}$ for $k = 1, \cdots, n$, we have $g(x) = 0$ for $x \notin [a'', b'']$. Also the set of open intervals $I_i \cap I'_j$, $i = 1, \cdots, r$; $j = 1, \cdots, s$ form a partition of $[a'', b'']$. If $c_i \neq 0$, then $I_i \subset [a'', b'']$ and $v(I_i) = \sum_{j=1}^{s} v(I_i \cap I'_j)$ so that

$$\sum_{i=1}^{r} c_i v(I_i) = \sum_{i=1}^{r} \sum_{j=1}^{s} c_i v(I_i \cap I'_j).$$

Similarly,

$$\sum_{j=1}^{s} c'_j v(I'_j) = \sum_{i=1}^{r} \sum_{j=1}^{s} c'_j v(I_i \cap I'_j).$$

If $I_i \cap I'_j \neq \varnothing$, then, for each $x \in I_i \cap I'_j$, $c_i = g(x) = c'_j$. Therefore,

$$\sum_{i=1}^{r} c_i v(I_i) = \sum_{j=1}^{s} c'_j v(I'_j).$$

Note that the values of a step function g on the faces of the open intervals I_i do not enter the computation of the integral of g. Any discontinuities of g will occur on these faces.

The following result is an immediate consequence of the definition of the integral of a step function.

3·3 Proposition. *If g is a step function such that $g(x) \geq 0$ for all x, then* $\int g \geq 0$.

For any set E in $\mathbf{R}^n$ the **characteristic function** χ_E of E is defined as follows:

$$\chi_{E(x)} = \begin{cases} 1 & \text{for } x \in E \\ 0 & \text{for } x \notin E. \end{cases}$$

If E is an interval I, then χ_I is a step function and

$$\int \chi_I = v(I).$$

We turn now to a discussion of the algebraic structure of the set $\mathscr{S}$ of step functions on $\mathbf{R}^n$.

3·4 Proposition. *The set $\mathscr{S}$ of step functions is a real linear space and the integral is a linear functional on $\mathscr{S}$.*

Proof. Let g and h be any elements in $\mathscr{S}$ and α and β be any real numbers. We can take representations of g and h in which $[a, b]$ is the basic interval for both; that is, $g(x) = 0$ and $h(x) = 0$ for $x \notin [a, b]$. Suppose that $\{I_i : i = 1, \cdots, r\}$ and $\{I'_j : j = 1, \cdots, s\}$ are partitions of $[a, b]$ such that g is constant on I_i and h is constant on I'_j. Then g and h are both constant on intervals of the partition

$$\{I_i \cap I'_j : i = 1, \ldots, r; j = 1, \ldots, s\}.$$

If

$$g(x) = c_{ij} = c_i \quad \text{and} \quad h(x) = d_{ij} = d_j \quad \text{for } x \in I_i \cap I'_j,$$

then $(\alpha g + \beta h)(x) = \alpha c_{ij} + \beta d_{ij}$ for $x \in I_i \cap I'_j$ and $(\alpha g + \beta h)(x) = 0$ for $x \notin [a, b]$. Therefore, $\alpha g + \beta h \in \mathscr{S}$ and this shows that $\mathscr{S}$ is a subspace of the real linear space of all real-valued functions on $\mathbf{R}^n$. Also,

$$\begin{aligned} \int (\alpha g + \beta h) &= \sum_{i=1}^{r} \sum_{j=1}^{s} (\alpha c_{ij} + \beta d_{ij}) v(I_i \cap I'_j) \\ &= \alpha \sum_{i=1}^{r} c_i v(I_i) + \beta \sum_{j=1}^{s} d_j v(I'_j) \\ &= \alpha \int g + \beta \int h. \blacksquare \end{aligned}$$

If for g and h in $\mathscr{S}$ we define $g \leq h$ to mean $g(x) \leq h(x)$ for all $x \in \mathbf{R}^n$, then $\mathscr{S}$ is a partially ordered set. A **lattice** is a partially ordered set in which any two elements have a least upper bound and a greatest lower bound. The least upper bound of the elements a and b is denoted by $a \vee b$ (a join b) and the greatest lower bound by $a \wedge b$ (a meet b). This terminology is related to union and interesection. In the partially ordered set $\mathscr{P}(X)$ of subsets of X ordered by inclusion the least upper bound of sets A and B is $A \cup B$ and the greatest lower bound is $A \cap B$.

We will now show that the set $\mathscr{S}$ of step functions is a lattice. Clearly, if $g \in \mathscr{S}$, then $|g| \in \mathscr{S}$. If for g and h in $\mathscr{S}$ we define

$$(g \wedge h)(x) = \min\{g(x), h(x)\} \quad \text{and} \quad (g \vee h)(x) = \max\{g(x), h(x)\},$$

then $g \wedge h$ and $g \vee h$ are the greatest lower bound and least upper bound of g and h, respectively. Also, the following formulas show that $g \wedge h$ and $g \vee h$ are in $\mathscr{S}$:

$$g \wedge h = \tfrac{1}{2}(g + h - |g - h|) \quad \text{and} \quad g \vee h = \tfrac{1}{2}(g + h + |g - h|).$$

Thus, we have proved:

3 · 5 Proposition. *The set $\mathscr{S}$ of step functions is a lattice.*

A set which is both a linear space and a lattice is called a **vector lattice**. The set $\mathscr{S}$ is a vector lattice. If $g \in \mathscr{S}$ and we define

$$g^+ = g \vee 0 \quad \text{and} \quad g^- = -(g \wedge 0),$$

then g^+ and g^- belong to $\mathscr{S}$. Observe that $g^+ \geq 0$ and $g^- \geq 0$ and that the following formulas hold:

$$g^+ = \tfrac{1}{2}(|g| + g), \qquad g^- = \tfrac{1}{2}(|g| - g),$$
$$g = g^+ - g^-, \qquad |g| = g^+ + g^-.$$

PROBLEMS

1. If g and h are step functions on $\mathbf{R}^n$ and $g \leq h$, show that $\int g \leq \int h$.

2. Let g be a step function on $\mathbf{R}$ with representation $g(x) = c_k$ for $x \in (x_{k-1}, x_k)$, $k = 1, \cdots, r$, and $g(x) = 0$ for $x \notin [x_0, x_r]$.

a. Express g as a linear combination of characteristic functions.

b. If $(x_0, x_r) \subset [a, b]$, determine the Riemann integral $\int_a^b g$.

3. Show that the product of two step functions is a step function.

4. If g is a step function on $\mathbf{R}^n$ and for $c \in \mathbf{R}^n$ the function φ from $\mathbf{R}^n$ into $\mathbf{R}^n$ is defined by $\varphi(x) = x - c$, show that $g \circ \varphi$ is a step function on $\mathbf{R}^n$ and

$$\int g \circ \varphi = \int g.$$

5. If $\{I_1, \cdots, I_r\}$ is a set of open intervals which are pairwise disjoint and $I_i \subset \bigcup_{j=1}^s J_j$ $(i = 1, \cdots, r)$ where each J_j is an open interval, show that

$$\sum_{i=1}^r v(I_i) \leq \sum_{j=1}^s v(J_j).$$

6. Show that the set of continuous functions from a metric space X into $\mathbf{R}$ is a vector lattice.

7. If f, g, and h are in $\mathcal{S}$, show that:

a. $f \wedge g = -((-f) \vee (-g))$ and $f \vee g = -((-f) \wedge (-g))$

b. $f \vee (g \wedge h) = (f \vee g) \wedge (f \vee h)$

$(f \vee g) \wedge h = (f \wedge h) \vee (g \wedge h)$

8. If $\lim g_k = g$ and $\lim h_k = h$, show that

$$\lim(g_k \wedge h_k) = g \wedge h \quad \text{and} \quad \lim(g_k \vee h_k) = g \vee h.$$

9. Let k and m be positive integers and let $\mathbf{R}^n = \mathbf{R}^k \times \mathbf{R}^m$. If $x = (x^1, \cdots, x^k) \in \mathbf{R}^k$ and $y = (y^1, \cdots, y^m) \in \mathbf{R}^m$, let $(x, y) = (x^1, \cdots, x^k, y^1, \cdots, y^m) \in \mathbf{R}^n$. Suppose $I^k = (a_1, b_1)$ and $I^m = (a_2, b_2)$ are open intervals in $\mathbf{R}^k$ and $\mathbf{R}^m$, respectively, and let $I^n = (a, b)$ where $a = (a_1, a_2)$, $b = (b_1, b_2)$; that is, $I^n = I^k \times I^m$. Show that

a. $\chi_{I^n}(x, y) = \chi_{I^k}(x)\chi_{I^m}(y)$

b. $v_n(I^n) = v_k(I^k)v_m(I^m)$ where the subscript on v denotes the dimension of the space in which volume is taken.

10. If g and h are step functions in $\mathbf{R}^k$ and $\mathbf{R}^m$, respectively, and $f(x, y) = g(x)h(y)$ (see Problem 9), show that f is a step function in $\mathbf{R}^n = \mathbf{R}^k \times \mathbf{R}^m$ and

$$\int f = \int g \cdot \int h.$$

4 · SETS OF MEASURE ZERO

As we have seen, the values of a step function on the faces of the open intervals of a representation of the function do not affect the integral of the function. For the Lebesgue integral there are also certain sets which are negligible. These are the sets of measure zero which we now define.

4 · 1 Definition. *A set E in $\mathbf{R}^n$ is a **set of measure zero** if for each $\varepsilon > 0$ there is a countable collection $\{I_k\}$ of open intervals such that*

$$E \subset \bigcup_{k=1}^{\infty} I_k \quad \text{and} \quad \sum_{k=1}^{\infty} v(I_k) \leq \varepsilon.$$

If a proposition $P(x)$ holds for all x except for a set of measure zero, we say that $P(x)$ holds **almost everywhere** (abbreviated: a.e.) or $P(x)$ holds **for almost all** x (abbreviated: for a.a.x).

Clearly, a countable union of sets of measure zero is a set of measure zero. Also, it is easy to see that an $(n-1)$-dimensional face of an n-dimensional interval is a set of measure zero in $\mathbf{R}^n$. Thus, the set of points of discontinuity of a step function on $\mathbf{R}^n$ is a set of measure zero.

Sets which are negligible with respect to Lebesgue integration can also be characterized in terms of step functions.

4 · 2 Definition. *A set E in $\mathbf{R}^n$ is a **null set** if there is a nondecreasing sequence (g_k) of nonnegative step functions such that $(g_k(x))$ diverges to ∞ for each $x \in E$ while $(\int g_k)$ converges.*

4 · 3 Proposition. *A set E in $\mathbf{R}^n$ is a null set if and only if it is of measure zero.*

PROOF. Assume E is a null set; that is, there exists a nondecreasing sequence (g_k) of nonnegative step functions such that $(g_k(x))$ diverges to ∞ for each $x \in E$ and $(\int g_k)$ converges. Let $\lim \int g_k = c$. Take $\varepsilon > 0$. For each step function g_k take a representation with basic interval $[a_k, b_k]$ and open subintervals I_i^k, $i = 1, \cdots, r_k$, on which g_k is constant. We may assume that the representations are chosen so that each interval I_i^{k+1} which intersects $[a_k, b_k]$ is contained in an interval I_j^k. The set E' of points which lie on a face of some I_i^k, $k = 1, 2, \cdots$, is a set of measure zero (Problem 2). Let $J_1, \cdots, J_{s_1}$ be the open intervals in $\{I_i^1\}$ on which $g_1 > c/\varepsilon$. Then

$$\sum_{i=1}^{s_1} \frac{c}{\varepsilon} v(J_i) < \int g_1 \leq c \quad \text{and, hence,} \quad \sum_{i=1}^{s_1} v(J_i) < \varepsilon.$$

The function g_2 will have values greater than c/ε on $J_1, \cdots, J_{s_1}$ and perhaps on some intervals $J_{s_1+1}, \cdots, J_{s_2}$ in $\{I_i^2\}$ which are disjoint from the previously chosen J-intervals. Then

$$\sum_{i=1}^{s_2} \frac{c}{\varepsilon} v(J_i) < \int g_2 \leq c \quad \text{and, hence,} \quad \sum_{i=1}^{s_2} v(J_i) < \varepsilon.$$

Continuing in this way we obtain a countable collection $\{J_i\}$ of open intervals which covers $E \sim E'$. Also, if $m \leq s_k$, then

$$\sum_{i=1}^{m} v(J_i) \leq \sum_{i=1}^{s_k} v(J_i) < \frac{\varepsilon}{c} \int g_k \leq \varepsilon$$

and, thus,

$$\sum v(J_i) \leq \varepsilon.$$

Therefore, $E \sim E'$ is a set of measure zero and, hence, E is of measure zero.

Now assume that E is of measure zero. For each positive integer, k, let $\{I_i^k\}$ be a countable covering of E by open intervals such that $\sum_{i=1}^{\infty} v(I_i^k) < 1/2^k$. The set $\{I_i^k : k = 1, 2, \cdots; i = 1, 2, \cdots\}$ is a countable collection and, hence, can be enumerated as $\{I_j : j = 1, 2, \cdots\}$. Let χ_j be the characteristic function of I_j. Define $g_m = \sum_{j=1}^{m} \chi_j$. Then, (g_m) is a nondecreasing sequence of nonnegative step functions. Also, for $x \in E$, $(g_m(x))$ diverges to ∞ but

$$\lim \int g_m = \lim \sum_{j=1}^{m} v(I_j) \leq \sum_{k=1}^{\infty} \frac{1}{2^k} = 1. \quad \blacksquare$$

4·4 Proposition. *If* (h_k) *is a nonincreasing sequence of nonnegative step functions on* $\mathbf{R}^n$ *and* $\lim \int h_k = 0$, *then*

$$\lim h_k(x) = 0 \quad \textit{for a.a.x.}$$

PROOF. Since, for each x, $(h_k(x))$ is a nonincreasing sequence of nonnegative numbers, $\lim h_k(x)$ exists. Let $\lim h_k(x) = h(x)$ and $E = \{x : h(x) \neq 0\}$. Then, $E = \bigcup_{m=1}^{\infty} E_m$ where $E_m = \{x : h(x) \geq 1/m\}$. Thus, if each E_m is a null set, then E is a null set. For a fixed positive integer m, (mh_k) is a nonincreasing sequence of nonnegative step functions. Also,

$$mh_k(x) \geq mh(x) \geq 1 \quad \text{for all } x \in E_m \text{ and all } k.$$

Since $\lim \int mh_k = 0$ we can choose a subsequence (mh_{k_i}) of (mh_k) such that $\int mh_{k_i} \leq \dfrac{1}{i^2}$. Let $g_j = \sum_{i=1}^{j} mh_{k_i}$. Then (g_j) is a nondecreasing sequence of nonnegative step functions such that $(g_j(x))$ diverges to ∞ for each $x \in E_m$ and $(\int g_j)$ converges. Thus, E_m is a null set. ∎

The converse of this proposition also holds.

4·5 Proposition. *If* (h_k) *is a nonincreasing sequence of nonnegative step functions and* $\lim h_k(x) = 0$ *for a.a.x, then*

$$\lim \int h_k = 0.$$

PROOF. Suppose h_1 is zero outside $[a, b]$. Then, for all k, h_k is zero outside $[a, b]$. Let E_1 be the set of measure zero on which $\lim h_k(x) \neq 0$, and let E_2 be the set of points where any of the step functions h_k is discontinuous. The set E_2 is also of measure zero and, hence, $E = E_1 \cup E_2$ is of measure zero. Take $\varepsilon > 0$. There exists a countable collection $\{I_i\}$ of open intervals such that $E \subset \bigcup I_i$ and $\sum v(I_i) < \varepsilon$. Take $x \in [a, b] \sim E$. Since $\lim h_k(x) = 0$ there exists a positive integer $k(x)$ such that $h_{k(x)}(x) < \varepsilon$. There also exists an open interval $J(x)$ containing x such that $h_{k(x)}$ is constant on $J(x)$. Then, $\{I_i\} \cup \{J(x) : x \in [a, b] \sim E\}$ is an open covering of the compact set $[a, b]$ and, hence, has a finite subcover $\{I_1, \cdots, I_r\} \cup \{J(x_1), \cdots, J(x_s)\}$.

Let $k_0 = \max\{k(x_1), \cdots, k(x_s)\}$ and take $k \geq k_0$. Suppose $\{I_j^k : j = 1, \cdots, t\}$ is a partition of $[a, b]$ such that $h_k(x) = c_j^k$ for $x \in I_j^k$. If $I_j^k \cap J(x_i) \neq \varnothing$ for some $i = 1, \cdots, s$, then for each $y \in I_j^k \cap J(x_i)$

$$c_j^k = h_k(y) \leq h_{k(x_i)}(y) = h_{k(x_i)}(x_i) < \varepsilon.$$

Letting $\sum'$ denote the sum over such I_j^k we have

$$\sum_j{}' c_j^k v(I_j^k) < \varepsilon v([a, b]). \tag{4·6}$$

If $I_j^k \cap J(x_i) = \varnothing$ for all $i = 1, \cdots, s$, then $I_j^k \subset \bigcup_{i=1}^{r} I_i$. Letting $\sum''$ denote the sum over such intervals I_j^k we have

$$(4\cdot 7) \qquad \sum_j{}'' c_j^k v(I_j^k) \le M \sum_{i=1}^{r} v(I_i) < M\varepsilon,$$

where M is an upper bound for h_1 and therefore for all h_k. Problem 5, p. 113, has been used to obtain both $4\cdot 6$ and $4\cdot 7$. Combining $4\cdot 6$ and $4\cdot 7$, for $k \ge k_0$ we have

$$\int h_k = \sum_{j=1}^{t} c_j^k v(I_j^k) < [v([a, b]) + M]\varepsilon$$

and, therefore,

$$\lim \int h_k = 0. \ \blacksquare$$

PROBLEMS

1. Show that a countable set in $\mathbf{R}^n$ is of measure zero.

2. Show that an $(n-1)$-dimensional face E of an n-dimensional interval is a set of measure zero in $\mathbf{R}^n$.

3. If (a, b) is an open interval in $\mathbf{R}^n$ with $a^i < b^i$ for $i = 1, \cdots, n$ show that (a, b) is not of measure zero.

4. Show that the Cantor set (p. 20) which is an uncountable set in $\mathbf{R}$ is a set of measure zero in $\mathbf{R}$.

5. If E is the set of points in $[0, 1]$ which have no 4 in their decimal representation, show that E is a set of measure zero.

6. Show that E is a null set if and only if there exists a sequence (h_k) of step functions such that $\sum h_k(x)$ diverges for each $x \in E$ and $\sum \int |h_k|$ converges.

7. If E is a set of measure zero in $\mathbf{R}^k$, show that

a. $E \times \mathbf{R}$ is of measure zero in $\mathbf{R}^{k+1}$

b. $E \times \mathbf{R}^m$ is of measure zero in $\mathbf{R}^n = \mathbf{R}^k \times \mathbf{R}^m$.

5 · THE RIEMANN INTEGRAL CONTINUED

Let f be a bounded real-valued function on the interval $[a, b]$ in $\mathbf{R}^n$, $P = \{I_i : i = 1, \cdots, r\}$ be a partition of $[a, b]$, and

$$m_i = \inf\{f(x) : x \in \bar{I}_i\} \quad \text{and} \quad M_i = \sup\{f(x) : x \in \bar{I}_i\}.$$

Define the step functions g and G as follows:

$$g(x) = \begin{cases} m_i & \text{for } x \in I_i \\ f(x) & \text{for } x \in \bar{I}_i \sim I_i \\ 0 & \text{for } x \notin [a, b] \end{cases} \quad \text{and} \quad G(x) = \begin{cases} M_i & \text{for } x \in I_i \\ f(x) & \text{for } x \in \bar{I}_i \sim I_i \\ 0 & \text{for } x \notin [a, b]. \end{cases}$$

Then

$$\int g = \sum_{i=1}^{r} m_i v(I_i) = L(f, P) \quad \text{and} \quad \int G = \sum_{i=1}^{r} M_i v(I_i) = U(f, P).$$

Let (P_k) be a sequence of partitions of $[a, b]$ such that $P_k \subset P_{k+1}$, $\lim |P_k| = 0$, and

$$\lim L(f, P_k) = \underline{\int_a^b} f \quad \text{and} \quad \lim U(f, P_k) = \overline{\int_a^b} f.$$

The existence of such a sequence was shown in Proposition 2 · 1. If g_k and G_k are the step functions corresponding to P_k according to the above prescription, then the sequences (g_k) and (G_k) are nondecreasing and nonincreasing, respectively, and

$$\lim \int g_k = \underline{\int_a^b} f \quad \text{and} \quad \lim \int G_k = \overline{\int_a^b} f.$$

Also,

$$g_k(x) \le f(x) \le G_k(x) \quad \text{for all } k \text{ and for all } x \in [a, b].$$

5 · 1 Theorem. *If f is Riemann integrable on $[a, b]$, then f is continuous almost everywhere on $[a, b]$.*

PROOF. Let (P_k) be a sequence of partitions of $[a, b]$ such that $P_k \subset P_{k+1}$, $\lim |P_k| = 0$, and

$$\lim \int g_k = \underline{\int_a^b} f \quad \text{and} \quad \lim \int G_k = \overline{\int_a^b} f.$$

If $h_k = G_k - g_k$, then (h_k) is a nonincreasing sequence of nonnegative step functions such that

$$\lim \int h_k = \lim \int G_k - \lim \int g_k = \overline{\int_a^b} f - \underline{\int_a^b} f = 0.$$

Then, by Proposition 4 · 4, $\lim h_k(x) = 0$ for a.a.x; that is,

$$\lim g_k(x) = f(x) = \lim G_k(x) \quad \text{for all } x \in [a, b] \sim E_1$$

where E_1 is of measure zero. If E_2 is the set of points on the faces of the intervals in any of the partitions P_k, then E_2 is of measure zero and, hence, $E = E_1 \cup E_2$ is of measure zero.

We will show that f is continuous on $[a, b] \sim E$. Take $x \in [a, b] \sim E$ and $\varepsilon > 0$. There exists a positive integer k such that

$$G_k(x) - g_k(x) < \varepsilon.$$

Since x is in some open interval I_i^k in P_k, there exists an open ball $B(x; \delta)$ such that

$$B(x; \delta) \subset I_i^k.$$

For all $y \in B(x; \delta)$

$$g_k(x) = g_k(y) \le f(y) \le G_k(y) = G_k(x).$$

In particular,

$$g_k(x) \leq f(x) \leq G_k(x)$$

and, therefore,

$$|f(x) - f(y)| \leq G_k(x) - g_k(x) < \varepsilon. \blacksquare$$

The converse of this result also holds.

5 · 2 Theorem. *If f is bounded and continuous almost everywhere on $[a, b]$, then f is Riemann integrable on $[a, b]$.*

PROOF. Suppose f is continuous on $[a, b] \sim E_1$ where E_1 is of measure zero. Let (P_k) be a sequence of partitions of $[a, b]$ such that $P_k \subset P_{k+1}$, $\lim |P_k| = 0$, and

$$\lim \int g_k = \underline{\int_a^b} f \quad \text{and} \quad \lim \int G_k = \overline{\int_a^b} f.$$

If E_2 is the set of points on the faces of intervals in any P_k, then E_2 is of measure zero and, hence, $E = E_1 \cup E_2$ is of measure zero. Take $x \in [a, b] \sim E$. For each $\varepsilon > 0$ there exists a $\delta > 0$ such that

$$|f(x) - f(y)| < \varepsilon/2 \quad \text{whenever } y \in B(x; \delta) \cap [a, b].$$

Since $\lim |P_k| = 0$ there exists a positive integer k_0 such that $|P_k| < \delta$ whenever $k \geq k_0$. Thus, for each $k \geq k_0$, if x is in the interval I_i^k in P_k then

$$h_k(x) = G_k(x) - g_k(x) = M_i - m_i \leq \varepsilon.$$

This shows that $\lim h_k(x) = 0$ for a.a.x and, therefore, by Proposition 4 · 5

$$\lim \int h_k = 0;$$

that is,

$$\underline{\int_a^b} f = \lim \int g_k = \lim \int G_k = \overline{\int_a^b} f. \blacksquare$$

If E is a bounded set in $\mathbf{R}^n$, say $E \subset [a, b]$, and f is a bounded real-valued function on E, then f is said to be Riemann integrable on E if $f\chi_E$ is Riemann integrable on $[a, b]$ and, in this case, we define

$$\int_E f = \int_a^b f\chi_E.$$

PROBLEMS

1. If f is defined on $[a, b]$ and (f_k) and (F_k) are, respectively, a nondecreasing and a nonincreasing sequence of step functions such that

$$f_k(x) \leq f(x) \leq F_k(x) \quad \text{for all } k \text{ and all } x \in [a, b]$$

and

$$\lim f_k(x) = f(x) = \lim F_k(x) \quad \text{for a.a.} x \in [a, b],$$

show that f is Riemann integrable on $[a, b]$.

2. If f and g are Riemann integrable on $[a, b]$, show that fg is Riemann integrable on $[a, b]$.

3. Let f be the function defined on $[0, 1]$ by the rule

$$f(x) = \begin{cases} 1/q & \text{when } x = p/q \text{ and } p \text{ and } q \text{ are relatively prime integers} \\ 0 & \text{when } x \text{ is irrational.} \end{cases}$$

Show that f is Riemann integrable on $[0, 1]$.

4. If f is the function defined in Problem 3 and

$$g(x, y) = f(x)f(y) \quad \text{for } (x, y) \in [(0, 0), (1, 1)]$$

is g Riemann integrable on $[(0, 0), (1, 1)]$?

5. If f is continuous on the interval $[a, b]$ in $\mathbf{R}$, show that the graph of f is a set of measure zero in $\mathbf{R}^2$.

6. If f is a bounded function which is continuous a.e. on the open ball $B(0; 1)$ in $\mathbf{R}^2$, show that $\int_{B(0;1)} f$ exists.

6 · EXTENSION OF THE INTEGRAL OF STEP FUNCTIONS

The Lebesgue integral will be developed from the integral of step functions. This development will be accomplished in two stages and the material in this section represents the first stage.

6 · 1 Definition. *Let $\mathscr{P}$ be the set of real-valued functions g defined on $\mathbf{R}^n$ for which there is a nondecreasing sequence (g_k) of step functions such that* $\lim g_k(x) = g(x)$ *for a.a.x and the sequence $(\int g_k)$ is bounded. For $g \in \mathscr{P}$, the integral is defined as follows:*

$$\int g = \lim \int g_k .$$

Some observations on this definition are in order. First, since $(\int g_k)$ is a bounded nondecreasing sequence of real numbers, it converges. To show that this is a proper definition of $\int g$ we must show that any other sequence with the properties of (g_k) would give the same result. This will follow from:

6 · 2 Proposition. *If (g_k) and (h_k) are nondecreasing sequences of step functions such that* $\lim g_k(x) \leq \lim h_k(x)$ *for a.a.x and $(\int g_k)$ and $(\int h_k)$ are bounded, then*

$$\lim \int g_k \leq \lim \int h_k .$$

PROOF. For a fixed positive integer i, $(g_i - h_k)$ is a nonincreasing sequence of step functions. And, $(g_i - h_k)^+$ is a nonincreasing sequence of nonnegative step functions such that $\lim_{k\to\infty} (g_i(x) - h_k(x))^+ = 0$ for a.a.x. Then, by Proposition 4 · 5,

$$\lim_{k\to\infty} \int (g_i - h_k)^+ = 0.$$

Since

$$g_i - h_k \leq (g_i - h_k)^+,$$

$$\lim_{k\to\infty} \int (g_i - h_k) \leq 0;$$

that is,

$$\int g_i \leq \lim_{k\to\infty} \int h_k.$$

This inequality holds for all positive integers i and, hence,

$$\lim_{i\to\infty} \int g_i \leq \lim_{k\to\infty} \int h_k. \blacksquare$$

6 · 3 Corollary. *If (g_k) and (h_k) are nondecreasing sequences of step functions such that $(\int g_k)$ and $(\int h_k)$ are bounded and* $\lim g_k(x) = g(x) = \lim h_k(x)$ *for a.a.x, then*

$$\lim \int g_k = \lim \int h_k$$

and the integral of g is this common limit.

6 · 4 Corollary. *If $g, h \in \bar{\mathscr{S}}$ and $g(x) \leq h(x)$ for a.a.x, then*

$$\int g \leq \int h.$$

The set $\mathscr{S}$ of step functions on $\mathbf{R}^n$ is a subset of the set $\bar{\mathscr{S}}$ and if $g \in \mathscr{S}$, then its integral as an element of $\bar{\mathscr{S}}$ agrees with its integral as an element of $\mathscr{S}$. Also, if $g \in \bar{\mathscr{S}}$ and $h = g$ almost everywhere, then $h \in \bar{\mathscr{S}}$ and $\int h = \int g$.

Consider the Dirichlet function g defined on $\mathbf{R}$ as follows:

$$g(x) = \begin{cases} 1 & \text{for } x \in [0, 1] \text{ and } x \text{ rational} \\ 0 & \text{otherwise.} \end{cases}$$

If $\{r_1, r_2, \cdots\}$ is an enumeration of the rational numbers in [0, 1] and

$$g_k(x) = \begin{cases} 1 & \text{for } x \in \{r_1, \cdots, r_k\} \\ 0 & \text{for } x \in \mathbf{R} \sim \{r_1, \cdots, r_k\}, \end{cases}$$

then (g_k) is a nondecreasing sequence of step functions such that $\lim g_k(x) = g(x)$ for all x. Also, for all k, $\int g_k = 0$. Thus, $g \in \bar{\mathscr{S}}$ and $\int g = 0$.

In general if $g \in \mathcal{P}$ we cannot say that $-g \in \mathcal{P}$. Although $\mathcal{P}$ is not a linear space, we do have the following partial result in that direction.

6 · 5 Proposition. *If g and h are in $\mathcal{P}$ and α is a nonnegative real number, then $g + h$ and αg are in $\mathcal{P}$ and*

$$\int g + h = \int g + \int h \quad \textit{and} \quad \int \alpha g = \alpha \int g.$$

The proof is left to the reader (Problem 1).

6 · 6 Proposition. *$\mathcal{P}$ is a lattice.*

PROOF. Let g and h be in $\mathcal{P}$ and let (g_k) and (h_k) be nondecreasing sequences of step functions such that $(\int g_k)$ and $(\int h_k)$ are bounded and (g_k) and (h_k) converge almost everywhere to g and h, respectively.

We will prove that $g \wedge h \in \mathcal{P}$. Clearly, $(g_k \wedge h_k)$ is a nondecreasing sequence of step functions. Since $g_k(x) \wedge h_k(x) \leq g_k(x)$, $\int(g_k \wedge h_k) \leq \int g_k$ and, hence, $(\int(g_k \wedge h_k))$ is bounded. Since $\lim g_k(x) = g(x)$ and $\lim h_k(x) = h(x)$ for a.a.x, $\lim(g_k(x) \wedge h_k(x)) = g(x) \wedge h(x)$ for a.a.x (Problem 8, p. 114) and this shows that $g \wedge h \in \mathcal{P}$.

The proof that $g \vee h \in \mathcal{P}$ is similar, using the inequality

$$g_k \vee h_k \leq g_k + h_k + |g_1| + |h_1|$$

to show that $(\int(g_k \vee h_k))$ is bounded.∎

The following is the basic convergence result for $\mathcal{P}$.

6 · 7 Proposition. *Let (g_k) be a nondecreasing sequence of functions in $\mathcal{P}$ such that $(\int g_k)$ is bounded. Then $(g_k(x))$ converges for a.a.x and if $g(x) = \lim g_k(x)$ for a.a.x, then $g \in \mathcal{P}$ and*

$$\int g = \lim \int g_k.$$

PROOF. For each positive integer k let $(g_{kj})_{j=1}^{\infty}$ be a nondecreasing sequence of step functions such that $(\int g_{kj})_{j=1}^{\infty}$ is bounded and $\lim_{j\to\infty} g_{kj}(x) = g_k(x)$ for a.a.x. If we define

$$h_m = g_{1m} \vee g_{2m} \vee \cdots \vee g_{mm},$$

then (h_m) is a nondecreasing sequence of step functions. Since, for all j, $g_{kj}(x) \leq g_k(x)$ for a.a.x,

$$h_m(x) \leq g_1(x) \vee g_2(x) \vee \cdots \vee g_m(x) = g_m(x) \quad \text{for a.a.}x \tag{6 · 8}$$

and, therefore,

$$\int h_m \leq \int g_m.$$

This shows that $(\int h_m)$ is bounded and, hence, converges. From the definition of a null set, (h_m) must then converge almost everywhere. If we let h be a function such that $h(x) = \lim h_m(x)$ for a.a.x, then $h \in \mathcal{P}$ and $\int h = \lim \int h_m$.

For each k and for $m \geq k$

$$g_{km}(x) \leq h_m(x) \quad \text{for all } x$$

and, therefore, taking the limit with respect to m, we obtain

$$g_k(x) \leq h(x) \quad \text{for a.a.}x.$$

This shows that $(g_k(x))$ converges for a.a.x. Also, if $g(x) = \lim g_k(x)$ for a.a.x, we have $g(x) \leq h(x)$ for a.a.x. But $6 \cdot 8$ implies $h(x) \leq g(x)$ for a.a.x. Thus, $g(x) = h(x)$ for a.a.x and, therefore, $g \in \mathscr{P}$ and $\int g = \int h$.

Since $g_k(x) \leq g(x)$ for a.a.x, Corollary $6 \cdot 4$ implies that

$$\int h_k \leq \int g_k \leq \int g \quad \text{for all } k$$

and, therefore,

$$\int h \leq \lim \int g_k \leq \int g.$$

Thus,

$$\int g = \lim \int g_k. \ \blacksquare$$

The following result is an immediate consequence of Proposition $6 \cdot 7$ and is sometimes more convenient to apply.

6 · 9 Corollary. *Let (h_k) be a sequence of nonnegative functions in $\mathscr{P}$ such that $(\int \sum_{k=1}^{m} h_k)$ is bounded. Then $\sum_{k=1}^{\infty} h_k(x)$ converges for a.a.x and if $g(x) = \sum_{k=1}^{\infty} h_k(x)$ for a.a.x then $g \in \mathscr{P}$ and*

$$\int g = \sum_{k=1}^{\infty} \int h_k.$$

PROBLEMS

1. Prove Proposition $6 \cdot 5$.

2. Prove Corollary $6 \cdot 9$.

3. If f is a function from $\mathbf{R}^n$ into $\mathbf{R}$, the closure of the set $\{x : f(x) \neq 0\}$ is called the **support** of f. Show that if f is a continuous function which has compact support then $f \in \mathscr{P}$.

4. a. Let $I = (a, b)$ be an open interval in $\mathbf{R}^n$ where $a = (a^1, \cdots, a^n)$ and $b = (b^1, \cdots, b^n)$. For a positive integer k such that $2/k < \min\{b^i - a^i : i = 1, \cdots, n\}$ and for each $t \in [-1/k, 1/k]$ let $I_t = (a_t, b_t)$ where $a_t = (a^1 + t, \cdots, a^n + t)$ and $b_t = (b^1 - t, \cdots, b^n - t)$ and define

$$g_k(x) = \begin{cases} -1/k & \text{if } x \notin I_{-1/k} \\ \sup\{t : x \in I_t\} & \text{if } x \in I_{-1/k}. \end{cases}$$

Show that: $I_r \supset I_s$ if $r < s$, $g_k(x) = 1/k$ for $x \in I_{1/k}$, and g_k is continuous.

b. If $f_k(x) = (k/2)(g_k(x) + 1/k)$, show that (f_k) is a sequence of continuous functions such that

$$\lim f_k(x) = \chi_I(x) \quad \text{for a.a.}x.$$

c. If h is a step function on $\mathbf{R}^n$, show that there is a sequence (f_k) of continuous functions such that

$$\lim f_k(x) = h(x) \quad \text{for a.a.}x.$$

7 · THE LEBESGUE INTEGRAL

Since the difference of two functions in $\tilde{\mathscr{S}}$ need not be in $\tilde{\mathscr{S}}$, this set is not a linear space. We now define a set which contains $\tilde{\mathscr{S}}$ and is also a linear space.

7 · 1 Definition. *Let $\mathscr{L}$ be the set of real-valued functions f such that for some g and h in $\tilde{\mathscr{S}}$*

$$f = g - h \quad a.e.$$

*The set $\mathscr{L}$ is called the set of **Lebesgue integrable functions** on $\mathbf{R}^n$ and the **Lebesgue integral** of f is defined as follows:*

$$\int f = \int g - \int h.$$

To justify the definition of the integral, suppose $f = g - h$ a.e. and $f = g_1 - h_1$ a.e. where g, h, g_1, and h_1 are in $\tilde{\mathscr{S}}$. Then

$$g + h_1 = g_1 + h \quad \text{a.e.}$$

and, therefore,

$$\int g + \int h_1 = \int (g + h_1) = \int (g_1 + h) = \int g_1 + \int h;$$

that is,

$$\int g - \int h = \int g_1 - \int h_1.$$

Note that if $f_1 \in \mathscr{L}$ and $f_2 = f_1$ a.e., then $f_2 \in \mathscr{L}$ and $\int f_2 = \int f_1$. Also, $\tilde{\mathscr{S}} \subset \mathscr{L}$ and if $f \in \tilde{\mathscr{S}}$ then its Lebesgue integral agrees with its integral as an element of $\tilde{\mathscr{S}}$.

Before comparing the Lebesgue integral with the Riemann integral, we define the Lebesgue integral over a set E in $\mathbf{R}^n$. If f is a function on $\mathbf{R}^n$ and E is a set in $\mathbf{R}^n$ such that $f\chi_E \in \mathscr{L}$, then the integral of f over E is defined to be:

$$\int_E f = \int f \chi_E .$$

If E is the interval $[a, b]$, we usually write $\int_a^b f$ instead of $\int_{[a,b]} f$. Using "R" to denote the Riemann integral, we recall that if E is a bounded set in $\mathbf{R}^n$,

$[a, b]$ is an interval containing E, and $f\chi_E$ is Riemann integrable on $[a, b]$, then

$$\mathrm{R}\int_E f = \mathrm{R}\int_a^b f\chi_E .$$

7 · 2 Theorem. *If f is Riemann integrable on $[a, b]$, then it is Lebesgue integrable on $[a, b]$ and*

$$\int_a^b f = \mathrm{R}\int_a^b f.$$

PROOF. In the proof of Theorem 5 · 1 we showed that if f is Riemann integrable on $[a, b]$, then there is a nondecreasing sequence (g_k) of step functions such that

$$\lim g_k(x) = f(x)\chi_{[a,b]}(x) \quad \text{for a.a.}x \quad \text{and} \quad \lim \int g_k = \mathrm{R}\int_a^b f.$$

This shows that $f\chi_{[a,b]} \in \mathscr{P}$ and, hence, $f\chi_{[a,b]} \in \mathscr{L}$ and also

$$\lim \int g_k = \int f\chi_{[a,b]} = \int_a^b f.$$

Thus,

$$\int_a^b f = \mathrm{R}\int_a^b f. \blacksquare$$

More generally, if f is Riemann integrable on a bounded set $E \subset [a, b]$, then

$$\mathrm{R}\int_E f = \mathrm{R}\int_a^b f\chi_E = \int_a^b f\chi_E = \int f\chi_E\chi_{[a,b]} = \int f\chi_E = \int_E f.$$

Thus, if f is Riemann integrable on a bounded set E, it is Lebesgue integrable on E and the two integrals are the same.

We turn now to a consideration of the basic properties of $\mathscr{L}$ and the integral on $\mathscr{L}$.

7 · 3 Theorem. *$\mathscr{L}$ is a linear space and the integral is a linear functional on $\mathscr{L}$; that is, if $f, g \in \mathscr{L}$ and $\alpha \in \mathbf{R}$, then $f + g$ and αf belong to $\mathscr{L}$ and*

$$\int f + g = \int f + \int g \quad \textit{and} \quad \int \alpha f = \alpha \int f.$$

The proof is left to the reader (Problem 3).

7 · 4 Theorem. *$\mathscr{L}$ is a lattice.*

PROOF. Take $f, g \in \mathscr{L}$ and suppose $f = f_1 - f_2$ a.e. where $f_1, f_2 \in \mathscr{P}$. Then $f_1 \wedge f_2$ and $f_1 \vee f_2$ are in $\mathscr{P}$ since $\mathscr{P}$ is a lattice (Proposition 6 · 6). Since

$$f_1 \wedge f_2 = \tfrac{1}{2}(f_1 + f_2 - |f_1 - f_2|)$$

and

$$f_1 \vee f_2 = \tfrac{1}{2}(f_1 + f_2 + |f_1 - f_2|),$$

we have

$$|f| = |f_1 - f_2| = f_1 \vee f_2 - f_1 \wedge f_2 \quad \text{a.e.}$$

and, hence, $|f| \in \mathscr{L}$. Using this result and Theorem 7 · 3, we have

$$f \wedge g = \tfrac{1}{2}(f + g - |f - g|) \in \mathscr{L} \quad \text{and} \quad f \vee g = \tfrac{1}{2}(f + g + |f - g|) \in \mathscr{L}. \ \blacksquare$$

Theorems 7 · 3 and 7 · 4 imply that $\mathscr{L}$ is a vector lattice. Note that in the proof of 7 · 4 we have shown that if $f \in \mathscr{L}$ then $|f| \in \mathscr{L}$. Also, it is evident that $f \in \mathscr{L}$ if and only if f^+ and f^- are in $\mathscr{L}$.

7 · 5 Theorem. *If $f \in \mathscr{L}$ and $f \geq 0$ a.e., then $\int f \geq 0$.*

PROOF. Let $f = g - h$ a.e. where g, $h \in \mathscr{P}$. Then $f \geq 0$ a.e. implies that $g \geq h$ a.e. and, hence, by Corollary 6 · 4

$$\int g \geq \int h.$$

Thus,

$$\int f = \int g - \int h \geq 0. \ \blacksquare$$

Using Theorems 7 · 3 and 7 · 5 we easily obtain: if f, $g \in \mathscr{L}$

$$f \leq g \quad \text{a.e.} \quad \text{implies} \quad \int f \leq \int g.$$

Also, if $f \in \mathscr{L}$, then

$$\left| \int f \right| \leq \int |f|.$$

If $\mathscr{L}$ is the set of Lebesgue integrable functions from **R** into **R** we now show that the constant function 1 does not belong to $\mathscr{L}$. For any positive integer k define

$$g_k(x) = \begin{cases} 1, & x \in [-k, k] \\ 0, & x \notin [-k, k]. \end{cases}$$

Then g_k is a step function and, hence, $g_k \in \mathscr{L}$. Also, $g_k \leq 1$ for all k. Then if $1 \in \mathscr{L}$,

$$\int g_k \leq \int 1 \quad \text{for all } k.$$

But the sequence $(\int g_k)$ is the sequence $(2k)$ which is not bounded and, therefore, $1 \notin \mathscr{L}$.

PROBLEMS

1. For any positive integer k let f_k be the function from **R** into **R** defined by

$$f_k(x) = \begin{cases} 2^k, & x \in [2^{-k}, 2^{-k+1}] \\ 0 & \text{otherwise.} \end{cases}$$

If $f(x) = \lim f_k(x)$, show that

$$\int f \neq \lim \int f_k.$$

2. If f is the function from **R** into **R** defined by

$$f(x) = \begin{cases} 0, & x \notin [0, 1] \\ 1, & x \in [0, 1] \text{ and } x \text{ rational} \\ -1, & x \in [0, 1] \text{ and } x \text{ irrational,} \end{cases}$$

show that $f \in \mathscr{L}$.

3. Prove Theorem 7 · 3

4. If $f \in \mathscr{L}$, show that there is a sequence (f_k) of continuous functions such that

$$\lim f_k(x) = f(x) \quad \text{for a.a.} x$$

(c.f. Problem 4d, p. 124).

5. Let (a_k) be a bounded sequence of real numbers and define

$$b_{kj} = a_k \wedge \cdots \wedge a_j \quad \text{and} \quad c_{kj} = a_k \vee \cdots \vee a_j \quad \text{for } j \geq k.$$

Show that

$$\underline{\lim}\, a_k = \lim_{k \to \infty} \left(\lim_{j \to \infty} b_{kj} \right) = \sup_k \left(\inf_{j \geq k} (a_k \wedge \cdots \wedge a_j) \right)$$

and

$$\overline{\lim}\, a_k = \lim_{k \to \infty} \left(\lim_{j \to \infty} c_{kj} \right) = \inf_k \left(\sup_{j \geq k} (a_k \vee \cdots \vee a_j) \right).$$

6. If $f \in \mathscr{L}$, show that

$$\int f(x)\, dx = \int f(-x)\, dx.$$

7. If $f \in \mathscr{L}$, then there exists a sequence (f_k) of step functions such that $\lim f_k(x) = f(x)$ for a.a.x and

$$\lim_{k \to \infty} \int |f - f_k| = 0.$$

8. If f_1 and f_2 are Riemann integrable on $[a, b]$ and $f_1 = f_2$ a.e. on $[a, b]$, show that

$$\mathrm{R} \int_a^b f_1 = \mathrm{R} \int_a^b f_2.$$

9. If f is the function defined on $[0, 1]$ by the rule

$$f(x) = \begin{cases} 1/q & \text{if } x = p/q \text{ and } p \text{ and } q \text{ are relatively prime positive integers} \\ 0 & \text{if } x \text{ is irrational,} \end{cases}$$

show that $R\int_0^1 f = 0$ (c.f. Problem 3, p. 120).

8 · SOME CONVERGENCE THEOREMS

Before proving the basic convergence theorem for functions in $\mathscr{L}$ which is due to the Italian mathematician Beppo Levi (1875–1961), we observe that a nonnegative function $f \in \mathscr{L}$ can be represented as the difference of two functions g', $h' \in \mathscr{P}$ which are almost everywhere nonnegative and are such that $\int h' < \varepsilon$. Suppose f is a nonnegative function in $\mathscr{L}$ and $f = g - h$ a.e. where g, $h \in \mathscr{P}$. Since $h \in \mathscr{P}$ there is a nondecreasing sequence (h_k) of step functions such that $\lim h_k(x) = h(x)$ for a.a.x and $\lim \int h_k = \int h$. Corresponding to $\varepsilon > 0$ take k such that

$$0 \leq \int h - \int h_k < \varepsilon.$$

Now $-h_k \in \mathscr{S} \subset \mathscr{P}$ so that $g' = g + (-h_k)$ and $h' = h + (-h_k)$ are in $\mathscr{P}$ and

$$f = g' - h' \quad \text{a.e.}$$

where $h' \geq 0$ a.e. and $0 \leq \int h' < \varepsilon$. Also since $g' = h' + f$ a.e., $g' \geq 0$ a.e.

8 · 1 Theorem (Levi). *Let (f_k) be a sequence of a.e. nonnegative functions in $\mathscr{L}$ such that the sequence $(\int \sum_{k=1}^m f_k)$ is bounded. Then $\sum_{k=1}^\infty f_k(x)$ converges for a.a.x and if $f(x) = \sum_{k=1}^\infty f_k(x)$ for a.a.x, then $f \in \mathscr{L}$ and*

$$\int f = \sum_{k=1}^{\infty} \int f_k.$$

PROOF. Suppose $\int \sum_{k=1}^m f_k \leq c$ for all m. For each k, f_k has a representation: $f_k = g_k - h_k$ a.e. where g_k and h_k are nonnegative functions in $\mathscr{P}$ and $\int h_k < 1/2^k$. Since (h_k) is a sequence of nonnegative functions in $\mathscr{P}$ such that $\int \sum_{k=1}^m h_k \leq 1$ for all m, Corollary 6 · 9 implies that $\sum_{k=1}^\infty h_k(x)$ converges for a.a.x and if $h(x) = \sum_{k=1}^\infty h_k(x)$ for a.a.x, then $h \in \mathscr{P}$ and $\int h = \sum_{k=1}^\infty \int h_k$. Also, (g_k) is a sequence of nonnegative functions in $\mathscr{P}$ such that for all m

$$\int \sum_{k=1}^{m} g_k = \int \sum_{k=1}^{m} f_k + \int \sum_{k=1}^{m} h_k \leq c + 1.$$

Then, $\sum_{k=1}^\infty g_k(x)$ converges for a.a.x and if $g(x) = \sum_{k=1}^\infty g_k(x)$ for a.a.x, then $g \in \mathscr{P}$ and $\int g = \sum_{k=1}^\infty \int g_k$. Thus $\sum_{k=1}^\infty f_k(x)$ converges for a.a.x and if $f(x) = \sum_{k=1}^\infty f_k(x)$ for a.a.x, then $f = g - h$ a.e. Therefore, $f \in \mathscr{L}$ and

$$\int f = \int g - \int h = \sum_{k=1}^{\infty} \int g_k - \sum_{k=1}^{\infty} \int h_k = \sum_{k=1}^{\infty} \int f_k. \quad \blacksquare$$

We now state the corresponding result for sequences. The proof is an immediate consequence of Theorem 8 · 1 and is omitted (Problem 1).

8 · 2 Corollary (Monotone Convergence Theorem). *Let (g_k) be an a.e. nondecreasing sequence of functions in $\mathscr{L}$ such that $(\int g_k)$ is bounded. Then $(g_k(x))$ converges for a.a.x and if $g(x) = \lim g_k(x)$ for a.a.x, then $g \in \mathscr{L}$ and*

$$\int g = \lim \int g_k.$$

In Corollary 8 · 2 the same conclusion holds if (g_k) is an a.e. nonincreasing sequence (Problem 2). Problem 1, § 7, shows that the condition that the sequence (g_k) be monotone is needed in this result. Later in this section we will obtain another convergence theorem in which monotonicity is replaced by another condition on the sequence but first we obtain an interesting consequence of the Monotone Convergence Theorem.

From the definition of an integrable function it is clear that a set of measure zero is insignificant in Lebesgue integration; that is, if $f_1 \in \mathscr{L}$ and $f_2 = f_1$ a.e., then $f_2 \in \mathscr{L}$ and $\int f_2 = \int f_1$. We now show that if a set is insignificant with respect to integration, then it is of measure zero.

8 · 3 Proposition. *If $f \in \mathscr{L}$ and $\int |f| = 0$, then $f = 0$ a.e.*

Proof. Let $g_k = k|f|$ for each positive integer k. Then (g_k) is a nondecreasing sequence of functions in $\mathscr{L}$ such that $(\int g_k)$ is bounded. Therefore, by the Monotone Convergence Theorem $(g_k(x))$ converges for a.a.x. But $(k|f(x)|)$ converges only at those points x such that $f(x) = 0$. ▮

Thus, a set E is of measure zero if and only if $\chi_E \in \mathscr{L}$ and $\int \chi_E = 0$.

Before obtaining another convergence theorem we will prove a lemma due to the French mathematician Pierre Fatou (1878–1929). Fatou's lemma is not only useful in proving the convergence theorem but is of interest in itself.

8 · 4 Lemma (Fatou). *If (f_k) is a sequence of functions in $\mathscr{L}$ and g is a function in $\mathscr{L}$ such that, for all k, $f_k(x) \geq g(x)$ for a.a.x and $\underline{\lim} \int f_k < \infty$, then $\underline{\lim} f_k(x)$ is finite for a.a.x and, if $f(x) = \underline{\lim} f_k(x)$ for a.a.x, $f \in \mathscr{L}$ and*

$$\int f \leq \underline{\lim} \int f_k.$$

Proof. Note that $\underline{\lim} f_k(x) = \lim_k (\lim_j (f_k(x) \wedge \cdots \wedge f_j(x)))$ for those values of x where these limits exist (Problem 5, p. 127). Let $h_{kj} = f_k \wedge \cdots \wedge f_j$ for $j \geq k$. Then, for fixed k, (h_{kj}) is a nonincreasing sequence of functions in $\mathscr{L}$ such that $h_{kj} \geq g$ a.e. Therefore, the sequence $(\int h_{kj})$ is bounded below by $\int g$ and, hence, by the Monotone Convergence Theorem $(h_{kj}(x))$ converges for a.a.x and if $h_k(x) = \lim_{j \to \infty} h_{kj}(x)$ for a.a.x then $h_k \in \mathscr{L}$ and $\int h_k = \lim_{j \to \infty} \int h_{kj}$. Since, for all $j \geq k$, $h_{kj} \leq f_k$ we have $h_k(x) \leq f_k(x)$ for a.a.x and, therefore, $\int h_k \leq \int f_k$. Since $h_{kj} = f_k \wedge h_{k+1,j} \leq h_{k+1,j}$ for all j and k, the sequence (h_k) is an a.e. nondecreasing sequence of functions in $\mathscr{L}$. Thus $(\int h_k)$ is a nondecreasing sequence of real numbers such that

$$\lim \int h_k = \underline{\lim} \int h_k \leq \underline{\lim} \int f_k < \infty.$$

Again, by the Monotone Convergence Theorem, $(h_k(x))$ converges for a.a.x and if $h(x) = \lim h_k(x)$ for a.a.x, then $h \in \mathscr{L}$ and

$$\int h = \lim \int h_k \leq \underline{\lim} \int f_k.$$

Since

$$f(x) = \underline{\lim} f_k(x) = \lim h_k(x) = h(x) \quad \text{for a.a.}x,$$

$$\int f = \int h \leq \underline{\lim} \int f_k. \blacksquare$$

Using the fact that $\overline{\lim}(f_k) = -\underline{\lim}(-f_k)$, we obtain a result similar to Fatou's Lemma for the upper limit.

8 · 5 Lemma. *If (f_k) is a sequence of functions in $\mathscr{L}$ and g is a function in $\mathscr{L}$ such that, for all k, $f_k(x) \leq g(x)$ for a.a.x and $\overline{\lim} \int f_k > -\infty$, then $\overline{\lim} f_k(x)$ is finite for a.a.x and, if $f(x) = \overline{\lim} f_k(x)$ for a.a.x, $f \in \mathscr{L}$ and*

$$\int f \geq \overline{\lim} \int f_k.$$

From the two previous lemmas we obtain the following important convergence theorem.

8 · 6 Theorem (Lebesgue Dominated Convergence). *Let (f_k) be a sequence of functions in $\mathscr{L}$ such that $(f_k(x))$ converges for a.a.x and, for all k, $|f_k| \leq g$ a.e. for some function $g \in \mathscr{L}$. If $f(x) = \lim f_k(x)$ for a.a.x, then $f \in \mathscr{L}$ and*

$$\int f = \lim \int f_k.$$

PROOF. Since, for all k, $-g(x) \leq f_k(x) \leq g(x)$ for a.a.x, we have

$$-\int g \leq \int f_k \leq \int g$$

and, therefore

$$\overline{\lim} \int f_k \geq -\int g > -\infty \quad \text{and} \quad \underline{\lim} \int f_k \leq \int g < \infty.$$

Thus by Lemmas 8 · 4 and 8 · 5, if

$$f(x) = \underline{\lim} f_k(x) = \lim f_k(x) = \overline{\lim} f_k(x) \quad \text{for a.a.}x,$$

then $f \in \mathscr{L}$ and

$$\overline{\lim} \int f_k \leq \int f \leq \underline{\lim} \int f_k;$$

that is,

$$\int f = \lim \int f_k. \blacksquare$$

PROBLEMS

1. Prove Corollary 8 · 2.

2. Show that Corollary 8 · 2 holds if nondecreasing is replaced by nonincreasing.

3. Let (f_k) be a sequence of functions in $\mathscr{L}$ such that $(\int \sum_{k=1}^{m} |f_k|)$ is bounded. Show that $\sum_{k=1}^{\infty} f_k(x)$ converges for a.a.x and if $f(x) = \sum_{k=1}^{\infty} f_k(x)$ for a.a.x, then $f \in \mathscr{L}$ and

$$\int f = \sum_{k=1}^{\infty} \int f_k.$$

4. If f is the function from **R** into **R** defined by

$$f(x) = \begin{cases} 0, & x \leq 1 \\ 1/x, & x > 1, \end{cases}$$

show that f is not Lebesgue integrable.

5. If f is the function from **R** into **R** defined by

$$f(x) = \begin{cases} 0, & x \leq 0 \quad \text{and} \quad x > 1 \\ 1/\sqrt{x}, & x \in (0, 1], \end{cases}$$

show that f is Lebesgue integrable.

6. If f is a nonnegative real-valued function on **R** such that the improper Riemann integral $\int_{-\infty}^{\infty} f$ exists, show that f is Lebesgue integrable and

$$\int f = \int_{-\infty}^{\infty} f.$$

7. If f is a nonnegative real-valued function which is Riemann integrable on any bounded interval in **R** but the improper Riemann integral $\int_{-\infty}^{\infty} f$ diverges, show that f is not Lebesgue integrable on **R**.

8. Let f be the function from **R** into **R** defined by

$$f(x) = \begin{cases} 0, & x < 1 \\ (-1)^m \dfrac{1}{m}, & x \in [m, m+1), \quad m = 1, 2, \cdots. \end{cases}$$

Show that f is not Lebesgue integrable but that the improper Riemann integral $\int_{-\infty}^{\infty} f$ does exist.

9. For each positive real number r let $f_r \in \mathscr{L}$. If there exists some function $g \in \mathscr{L}$ such that for all $r > 0$, $|f_r| \leq g$ a.e. and

$$\lim_{r \to \infty} f_r(x) = f(x) \quad \text{for a.a.}x,$$

show that $f \in \mathscr{L}$ and

$$\lim_{r \to \infty} \int f_r = \int f.$$

9 · FUBINI'S THEOREM

In the theory of Riemann integration, under certain circumstances, a multiple integral is equal to an iterated integral and this provides a method for evaluating multiple integrals. In the Lebesgue theory this result has a particularly simple form and is one of the nice aspects of this theory.

For $n \geq 2$ let $\mathbf{R}^n = \mathbf{R}^k \times \mathbf{R}^m$ and denote a point in $\mathbf{R}^n$ by (x, y) where $x \in \mathbf{R}^k$ and $y \in \mathbf{R}^m$. If f is a function defined on $\mathbf{R}^n$, then for any $x \in \mathbf{R}^k$ let $f(x, \cdot)$ denote the function on $\mathbf{R}^m$ whose value at y is $f(x, y)$. Similarly, for $y \in \mathbf{R}^m$, $f(\cdot, y)$ is the function on $\mathbf{R}^k$ whose value at x is $f(x, y)$.

9 · 1 Definition. *If, for a.a. $x \in \mathbf{R}^k$, $f(x, \cdot)$ is integrable and if g is an integrable function on $\mathbf{R}^k$ such that*

$$g(x) = \int_{\mathbf{R}^m} f(x, \cdot) \quad \textit{for a.a. } x \in \mathbf{R}^k,$$

then we say that the iterated integral of f over $\mathbf{R}^k$ and $\mathbf{R}^m$ exists and this iterated integral is defined as follows:

$$\int_{\mathbf{R}^k} \int_{\mathbf{R}^m} f(x, y)\, dy\, dx = \int_{\mathbf{R}^k} g.$$

In a similar fashion we define the iterated integral $\int_{\mathbf{R}^m} \int_{\mathbf{R}^k} f(x, y)\, dx\, dy$.

Fubini's Theorem states that if f is integrable on $\mathbf{R}^n$, then both the iterated integrals given above exist and are equal to the integral of f on $\mathbf{R}^n$. We will prove this theorem in a sequence of steps.

9 · 2. Lemma. *If I is an open interval in $\mathbf{R}^n$, then both the iterated integrals of χ_I over $\mathbf{R}^k$ and $\mathbf{R}^m$ exist and*

$$\int_{\mathbf{R}^n} \chi_I = \int_{\mathbf{R}^k} \int_{\mathbf{R}^m} \chi_I(x, y)\, dy\, dx = \int_{\mathbf{R}^m} \int_{\mathbf{R}^k} \chi_I(x, y)\, dx\, dy.$$

PROOF. Let $I = I_k \times I_m$ where $I_k \subset \mathbf{R}^k$ and $I_m \subset \mathbf{R}^m$. Since

$$\chi_I(x, y) = \chi_{I_k}(x)\chi_{I_m}(y),$$

$$\int_{\mathbf{R}^m} \chi_I(x, \cdot) = \chi_{I_k}(x) v_m(I_m) \quad \text{for all } x \in \mathbf{R}^k$$

and

$$\int_{\mathbf{R}^k} \int_{\mathbf{R}^m} \chi_I(x, y)\, dy\, dx = v_m(I_m) \int_{\mathbf{R}^k} \chi_{I_k}(x)\, dx = v_m(I_m) v_k(I_k)$$

$$= v_n(I) = \int_{\mathbf{R}^n} \chi_I .$$

The other equality follows in similar fashion. ▮

9 · 3 Corollary. *If E is a set of measure zero in $\mathbf{R}^n$, then for a.a.$x \in \mathbf{R}^k$ the set $\{y : (x, y) \in E\}$ is of measure zero in $\mathbf{R}^m$.*

PROOF. In the proof of Proposition 4 · 3 we showed that a set E of measure zero can be covered by a countable collection $\{I_j\}$ of open intervals such that each point of E is in infinitely many of these intervals and $\sum_{j=1}^{\infty} v(I_j) \leq 1$. If we let χ_j denote the characteristic function of I_j, then

$$\sum_{j=1}^{\infty} \int_{\mathbf{R}^n} \chi_j \leq 1.$$

By Lemma 9 · 2

$$\int_{\mathbf{R}^n} \chi_j = \int_{\mathbf{R}^k} \int_{\mathbf{R}^m} \chi_j(x, y)\, dy\, dx.$$

Then, for a.a.$x \in \mathbf{R}^k$, $(\int_{\mathbf{R}^m} \chi_j(x, \cdot))$ is a sequence of values of a.e. nonnegative functions in $\mathscr{L}$ such that $(\int_{\mathbf{R}^k} \sum_{j=1}^{i} \int_{\mathbf{R}^m} \chi_j(x, y) dy\, dx)$ is bounded. Levi's Theorem 8 · 1 implies that $\sum_{j=1}^{\infty} \int_{\mathbf{R}^m} \chi_j(x, \cdot)$ converges for a.a.x. If $\sum_{j=1}^{\infty} \int_{\mathbf{R}^m} \chi_j(x_0, \cdot)$ converges then Levi's Theorem implies that $\sum_{j=1}^{\infty} \chi_j(x_0, y)$ converges for a.a.y. Since $(x_0, y) \in E$ implies that $\sum_{j=1}^{\infty} \chi_j(x_0, y)$ diverges, we see that $\{y: (x_0, y) \in E\}$ is of measure zero in $\mathbf{R}^m$. ▮

9 · 4 Lemma. *If g is a step function on $\mathbf{R}^n$, then the iterated integrals of g over $\mathbf{R}^k$ and $\mathbf{R}^m$ exist and*

$$\int_{\mathbf{R}^n} g = \int_{\mathbf{R}^k} \int_{\mathbf{R}^m} g(x, y)\, dy\, dx = \int_{\mathbf{R}^m} \int_{\mathbf{R}^k} g(x, y)\, dx\, dy.$$

PROOF. We give the proof for one of the iterated integrals only; that for the other iterated integral is similar. Let g have a representation consisting of $[a, b]$ and a partition $\{I_1, \cdots, I_r\}$ of $[a, b]$ such that

$$g(x, y) = c_j \quad \text{for } x, y \in I_j \quad \text{and } g(x, y) = 0 \quad \text{for } x, y \notin [a, b].$$

Then

$$g = \sum_{j=1}^{r} c_j \chi_{I_j} \quad \text{a.e. on } \mathbf{R}^n$$

and, for a.a.$x \in \mathbf{R}^k$,

$$g(x, \cdot) = \sum_{j=1}^{r} c_j \chi_{I_j}(x, \cdot) \quad \text{a.e. on } \mathbf{R}^m.$$

Thus,

$$\begin{aligned}\int_{\mathbf{R}^k} \int_{\mathbf{R}^m} g(x, y)\, dy\, dx &= \sum_{j=1}^{r} c_j \int_{\mathbf{R}^k} \int_{\mathbf{R}^m} \chi_{I_j}(x, y)\, dy\, dx \\ &= \sum_{j=1}^{r} c_j \int_{\mathbf{R}^n} \chi_{I_j} = \int_{\mathbf{R}^n} g. \quad ▮\end{aligned}$$

9 · 5 Lemma. *If $g \in \mathscr{P}$ on $\mathbf{R}^n$, then the iterated integrals of g over $\mathbf{R}^k$ and $\mathbf{R}^m$ exist and*

$$\int_{\mathbf{R}^n} g = \int_{\mathbf{R}^k} \int_{\mathbf{R}^m} g(x, y)\, dy\, dx = \int_{\mathbf{R}^m} \int_{\mathbf{R}^k} g(x, y)\, dx\, dy.$$

PROOF. Again we prove this for only one of the iterated integrals. Let (g_i) be a nondecreasing sequence of step functions such that $(\int_{\mathbf{R}^n} g_i)$ is bounded and

$$\lim g_i(x, y) = g(x, y) \quad \text{for a.a.}(x, y) \in \mathbf{R}^n.$$

Then

$$\int_{\mathbf{R}^n} g = \lim \int_{\mathbf{R}^n} g_i = \lim \int_{\mathbf{R}^k} \int_{\mathbf{R}^m} g_i(x, y)\, dy\, dx.$$

Let h_i be an integrable function on $\mathbf{R}^k$ such that

$$h_i(x) = \int_{\mathbf{R}^m} g_i(x, \cdot) \quad \text{for a.a.} x \in \mathbf{R}^k.$$

Then (h_i) is an a.e. nondecreasing sequence of integrable functions on $\mathbf{R}^k$ such that $(\int_{\mathbf{R}^k} h_i)$ is bounded. Then, by the Monotone Convergence Theorem, (h_i) converges a.e. on $\mathbf{R}^k$ and, if

$$h(x) = \lim h_i(x) \quad \text{for a.a.} x \in \mathbf{R}^k,$$

h is integrable over $\mathbf{R}^k$ and

$$\int_{\mathbf{R}^k} h = \lim \int_{\mathbf{R}^k} h_i = \lim \int_{\mathbf{R}^k} \int_{\mathbf{R}^m} g_i(x, y)\, dy\, dx.$$

For a.a.$x \in \mathbf{R}^k$, $(g_i(x, \cdot))$ is a nondecreasing sequence of integrable functions on $\mathbf{R}^m$ such that

$$\lim g_i(x, y) = g(x, y) \quad \text{for a.a.} y \in \mathbf{R}^m$$

and $(\int_{\mathbf{R}^m} g_i(x, \cdot))$ is bounded. By the Monotone Convergence Theorem for a.a.$x \in \mathbf{R}^k$

$$\int_{\mathbf{R}^m} g(x, \cdot) = \lim \int_{\mathbf{R}^m} g_i(x, \cdot) = \lim h_i(x) = h(x).$$

Therefore,

$$\int_{\mathbf{R}^k} \int_{\mathbf{R}^m} g(x, y)\, dy\, dx = \int_{\mathbf{R}^k} h = \lim \int_{\mathbf{R}^k} \int_{\mathbf{R}^m} g_i(x, y)\, dy\, dx = \int_{\mathbf{R}^n} g. \quad \blacksquare$$

9 · 6 Theorem (Fubini). *If f is integrable on $\mathbf{R}^n = \mathbf{R}^k \times \mathbf{R}^m$, then the iterated integrals of f over $\mathbf{R}^k$ and $\mathbf{R}^m$ exist and*

$$\int_{\mathbf{R}^n} f = \int_{\mathbf{R}^k} \int_{\mathbf{R}^m} f(x, y)\, dy\, dx = \int_{\mathbf{R}^m} \int_{\mathbf{R}^k} f(x, y)\, dx\, dy.$$

PROOF. Let $f = g - h$ a.e. where g and h are in $\mathscr{P}$. Using Lemma 9 · 5 we have

$$\int_{\mathbf{R}^n} f = \int_{\mathbf{R}^n} g - \int_{\mathbf{R}^n} h = \int_{\mathbf{R}^k} \int_{\mathbf{R}^m} g(x, y)\, dy\, dx - \int_{\mathbf{R}^k} \int_{\mathbf{R}^m} h(x, y)\, dy\, dx$$
$$= \int_{\mathbf{R}^k} \int_{\mathbf{R}^m} f(x, y)\, dy\, dx$$

and

$$\int_{\mathbf{R}^n} f = \int_{\mathbf{R}^n} g - \int_{\mathbf{R}^n} h = \int_{\mathbf{R}^m} \int_{\mathbf{R}^k} g(x, y)\, dx\, dy - \int_{\mathbf{R}^m} \int_{\mathbf{R}^k} h(x, y)\, dx\, dy$$
$$= \int_{\mathbf{R}^m} \int_{\mathbf{R}^k} f(x, y)\, dx\, dy. \quad \blacksquare$$

Since $\mathbf{R}^n = \bigtimes_{i=1}^{n} \mathbf{R}$, Fubini's Theorem implies that if f is integrable on $\mathbf{R}^n$, then the n-fold iterated integrals of f over $\mathbf{R}$ exist and

$$\int_{\mathbf{R}^n} f = \int_{\mathbf{R}} \cdots \int_{\mathbf{R}} f(x^1, \cdots, x^n)\, dx^{p(1)} \cdots dx^{p(n)}$$

where p is any permutation of $1, \cdots, n$. Thus an integral over $\mathbf{R}^n$ can be evaluated by a succession of integrations over $\mathbf{R}$. This is one aspect of Fubini's Theorem but equally important is its use in changing the order of integration of iterated integrals.

PROBLEMS

1. If f is integrable over the interval $I = [a, b] \times [c, d]$ in $\mathbf{R}^2$, show that

$$\int_I f = \int_a^b \int_c^d f(x, y)\, dy\, dx = \int_c^d \int_a^b f(x, y)\, dx\, dy.$$

2. Let $I = [0, 1] \times [0, 1]$ and let

$$f(x, y) = \begin{cases} 0, & (x, y) = (0, 0) \\ \dfrac{x^2 - y^2}{(x^2 + y^2)^2}, & (x, y) \neq (0, 0). \end{cases}$$

Determine $\int_0^1 \int_0^1 f(x, y)\, dx\, dy$ and $\int_0^1 \int_0^1 f(x, y)\, dy\, dx$. Is f integrable over I?

3. Determine

$$\int_0^1 \int_1^2 \frac{1}{x^3} e^{y/x}\, dx\, dy.$$

10 · MEASURABLE FUNCTIONS

If in the Lebesgue Dominated Convergence Theorem we drop the hypothesis that the sequence (f_k) of functions in $\mathscr{L}$ is bounded by an integrable function, then we can no longer conclude that the limit function is integrable.

A function which is the limit a.e. of a sequence of functions in $\mathscr{L}$ will be said to be a measurable function.

Suppose (f_k) is a sequence of functions in $\mathscr{L}$ and $\lim f_k(x) = f(x)$ for a.a.x. Take $g \in \mathscr{L}$ such that $g \geq 0$ and let $h_k = (-g) \vee (f_k \wedge g)$. Then

$$h_k(x) = \begin{cases} -g(x) & \text{if } f_k(x) < -g(x) \\ f_k(x) & \text{if } -g(x) \leq f_k(x) \leq g(x) \\ g(x) & \text{if } f_k(x) > g(x); \end{cases}$$

that is, h_k is the function f_k cut off above and below by the integrable function g. Since (h_k) is a sequence of functions in $\mathscr{L}$ such that $|h_k| \leq g$ and

$$\lim h_k(x) = (-g(x)) \vee (f(x) \wedge g(x)) \quad \text{for a.a.}x,$$

by the Lebesgue Dominated Convergence Theorem $(-g) \vee (f \wedge g)$ is integrable.

10 · 1 Definition. *A function f from $\mathbf{R}^n$ into $\mathbf{R}$ is said to be* ***a measurable function*** *if, for each $g \in \mathscr{L}$ such that $g \geq 0$, $(-g) \vee (f \wedge g)$ is integrable. We denote the set of measurable functions by $\tilde{\mathscr{M}}$.*

Observe the following immediate consequences of this definition. An integrable function is measurable; that is, $\mathscr{L} \subset \tilde{\mathscr{M}}$. If $f_1 \in \tilde{\mathscr{M}}$ and $f_2 = f_1$ a.e., then $f_2 \in \tilde{\mathscr{M}}$. A measurable function which is bounded by an integrable function is integrable. For, if $f \in \tilde{\mathscr{M}}$ and $|f| \leq g \in \mathscr{L}$, then

$$f = (-g) \vee (f \wedge g) \in \mathscr{L}.$$

This observation allows us to rewrite the Lebesgue Dominated Convergence Theorem as follows:

10 · 2 Theorem (Lebesgue Dominated Convergence). *If (f_k) is a sequence of measurable functions such that, for all k, $|f_k| \leq g$ for some integrable function g and $f(x) = \lim f_k(x)$ for a.a.x, then f is integrable and*

$$\int f = \lim \int f_k.$$

The discussion preceding Definition 10 · 1 shows that the limit of a sequence of integrable functions is measurable. In fact, this property characterizes measurable functions in $\mathbf{R}^n$.

10 · 3 Proposition. *A function f from $\mathbf{R}^n$ to $\mathbf{R}$ is measurable if and only if there exists a sequence (f_k) of integrable functions such that $f(x) = \lim f_k(x)$ for a.a.x.*

PROOF. The "if" part has already been proved. Suppose now that f is measurable. For each positive integer k let $a_k = (-k, \cdots, -k)$ and $b_k = (k, \cdots, k)$ be points in $\mathbf{R}^n$ and define

$$g_k = k\chi_{[a_k,\, b_k]}.$$

Then g_k is a nonnegative integrable function and if

$$f_k = (-g_k) \vee (f \wedge g_k),$$

then $f_k \in \mathscr{L}$ and

$$f(x) = \lim f_k(x) \quad \text{for all } x. \blacksquare$$

From this characterization of measurable functions it is easy to see that continuous functions are measurable.

10 · 4 Proposition. *If f is continuous, then f is measurable.*

PROOF. For each positive integer k, let $a_k = (-k, \cdots, -k)$ and $b_k = (k, \cdots, k)$. Since $f\chi_{[a_k, b_k]}$ is continuous on (a_k, b_k), it is Riemann integrable on $[a_k, b_k]$ and, hence, is Lebesgue integrable on $\mathbf{R}^n$. Also,

$$f(x) = \lim_{k \to \infty} [f(x)\chi_{[a_k, b_k]}(x)] \quad \text{for all } x$$

and, therefore, f is measurable. ∎

We now show that $\mathscr{M}$ unlike $\mathscr{L}$ is closed with respect to a.e. pointwise convergence.

10 · 5 Proposition. *If (f_k) is a sequence of functions in $\mathscr{M}$ and $f(x) = \lim f_k(x)$ for a.a.x, then $f \in \mathscr{M}$.*

PROOF. Take $g \in \mathscr{L}$ such that $g \geq 0$ and let

$$h_k = (-g) \vee (f_k \wedge g).$$

Then, $h_k \in \mathscr{L}$, $|h_k| \leq g$, and

$$\lim h_k(x) = (-g(x)) \vee (f(x) \wedge g(x)) \quad \text{for a.a.} x.$$

Thus, by the Lebesgue Dominated Convergence Theorem 10 · 2 $(-g) \vee (f \wedge g) \in \mathscr{L}$ and, hence, $f \in \mathscr{M}$. ∎

10 · 6 Proposition. *$\mathscr{M}$ is a real linear space.*

PROOF. Take f, $g \in \mathscr{M}$ and $\alpha \in \mathbf{R}$. By Proposition 10 · 3, there exist sequences (f_k) and (g_k) of integrable functions such that

$$\lim f_k(x) = f(x) \quad \text{and} \quad \lim g_k(x) = g(x) \quad \text{for a.a.} x.$$

Then, $(f_k + g_k)$ and (αf_k) are sequences of integrable functions such that

$$\lim(f_k(x) + g_k(x)) = f(x) + g(x)$$

and

$$\lim \alpha f_k(x) = \alpha f(x) \quad \text{for a.a.} x$$

and, hence, $f + g$ and αf are in $\mathscr{M}$. ∎

The proof of the following result is similar to that of Proposition 10 · 6.

10 · 7 Proposition. *$\tilde{\mathcal{M}}$ is a lattice.*

Thus, $\tilde{\mathcal{M}}$ is a vector lattice. This implies that if $f \in \tilde{\mathcal{M}}$, then $f^+ = f \vee 0$, $f^- = -(f \wedge 0)$, and $|f| = f^+ + f^-$ are in $\tilde{\mathcal{M}}$.

10 · 8 Proposition. *If (f_k) is a sequence of functions in $\tilde{\mathcal{M}}$ and $g(x) = \inf_k f_k(x)$, $G(x) = \sup_k f_k(x)$, $h(x) = \varliminf f_k(x)$, and $H(x) = \varlimsup f_k(x)$ for a.a. x, then g, G, h, and H are in $\tilde{\mathcal{M}}$.*

PROOF. Since

$$g = \lim_{k\to\infty} \bigwedge_{j=1}^{k} f_j \quad \text{and} \quad G = \lim_{k\to\infty} \bigvee_{j=1}^{k} f_j \quad \text{a.e.},$$

and

$$h = \lim_{k\to\infty}\left(\lim_{j\to\infty}(f_k \wedge \cdots \wedge f_j)\right) \quad \text{and} \quad H = \lim_{k\to\infty}\left(\lim_{j\to\infty}(f_k \vee \cdots \vee f_j)\right) \quad \text{a.e.},$$

Propositions 10 · 7 and 10 · 5 imply that g, G, h, and H are in $\tilde{\mathcal{M}}$. ▮

We can use this result to show that $\tilde{\mathcal{M}}$ is an algebra.

10 · 9 Proposition. *$\tilde{\mathcal{M}}$ is an algebra.*

PROOF. First we show that if $f \in \tilde{\mathcal{M}}$, then $f^2 \in \tilde{\mathcal{M}}$. Let $\{r_k : k = 1, 2, \cdots\}$ be an enumeration of the rational numbers in $\mathbf{R}$. For each $x \in \mathbf{R}^n$

$$\inf_k (f(x) - r_k)^2 = 0;$$

that is,

$$f^2(x) = \sup_k (2r_k f(x) - r_k^2).$$

Since, for each k, $2r_k f - r_k^2 \in \tilde{\mathcal{M}}$, $f^2 \in \tilde{\mathcal{M}}$.

For any $f, g \in \tilde{\mathcal{M}}$

$$fg = \tfrac{1}{4}[(f+g)^2 - (f-g)^2]$$

and, therefore, $fg \in \tilde{\mathcal{M}}$. ▮

We know that if f is measurable and $|f|$ is integrable, then $f = (-|f|) \vee (f \wedge |f|)$ is integrable. The following result shows that if f is measurable and an iterated integral of $|f|$ exists, then f is integrable.

10 · 10 Theorem (Tonelli). *If $f \in \tilde{\mathcal{M}}$ and $\int_{\mathbf{R}} \cdots \int_{\mathbf{R}} |f(x^1, \cdots, x^n)|\, dx^{p(1)} \cdots dx^{p(n)}$ exists for some permutation p of $\{1, \cdots, n\}$, then f is integrable.*

PROOF. For each positive integer k let $a_k = (-k, \cdots, -k)$ and $b_k = (k, \cdots, k)$ and let $g_k = k\chi_{[a_k, b_k]}$. If

$$f_k = g_k \wedge |f|,$$

then (f_k) is a nondecreasing sequence of integrable functions such that

$$\lim f_k(x) = |f(x)| \quad \text{for all } x.$$

Using Fubini's Theorem (p. 134) we have

$$0 \le \int f_k = \int_{\mathbf{R}} \cdots \int_{\mathbf{R}} f_k(x^1, \cdots, x^n)\, dx^{p(1)} \cdots dx^{p(n)}$$

$$\le \int_{\mathbf{R}} \cdots \int_{\mathbf{R}} |f(x^1, \cdots, x^n)|\, dx^{p(1)} \cdots dx^{p(n)}.$$

Thus, $(\int f_k)$ is bounded and by the Monotone Convergence Theorem (p. 129) $|f|$ is integrable and, hence, f is integrable. ∎

PROBLEMS

1. If f is the function from **R** to **R** defined by

$$f(x) = \begin{cases} \dfrac{1}{x(x-1)}, & x \ne 0, 1 \\ 2, & x = 0, 1, \end{cases}$$

is f measurable?

2. If $f(x) = [x]$, the greatest integer not greater than x, is f measurable?

3. If $f \in \mathscr{L}$ and g is a bounded measurable function, show that $fg \in \mathscr{L}$.

4. Use Definition 10·1 to prove directly that $\mathscr{M}$ is a lattice.

5. If f is a nonnegative measurable function, show that there exists a nondecreasing sequence (f_k) of nonnegative integrable functions such that

$$\lim f_k(x) = f(x) \quad \text{for all } x.$$

6. If f is a differentiable function from **R** to **R**, show that f' is measurable.

7. Let k and m be positive integers and $\mathbf{R}^n = \mathbf{R}^k \times \mathbf{R}^m$. If g and h are integrable functions on $\mathbf{R}^k$ and $\mathbf{R}^m$, respectively, and if $f(x, y) = g(x)h(y)$, show that

a. f is measurable on $\mathbf{R}^n$

b. f is integrable on $\mathbf{R}^n$.

8. If g and h are measurable functions on $\mathbf{R}^k$ and $\mathbf{R}^m$, respectively, and if $f(x, y) = g(x)h(y)$, show that f is measurable on $\mathbf{R}^n = \mathbf{R}^k \times \mathbf{R}^m$.

9. Let f be a nonnegative measurable function and let $\{r_k : k = 1, 2, \cdots\}$ be an enumeration of the positive rational numbers. Show that

$$\sqrt{f(x)} = \tfrac{1}{2} \inf_k \left(\frac{1}{r_k} f(x) + r_k \right)$$

and, hence, $f^{1/2}$ is a measurable function.

10. If f is measurable and a.e. nonzero, show that $1/f$ is measurable.

11. Show that a function f is measurable if and only if there exists a sequence (f_m) of step functions such that $f(x) = \lim f_m(x)$ for a.a.x.

12. If for each $\varepsilon > 0$ there exist integrable functions g and h such that $g \leq f \leq h$ and $\int (h - g) < \varepsilon$, show that f is integrable.

Suggestion. Let $\varepsilon = 2^{-n}$ $(n = 1, 2, \cdots)$ and apply Levi's Theorem 8 · 1.

11 · COMPLEX-VALUED FUNCTIONS

The Lebesgue integral of a complex-valued function can be defined in terms of the integral of the real and imaginary parts of the function. The theory for complex-valued functions then follows readily from that for real-valued functions.

Suppose that f is a complex-valued function defined on $\mathbf{R}^n$ and let $f = f_1 + if_2$ where f_1 and f_2 are real-valued functions. We say that f is integrable if both f_1 and f_2 are integrable and in this case we define

$$\int f = \int f_1 + i \int f_2 .$$

We say that f is measurable if f_1 and f_2 are measurable.

If we let $\mathscr{L}$ and $\mathscr{M}$ denote, respectively, the set of complex-valued integrable and measurable functions on $\mathbf{R}^n$, then the following results are easily proved.

11 · 1 Proposition. *$\mathscr{L}$ is a complex linear space and the integral is a linear functional on $\mathscr{L}$.*

11 · 2 Proposition. *$\mathscr{M}$ is an algebra over the complex field.*

11 · 3 Proposition. *If $f \in \mathscr{L}$, then $|f| \in \mathscr{L}$ and*

$$\left| \int f \right| \leq \int |f|.$$

PROOF. Since f_1 and f_2 are real-valued measurable functions, $|f| = [f_1{}^2 + f_2{}^2]^{1/2}$ is a real-valued measurable function. Then using the facts that $|f_1|$ and $|f_2|$ are real-valued integrable functions and $|f| \leq |f_1| + |f_2|$, we see that $|f|$ is integrable.

Let

$$\int f = \left| \int f \right| e^{i\varphi}.$$

If $e^{-i\varphi} f = g_1 + ig_2$ where g_1 and g_2 are real-valued functions on $\mathbf{R}^n$, then

$$\int g_1 + i \int g_2 = e^{-i\varphi} \int f = \left| \int f \right| \geq 0$$

so that $\int g_2 = 0$. Since $g_1 \le |f|$,

$$\left|\int f\right| = \int g_1 \le \int |f|. \quad \blacksquare$$

11 · 4 Proposition. *If $f \in \tilde{\mathcal{M}}$ and $|f| \le g$ where g is a real-valued integrable function, then $f \in \mathcal{L}$.*

PROOF. The real-valued functions f_1 and f_2 are measurable functions with $|f_1| \le g$ and $|f_2| \le g$ and, hence, f_1 and f_2 are integrable. $\blacksquare$

11 · 5 Theorem. (Lebesgue Dominated Convergence). *If (f_k) is a sequence of complex-valued measurable functions such that, for all k, $|f_k| \le g$ for some real-valued integrable function g and if $f(x) = \lim f_k(x)$ for a.a.x, then f is integrable and*

$$\int f = \lim \int f_k .$$

This theorem follows readily by applying the Lebesgue Dominated Convergence Theorem for real-valued functions to the real and imaginary parts of f_k.

Since integration of complex-valued functions is handled in terms of the real and imaginary parts of the functions, there is nothing essentially new here and we revert to discussing only real-valued functions in the remainder of this chapter.

PROBLEMS

1. Prove 11 · 1.

2. Prove 11 · 2.

3. If f is a complex-valued integrable function such that

$$\int |f| = 0,$$

show that $f = 0$ a.e.

4. Prove 11 · 5.

5. Let (f_k) be a sequence of complex-valued integrable functions such that $(\int \sum_{k=1}^{m} |f_k|)$ is bounded. Show that $\sum_{k=1}^{\infty} f_k(x)$ converges for a.a.x and if $f(x) = \sum_{k=1}^{\infty} f_k(x)$ for a.a.x, then $f \in \mathcal{L}$ and

$$\int f = \sum_{k=1}^{\infty} \int f_k .$$

6. Prove Tonelli's Theorem 10 · 10 for complex-valued functions.

7. Show that

$$\int_{-r}^{r} \int_{-\infty}^{\infty} e^{-y^2 + ixy}\, dy\, dx = 2 \int_{-\infty}^{\infty} \frac{\sin ry}{ye^{y^2}}\, dy.$$

12 · MEASURABLE SETS

12 · 1 Definition. *A subset E of $\mathbf{R}^n$ is **measurable** if χ_E is a measurable function; E is **integrable** if χ_E is an integrable function.*

We will denote the set of measurable subsets of $\mathbf{R}^n$ by $\mathscr{M}$.

Note that if a set is integrable, then it is measurable. Since the functions 0 and $\chi_{(a,b)}$ are integrable, the empty set and an open interval (a, b) in $\mathbf{R}^n$ are integrable sets. Although the function 1 is not integrable, it is measurable and, hence, $\mathbf{R}^n$ is a measurable set.

Recall that a collection $\mathscr{A}$ of sets in $\mathscr{P}(X)$ is called an algebra of sets if $X \in \mathscr{A}$ and if $A \cup B$ and $A \sim B$ are in $\mathscr{A}$ whenever A and B are in $\mathscr{A}$. An algebra $\mathscr{A}$ is called a **σ-algebra** if any countable union of sets in $\mathscr{A}$ is in $\mathscr{A}$.

12 · 2 Theorem. *The set of measurable subsets of $\mathbf{R}^n$ is a σ-algebra.*

PROOF. We have already established that $\mathbf{R}^n \in \mathscr{M}$. Suppose E and F are in $\mathscr{M}$; that is, χ_E and χ_F are measurable functions. Since

$$\chi_{E \cup F} = \chi_E \vee \chi_F \quad \text{and} \quad \chi_{E \sim F} = \chi_E - (\chi_E \wedge \chi_F),$$

$\chi_{E \cup F}$ and $\chi_{E \sim F}$ are measurable functions and, hence, $E \cup F \in \mathscr{M}$ and $E \sim F \in \mathscr{M}$. Thus, $\mathscr{M}$ is an algebra of sets. Now suppose $E_j \in \mathscr{M}$ for $j = 1, 2, \cdots$. Then, if $E = \bigcup_{j=1}^{\infty} E_j$,

$$\chi_E = \lim_{k \to \infty} \bigvee_{j=1}^{k} \chi_{E_j}.$$

Proposition 10 · 8 implies that χ_E is measurable and, hence, E is a measurable set. Thus, $\mathscr{M}$ is a σ-algebra. ∎

Note that, since $\mathscr{M}$ is a σ-algebra, any countable intersection of sets in $\mathscr{M}$ is also in $\mathscr{M}$.

We now show that any open or closed set in $\mathbf{R}^n$ is measurable. We know that any open interval (a, b) is measurable.

12 · 3 Proposition. *Any open set E in $\mathbf{R}^n$ is measurable.*

PROOF. If X is the set of n-tuples of real numbers and

$$d(x, y) = \max\{|x^k - y^k| : k = 1, \cdots, n\},$$

then (X, d) is a metric space which is homeomorphic to $\mathbf{R}^n$. Thus, E is open in (X, d) and (X, d) is separable. Note that the open balls in (X, d) are open intervals; that is,

$$B(x; r) = ((x^1 - r, \cdots, x^n - r), (x^1 + r, \cdots, x^n + r)).$$

Therefore, by Proposition 3 · 10, p. 63, the open set E is a countable union of open intervals and, hence, is measurable. ∎

A closed set in $\mathbf{R}^n$ is also measurable since it is the complement of an open set. In fact, if we define the set of **Borel sets** to be the smallest σ-algebra which contains all the open sets, then we can state that any Borel set is measurable. Since the set $\mathscr{B}$ of Borel sets is a σ-algebra it must contain any set which is a countable intersection of open sets; such a set is called a G_δ set. It must also contain any closed set and also any countable union of closed sets, called an F_σ set. The set of Borel sets must contain any countable union of G_δ sets, called a $G_{\delta\sigma}$ set, as well as any countable intersection of F_σ sets, called an $F_{\sigma\delta}$ set. Clearly, we can continue this indefinitely so that there is no problem in showing that there is an abundance of measurable sets in $\mathbf{R}^n$. As a matter of fact we may wonder if all subsets of $\mathbf{R}^n$ are measurable. However, using the Axiom of Choice we can show the existence of a nonmeasurable set in $\mathbf{R}^n$. This will be done in the next section.

The following proposition establishes a useful relationship between measurable sets and measurable functions.

12 · 4 Proposition. *A function f is measurable if and only if for all real numbers c the set $\{x : f(x) > c\}$ is measurable.*

PROOF. Suppose f is measurable. For each real number c let $E(c) = \{x : f(x) > c\}$ and for each positive integer k define

$$g_k = k\left[\left(f \wedge \left(c + \frac{1}{k}\right)\right) - (f \wedge c)\right].$$

Then, g_k is measurable and

$$g_k(x) = \begin{cases} 0 & \text{if } f(x) \le c \\ k[f(x) - c] & \text{if } c < f(x) \le c + 1/k \\ 1 & \text{if } f(x) > c + 1/k. \end{cases}$$

Since $\lim g_k(x) = \chi_{E(c)}(x)$ for all x, $\chi_{E(c)}$ is measurable (Proposition 10 · 5) and, hence, $E(c)$ is measurable.

Now suppose that, for all real numbers c, $E(c)$ is measurable. Then, for any real numbers c and d with $c < d$,

$$\{x : c < f(x) \le d\} = E(c) \sim E(d)$$

which is measurable. For each positive integer k, define

$$h_k(x) = \begin{cases} j/k & \text{if } (j-1)/k < f(x) \le j/k, \quad j = -k^2 + 1, \ldots, k^2 \\ 0 & \text{if } f(x) \le -k \text{ or } f(x) > k. \end{cases}$$

Let

$$E_{kj} = \{x : (j-1)/k < f(x) \le j/k\}.$$

Then

$$h_k(x) = \sum_{j=-k^2+1}^{k^2} \frac{j}{k} \chi_{E_{kj}}(x)$$

and, therefore, h_k is measurable since E_{kj} is measurable. For each x there exists a positive integer k_0 such that $-k_0 < f(x) \leq k_0$ and, hence,

$$0 \leq h_k(x) - f(x) < 1/k \quad \text{whenever } k \geq k_0.$$

Thus, $f(x) = \lim h_k(x)$ and, therefore, f is measurable. ▮

In Proposition 12 · 4 any of the sets $\{x : f(x) \geq c\}$, $\{x : f(x) < c\}$, and $\{x : f(x) \leq c\}$ could have been used in place of $\{x : f(x) > c\}$. For example, if $\{x : f(x) > c\}$ is measurable for all real numbers c, then

$$\{x : f(x) \geq c\} = \bigcap_{k=1}^{\infty} \left\{x : f(x) > c - \frac{1}{k}\right\}$$

is measurable for all real numbers c. Conversely, if $\{x : f(x) \geq c\}$ is measurable for all real numbers c, then

$$\{x : f(x) > c\} = \bigcup_{k=1}^{\infty} \left\{x : f(x) \geq c + \frac{1}{k}\right\}$$

is measurable for all real numbers c.

PROBLEMS

1. If E_1 and E_2 are measurable sets in $\mathbf{R}^k$ and $\mathbf{R}^m$, respectively, show that $E_1 \times E_2$ is measurable in $\mathbf{R}^n = \mathbf{R}^k \times \mathbf{R}^m$.

2. If f and g are measurable functions, show that the set $\{x : f(x) > g(x)\}$ is measurable.

3. Show that the sets $\{x : f(x) < c\}$ and $\{x : f(x) \leq c\}$ are measurable for all real numbers c if and only if the sets $\{x : f(x) > c\}$ are measurable for all real numbers c.

4. If, for some $p > 0$, $|f|^p$ is measurable, show that $|f|$ is measurable.

5. Show that a function f is measurable if and only if, for each open set $G \subset \mathbf{R}$, $f^{-1}(G)$ is a measurable set.

6. If g is continuous and f is measurable, show that $g \circ f$ is measurable.

13 · MEASURE

In this section we define a measure for measurable sets which is a generalization of the volume of an interval. The volume of an interval (a, b) is $\int \chi_{(a,b)}$.

13 · 1 Definition. *The (Lebesgue)* ***measure*** μ *is a nonnegative extended real-valued function defined on the set* $\mathscr{M}$ *of measurable sets as follows: if* $E \in \mathscr{M}$, *then*

$$\mu(E) = \begin{cases} \int \chi_E & \text{if } E \text{ is integrable} \\ \infty & \text{otherwise.} \end{cases}$$

Recall that we have shown that a set E is a set of measure zero or a null set if and only if $\int \chi_E = 0$. Thus, our terminology is consistent.

The following properties of the Lebesgue measure μ are easily obtained and the proofs of most of them are left to the reader. Assume that E and F are measurable sets.

(13 · 2) $\mu(\varnothing) = 0$

(13 · 3) $E \subset F \Rightarrow \mu(E) \leq \mu(F)$

(13 · 4) $\mu(E \cup F) \leq \mu(E) + \mu(F)$

(13 · 5) $E \cap F = \varnothing \Rightarrow \mu(E \cup F) = \mu(E) + \mu(F)$

(13 · 6) $E \subset F$ and $\mu(E) < \infty \Rightarrow \mu(F \sim E) = \mu(F) - \mu(E)$

Note that addition and subtraction of extended real numbers occur in 13 · 4–13 · 6. The following rules hold for these operations with the extended real number ∞: if $r \in \mathbf{R}$, then $\infty \pm r = \infty$ and $\infty + \infty = \infty$. The operation $\infty - \infty$ is not defined and is avoided.

PROOF OF 13 · 3. If F is not integrable, then $\mu(F) = \infty$ and $\mu(E) \leq \mu(F)$. If F is integrable, then, since $\chi_E \leq \chi_F$, E is integrable and

$$\mu(E) = \int \chi_E \leq \int \chi_F = \mu(F). \quad \blacksquare$$

PROOF OF 13 · 5. Suppose E is not integrable. Then $\mu(E) = \infty$. Since $E \subset E \cup F$, $\mu(E \cup F) = \infty$ and the result holds. If E and F are integrable, then, since $\chi_{E \cup F} = \chi_E + \chi_F$, $E \cup F$ is integrable and

$$\mu(E \cup F) = \int \chi_{E \cup F} = \int \chi_E + \int \chi_F = \mu(E) + \mu(F). \quad \blacksquare$$

A function which satisfies 13 · 5 is called **additive.** In general, if φ is an extended real-valued function defined on the sets of a class $\mathscr{S}$, then φ is said to be **finitely additive** if for any collection $\{E_1, \cdots, E_m\}$ of (pairwise) disjoint sets in $\mathscr{S}$ whose union is in $\mathscr{S}$ we have

$$\varphi\left(\bigcup_{k=1}^{m} E_k\right) = \sum_{k=1}^{m} \varphi(E_k).$$

An extended real-valued function φ defined on a class $\mathscr{S}$ is said to be **countably additive** if for any sequence (E_k) of pairwise disjoint sets in $\mathscr{S}$ whose union is in $\mathscr{S}$ we have

$$\varphi\left(\bigcup_{k=1}^{\infty} E_k\right) = \sum_{k=1}^{\infty} \varphi(E_k).$$

Note that in this definition $\sum_{k=1}^{\infty} \varphi(E_k)$ need not be finite; that is, it need not converge in the usual sense.

Using induction it is easy to show from 13 · 5 that μ is finitely additive.

A basic property of a measure is that it be countably additive and this we now prove.

13 · 7 Theorem. *The Lebesgue measure μ is countably additive.*

PROOF. Let (E_k) be a sequence of disjoint measurable sets and let $E = \bigcup_{k=1}^{\infty} E_k$. Then E is measurable. If, for some k, $\mu(E_k) = \infty$, then $\sum_{k=1}^{\infty} \mu(E_k) = \infty$. Also, since $E_k \subset E$, then $\mu(E) = \infty$ and therefore $\mu(E) = \sum_{k=1}^{\infty} \mu(E_k)$.

Suppose now, for all k, E_k is integrable. Then $\sum_{k=1}^{\infty} \mu(E_k)$ either converges or diverges to ∞. First, suppose $\sum_{k=1}^{\infty} \mu(E_k) = \infty$. Since, for all m,

$$\bigcup_{k=1}^{m} E_k \subset E,$$

$$\mu(E) \geq \mu\left(\bigcup_{k=1}^{m} E_k\right) = \sum_{k=1}^{m} \mu(E_k)$$

and, therefore,

$$\mu(E) = \infty = \sum_{k=1}^{\infty} \mu(E_k).$$

Next, suppose $\sum_{k=1}^{\infty} \mu(E_k)$ converges. Then, $(\int \sum_{k=1}^{m} \chi_{E_k})$ is bounded and, since $\chi_E = \sum_{k=1}^{\infty} \chi_{E_k}$, Levi's Theorem (8 · 1, p. 128) implies that E is integrable and

$$\mu(E) = \int \chi_E = \sum_{k=1}^{\infty} \int \chi_{E_k} = \sum_{k=1}^{\infty} \mu(E_k). \quad \blacksquare$$

If we drop the requirement that the sets be disjoint we obtain the following result whose proof is similar to that of Theorem 13 · 7 and is left to the reader.

13 · 8 Proposition. *If (E_k) is a sequence of measurable sets, then*

$$\mu\left(\bigcup_{k=1}^{\infty} E_k\right) \leq \sum_{k=1}^{\infty} \mu(E_k).$$

It is of interest to compute the Lebesgue measure of the Cantor set (p. 20). Recall that the Cantor set is obtained by deleting from [0, 1] its open middle third, then deleting the open middle third of the remaining subintervals and so on. We know that the cardinal number of this set is c, the cardinal number of the reals. Denoting the Cantor set by C, we have

$$C = [0, 1] \sim ((1/3, 2/3) \cup (1/9, 2/9) \cup (7/9, 8/9) \cup \cdots)$$

and, therefore,

$$\mu(C) = 1 - (1/3 + 2/9 + 4/27 + \cdots) = 1 - \tfrac{1}{3}\sum_{k=0}^{\infty} (\tfrac{2}{3})^k = 0.$$

Thus, even though the Cantor set has cardinal number c, it is insignificant with respect to Lebesgue integration.

As a generalization of the volume of an interval, we might expect that Lebesgue measure would be invariant under translations and we will show that such is the case. If E is a set in $\mathbf{R}^n$ and $c \in \mathbf{R}^n$, then the set $E + c = \{x + c : x \in E\}$ is a translation of E. Note that since

$$\chi_{E+c}(x) = \chi_E(x - c),$$

if we let $\varphi(x) = x - c$, then $\chi_{E+c} = \chi_E \circ \varphi$.

If g is a step function with representation

$$g(x) = \begin{cases} c_k & \text{for } x \in I_k, \quad k = 1, \cdots, r \\ 0 & \text{for } x \notin [a, b], \end{cases}$$

then

$$g \circ \varphi(x) = \begin{cases} c_k & \text{for } x \in I_k + c, \quad k = 1, \cdots, r \\ 0 & \text{for } x \notin [a + c, b + c]. \end{cases}$$

Thus $g \circ \varphi$ is a step function and

$$\int g \circ \varphi = \sum_{k=1}^{r} c_k v(I_k + c) = \sum_{k=1}^{r} c_k v(I_k) = \int g.$$

13 · 9 Proposition. *If E is a measurable set in $\mathbf{R}^n$ and $c \in \mathbf{R}^n$, then $E + c$ is measurable and*

$$\mu(E + c) = \mu(E).$$

PROOF. Let $\varphi(x) = x - c$. First, suppose that E is integrable. Then, there exist functions g and h and nondecreasing sequences (g_k) and (h_k) of step functions such that

$$\chi_E = g - h \quad \text{a.e.},$$

$$\lim g_k(x) = g(x) \quad \text{and} \quad \lim h_k(x) = h(x) \quad \text{for a.a.} x,$$

and

$$\mu(E) = \int \chi_E = \int g - \int h = \lim \int g_k - \lim \int h_k.$$

From the discussion preceding this proposition we see that $(g_k \circ \varphi)$ and $(h_k \circ \varphi)$ are nondecreasing sequences of step functions such that

$$\lim g_k \circ \varphi(x) = g \circ \varphi(x) \quad \text{and} \quad \lim h_k \circ \varphi(x) = h \circ \varphi(x) \quad \text{for a.a.} x$$

and $(\int g_k \circ \varphi)$ and $(\int h_k \circ \varphi)$ are bounded. Thus,

$$\int g \circ \varphi = \lim \int g_k \circ \varphi = \lim \int g_k = \int g$$

$$\int h \circ \varphi = \lim \int h_k \circ \varphi = \lim \int h_k = \int h$$

and, therefore, $\chi_E \circ \varphi$ is integrable and

$$\mu(E + c) = \int \chi_{E+c} = \int \chi_E \circ \varphi = \int g \circ \varphi - \int h \circ \varphi = \int g - \int h = \mu(E).$$

If E is measurable, then there exists a sequence (f_k) of integrable functions such that $\chi_E(x) = \lim f_k(x)$ for a.a.x. Then, $(f_k \circ \varphi)$ is a sequence of integrable functions such that $\chi_E \circ \varphi(x) = \lim f_k \circ \varphi(x)$ for a.a.x. Therefore, $\chi_{E+c} = \chi_E \circ \varphi$ is a measurable function and $E + c$ is a measurable set. From the first part of the proof it is clear that if $\mu(E) = \infty$ then $\mu(E + c) = \infty$. ∎

We will now show the existence of a nonmeasurable set in **R**. Define a relation R on $(0, 1)$ by the rule: xRy if $y - x$ is rational. It is easy to see that this is an equivalence relation and, thus, $(0, 1)$ is partitioned into equivalence classes:

$$R(a) = \{x \in (0, 1) : aRx\} = \{a + r : a + r \in (0, 1),\ r \text{ rational}\}.$$

Using the Axiom of Choice, form a set E by taking one element from each of the equivalence classes. We will show that E is not measurable.

If $\{r_1, r_2, \cdots\}$ is an enumeration of the rational numbers in the interval $(-1, 1)$, then it is easy to show that $\{E + r_i : i = 1, 2, \cdots\}$ is a collection of disjoint sets such that

$$(0, 1) \subset \bigcup_{i=1}^{\infty} (E + r_i) \subset (-1, 2).$$

Then, using the invariance of measure under translation, we obtain under the assumption that E is measurable:

$$1 = \mu(0, 1) \leq \mu\left(\bigcup_{i=1}^{\infty} (E + r_i)\right) = \sum_{i=1}^{\infty} \mu(E + r_i) \leq \mu(-1, 2) = 3$$

and, therefore,

$$1 \leq \sum_{i=1}^{\infty} \mu(E) \leq 3.$$

This is impossible since $\sum_{i=1}^{\infty} \mu(E) = 0$ if $\mu(E) = 0$ and $\sum_{i=1}^{\infty} \mu(E) = \infty$ if $\mu(E) > 0$. Thus, the assumption that E is measurable cannot hold.

PROBLEMS

1. Prove 13 · 2, 13 · 4, 13 · 6.

2. Prove 13 · 8.

3. If (E_k) is an increasing sequence of measurable sets, i.e., $E_k \subset E_{k+1}$, show that

$$\mu\left(\bigcup_{k=1}^{\infty} E_k\right) = \lim_{k\to\infty} \mu E_k.$$

4. If (E_k) is a decreasing sequence of measurable sets, i.e., $E_{k+1} \subset E_k$, and $\mu E_1 < \infty$, show that

$$\mu\left(\bigcap_{k=1}^{\infty} E_k\right) = \lim_{k\to\infty} \mu E_k.$$

5. If E_1 and E_2 are measurable, show that

$$\mu(E_1 \cup E_2) + \mu(E_1 \cap E_2) = \mu E_1 + \mu E_2 .$$

6. If E is a measurable set in $[a, b]$, show that for each $\varepsilon > 0$ there exists an open set G such that $E \subset G$ and $\mu(G \sim E) < \varepsilon$.

Suggestion. There exists a sequence (g_k) of step functions such that $\lim g_k(x) = \chi_E(x)$ for a.a.x and $g_k(x) = 0$ if $x \notin [a, b]$ or if g_k is discontinuous at x. For $n = 1, 2, \cdots$, let

$$h_n(x) = \begin{cases} 1 & \text{if } g_k(x) > 1/2 \text{ for some } k \geq n \\ 0 & \text{otherwise.} \end{cases}$$

Then h_n is the characteristic function of a set which is the countable union of open intervals in $[a, b]$ and the Monotone Convergence Theorem can be applied to the sequence (h_n) which converges to χ_E a.e.

7. If E is a measurable set in $\mathbf{R}^n$, show that for each $\varepsilon > 0$ there exists an open set G such that $E \subset G$ and $\mu(G \sim E) < \varepsilon$. Also, there exists a closed set F such that $F \subset E$ and $\mu(E \sim F) < \varepsilon$.

14 · THE LEBESGUE INTEGRAL OVER A MEASURABLE SET

If f is a function from $\mathbf{R}^n$ to $\mathbf{R}$ and E is a measurable set in $\mathbf{R}^n$, then f is **integrable over** E if $f\chi_E$ is integrable (over $\mathbf{R}^n$) and

$$\int_E f = \int f\chi_E .$$

Actually f need only be defined on E. If any extension of f to all of $\mathbf{R}^n$ is taken, then $f\chi_E$ will be defined on $\mathbf{R}^n$ and will be zero outside of E.

Observe that if f is integrable (over $\mathbf{R}^n$), then f is integrable over any measurable set E in $\mathbf{R}^n$. For, $f\chi_E$ is a measurable function bounded by the integrable function $|f|$ and, hence, $f\chi_E$ is integrable.

If we let $\mathscr{L}(E)$ denote the set of real-valued functions which are integrable over E, then it is a simple matter to check that $\mathscr{L}(E)$ is a real linear space and a lattice. Thus, $\mathscr{L}(E)$ has the same basic properties as $\mathscr{L}$. We can also define a function f to be measurable over E if $f\chi_E$ is measurable. Then with obvious modifications the whole theory of integration developed up to this point will apply to integration over E.

We now consider some special results for integration over a measurable subset of $\mathbf{R}^n$.

14 · 1 Proposition. *If $\{E_1, E_2, \cdots\}$ is a collection of disjoint measurable sets and f is integrable over $E = \bigcup_{k=1}^{\infty} E_k$, then f is integrable over each E_k and*

$$\int_E f = \sum_{k=1}^{\infty} \int_{E_k} f.$$

PROOF. Note that $\chi_E = \sum_{k=1}^{\infty} \chi_{E_k}$ and, therefore, $f\chi_E = \lim_{m\to\infty} \sum_{k=1}^{m} f\chi_{E_k}$. Since $f\chi_{E_k}$ is measurable and, for all m, $|\sum_{k=1}^{m} f\chi_{E_k}| \le |f|\chi_E$ which is integrable, by the Lebesgue Dominated Convergence Theorem (10 · 2) we have

$$\int_E f = \int f\chi_E = \lim \int \sum_{k=1}^{m} f\chi_{E_k} = \sum_{k=1}^{\infty} \int_{E_k} f. \quad \blacksquare$$

The converse of this proposition is not true unless we add the hypothesis that f is a.e. nonnegative on each E_k.

14 · 2 Proposition. *If $\{E_1, E_2, \cdots\}$ is a collection of disjoint measurable sets, f is a.e. nonnegative and integrable on each E_k, and $\sum_{k=1}^{\infty} \int_{E_k} f$ converges, then f is integrable on $E = \bigcup_{k=1}^{\infty} E_k$ and*

$$\int_E f = \sum_{k=1}^{\infty} \int_{E_k} f.$$

PROOF. Apply Levi's Theorem (8 · 1, p. 128). $\blacksquare$

If we integrate over a set E of finite measure, that is, an integrable set, we can obtain some results that are special to that case. The reader can easily supply the proofs for the following results.

14 · 3 Proposition. *If E is of measure zero, then any function f is integrable over E and $\int_E f = 0$.*

14 · 4 Proposition. *A constant function c is integrable over any set E of finite measure and*

$$\int_E c = c\mu(E).$$

14 · 5 Proposition. *If f is measurable and $m \le f(x) \le M$ for a.a.x in a set E of finite measure, then f is integrable over E and*

$$m\mu(E) \le \int_E f \le M\mu(E).$$

The Russian mathematician D. F. Egoroff (1869–1931) showed that if a sequence of measurable functions on a set E of finite measure converges to a measurable function on E then the convergence is almost uniform on E.

14 · 6 Theorem (Egoroff, 1911). *Let E be a set of finite measure and let f and $f_1, f_2, \cdots$ be measurable functions on E such that*

$$\lim f_k(x) = f(x) \quad \text{for a.a.} x \in E.$$

Then corresponding to each $\varepsilon > 0$ there is a measurable set B with $\mu(B) < \varepsilon$ such that (f_k) converges to f uniformly on $E \sim B$.

PROOF. Assume that $\lim f_k(x) = f(x)$ for all $x \in E \sim A$ where A is a set of measure zero. For each positive integer j, let

$$E_{j,k} = \{x \in E : |f_k(x) - f(x)| \geq 1/j\}$$

and

$$B_{j,m} = \bigcup_{k=m}^{\infty} E_{j,k} = \{x \in E : |f_k(x) - f(x)| \geq 1/j \text{ for some } k \geq m\}.$$

Then since f_k and f are measurable, $E_{j,k}$ and $B_{j,m}$ are measurable sets. Also, $B_{j,m+1} \subset B_{j,m}$ and $\bigcap_{m=1}^{\infty} B_{j,m} \subset A$. Therefore,

$$\lim_{m\to\infty} \mu(B_{j,m}) = \mu\left(\bigcap_{m=1}^{\infty} B_{j,m}\right) \leq \mu(A) = 0.$$

Thus, for each j, there exists an $m(j)$ such that

$$\mu(B_{j,m(j)}) < \varepsilon/2^j.$$

Let $B = \bigcup_{j=1}^{\infty} B_{j,m(j)}$. Then

$$\mu(B) \leq \sum_{j=1}^{\infty} \mu(B_{j,m(j)}) < \varepsilon.$$

Since

$$E \sim B = E \sim \bigcup_{j=1}^{\infty} B_{j,m(j)} = \bigcap_{j=1}^{\infty} (E \sim B_{j,m(j)}),$$

for each positive integer j there exists an $m(j)$ such that for all $x \in E \sim B$

$$|f_k(x) - f(x)| < 1/j \quad \text{whenever } k \geq m(j). \blacksquare$$

Using Egoroff's Theorem we can show that a function which is measurable on a measurable set is almost a continuous function. This result is due to the Russian mathematician N. N. Lusin (1883–1952).

14 · 7 Theorem (Lusin). *If f is a measurable function on a measurable set $E \subset \mathbf{R}^n$, then for each $\delta > 0$ there exists a closed set $F \subset E$ such that $\mu(E \sim F) < \delta$ and f is continuous on F.*

The proof of this theorem is left to the reader (Problems 9, 10). See also Problem 9, p. 194.

PROBLEMS

1. Prove 14 · 3.
2. Prove 14 · 4.
3. Prove 14 · 5.
4. Show that $\int_1^{\infty}(1/x^2)\,dx$ exists and find its value.

5. Show that if $E = \{(x, y) : x \in [a, b], y \in [g(x), h(x)]\}$ where g and h are continuous on $[a, b]$ and $g(x) \le h(x)$ for all $x \in [a, b]$, then for any integrable function f on $\mathbf{R}^2$

$$\int_E f = \int_a^b \int_{g(x)}^{h(x)} f(x, y)\, dy\, dx.$$

6. For each $x \in \mathbf{R}$ show that $\int_{-\infty}^{x} e^{-t^2}\, dt$ exists.

7. If f is integrable over $[a, b] \subset \mathbf{R}$, show that

$$\int_a^b f(x)\, dx = \int_{-b}^{-a} f(-x)\, dx.$$

8. Show that Egoroff's Theorem does not hold if $\mu(E) = \infty$.

Suggestion. Let $E = \mathbf{R}$ and $f_n = \chi_{[n, n+1]}$.

9. If f is a measurable function on the measurable set $E \subset \mathbf{R}^n$ with $\mu(E) < \infty$, show that for each $\delta > 0$ there exists a closed set $F \subset E$ such that $\mu(E \sim F) < \delta$ and f is continuous on F.

Suggestion. There exists a sequence (g_k) of step functions such that $\lim g_k(x) = f(x)$ for a.a. $x \in E$. Then, there exists a set A of measure zero such that, on $E \sim A$, (g_k) is a sequence of continuous functions which converges to f. By Egoroff's Theorem there exists a set B with $\mu(B) < \delta/2$ such that (g_k) converges uniformly to f on $E \sim (A \cup B) = C$. By Problem 7, p. 149, there exists a closed set $F \subset C$ such that $\mu(C \sim F) < \delta/2$.

10. Prove Lusin's Theorem 14 · 7.

Suggestion. Let $E_1 = E \cap B[0; 1]$ and $E_m = E \cap B[0; m] \sim B[0; m-1]$, $m = 2, 3, \cdots$. By Problem 9 there exists a closed set $F_m \subset E_m$ such that $\mu(E_m \sim F_m) < 2^{-m}\delta$ and f is continuous on F_m. Let $F = \bigcup_{m=1}^{\infty} F_m$.

15 · THE DANIELL INTEGRAL

In this chapter we have considered Lebesgue integration for real-valued functions defined on a subset of $\mathbf{R}^n$. However, the method of development, called the Daniell method after the British-American mathematician P. J. Daniell (1889–1946), is quite general and can be used to obtain other integrals. In outline, this development of an integral starts with a vector lattice $\mathscr{S}$ of real-valued functions defined on a nonempty set X and a linear functional, called an integral and so denoted, defined on $\mathscr{S}$ such that

(1) if $f \in \mathscr{S}$ and $f \ge 0$, then $\int f \ge 0$;

(2) if (f_k) is a nonincreasing sequence of nonnegative functions in $\mathscr{S}$ and $\lim f_k(x) = 0$ for all $x \in X$, then

$$\lim \int f_k = 0.$$

A set E in X is said to be a null set if there is a nondecreasing sequence (g_k) of nonnegative functions in $\mathscr{S}$ such that $(g_k(x))$ diverges to ∞ for each $x \in E$ and $(\int g_k)$ converges. Then we can prove:

15 · 1 Proposition. *The countable union of null sets is a null set.*

15 · 2 Proposition. *If* (f_k) *is a nonincreasing sequence of nonnegative functions in* $\mathscr{S}$ *and* $\lim f_k(x) = 0$ *for a.a.x, then*

$$\lim \int f_k = 0.$$

15 · 3 Proposition. *If* (g_k) *and* (h_k) *are nondecreasing sequences of functions in* $\mathscr{S}$ *such that*

$$\lim g_k(x) \leq \lim h_k(x) \quad \textit{for a.a.x}$$

and $(\int g_k)$ *and* $(\int h_k)$ *are bounded, then*

$$\lim \int g_k \leq \lim \int h_k.$$

Then we define $\bar{\mathscr{S}}$ to be the set of real-valued functions g defined on X such that there is a nondecreasing sequence (g_k) of functions in $\mathscr{S}$ such that $\lim g_k(x) = g(x)$ for a.a.x and $(\int g_k)$ is bounded. For $g \in \bar{\mathscr{S}}$ we define

$$\int g = \lim \int g_k.$$

In this chapter we considered in detail the case where $X = \mathbf{R}^n$, $\mathscr{S}$ is the set of step functions on $\mathbf{R}^n$, and an integral is defined on $\mathscr{S}$. We then showed that $\mathscr{S}$ had the properties stated above. Since the further development of integration depends only on these properties of $\mathscr{S}$, the general case is essentially the same as that of the Lebesgue integral on $\mathbf{R}^n$.

15 · 4 Proposition. *If g and h are in* $\bar{\mathscr{S}}$ *and* α *is a nonnegative real number, then* $g + h$ *and* αg *are in* $\bar{\mathscr{S}}$ *and*

$$\int g + h = \int g + \int h \quad \textit{and} \quad \int \alpha g = \alpha \int g.$$

15 · 5 Proposition. $\bar{\mathscr{S}}$ *is a lattice.*

15 · 6 Proposition. *If* $g, h \in \bar{\mathscr{S}}$ *and* $g(x) \leq h(x)$ *for a.a.x, then*

$$\int g \leq \int h.$$

15 · 7 Proposition. *Let* (g_n) *be a nondecreasing sequence of functions in* $\bar{\mathscr{S}}$ *such that* $(\int g_n)$ *is bounded. Then* $(g_n(x))$ *converges for a.a.x and if* $g(x) = \lim g_n(x)$ *for a.a.x then* $g \in \bar{\mathscr{S}}$ *and*

$$\int g = \lim \int g_n.$$

15 · 8 Corollary. *Let (h_k) be a sequence of nonnegative functions in $\mathscr{P}$ such that $(\int \sum_{k=1}^{n} h_k)$ is bounded. Then $\sum_{k=1}^{\infty} h_k(x)$ converges for a.a.x and if $g(x) = \sum_{k=1}^{\infty} h_k(x)$ for a.a.x, then $g \in \mathscr{P}$ and*

$$\int g = \sum_{k=1}^{\infty} \int h_k.$$

Let $\mathscr{L}$ be the set of real-valued functions f defined on X such that $f = g - h$ a.e. where g and h are in $\mathscr{P}$ and define

$$\int f = \int g - \int h.$$

15 · 9 Theorem. *$\mathscr{L}$ is a vector lattice and the integral is a linear functional on $\mathscr{L}$.*

15 · 10 Theorem. *If $f \in \mathscr{L}$ and $f \geq 0$ a.e., then $\int f \geq 0$.*

15 · 11 Theorem (Levi). *Let (f_k) be a sequence of functions in $\mathscr{L}$ which are nonnegative a.e. such that $(\int \sum_{k=1}^{n} f_k)$ is bounded. Then $\sum_{k=1}^{\infty} f_k(x)$ converges for a.a.x and if $f(x) = \sum_{k=1}^{\infty} f_k(x)$ for a.a.x, then $f \in \mathscr{L}$ and*

$$\int f = \sum_{k=1}^{\infty} \int f_k.$$

15 · 12 Corollary (Monotone Convergence). *Let (g_n) be an a.e. nondecreasing sequence of functions in $\mathscr{L}$ such that $(\int g_n)$ is bounded. Then $(g_n(x))$ converges for a.a.x and if $g(x) = \lim g_n(x)$ for a.a.x, then $g \in \mathscr{L}$ and*

$$\int g = \lim \int g_n.$$

15 · 13 Corollary. *If $f \in \mathscr{L}$ and $\int |f| = 0$, then $f = 0$ a.e.*

15 · 14 Lemma (Fatou). *If (f_n) is a sequence of functions in $\mathscr{L}$ and g is a function in $\mathscr{L}$ such that, for all n, $f_n(x) \geq g(x)$ for a.a.x and $\underline{\lim} \int f_n < \infty$, then $\underline{\lim}\, f_n(x)$ exists for a.a.x and, if $f(x) = \underline{\lim}\, f_n(x)$ for a.a.x, $f \in \mathscr{L}$ and*

$$\int f \leq \underline{\lim} \int f_n.$$

15 · 15 Theorem (Lebesgue Dominated Convergence). *Let (f_n) be a sequence of functions in $\mathscr{L}$ such that $(f_n(x))$ converges for a.a.x and, for all n, $|f_n| \leq g$ a.e. for some function $g \in \mathscr{L}$. If $f(x) = \lim f_n(x)$ for a.a.x, then $f \in \mathscr{L}$ and*

$$\int f = \lim \int f_n.$$

A real-valued function f defined on X is **measurable** if, for each $g \in \mathscr{L}$ such that $g \geq 0$,

$$(-g) \vee (f \wedge g) \in \mathscr{L}.$$

The set of measurable functions on X is denoted by $\tilde{\mathscr{M}}$.

15 · 16 Proposition. *$\tilde{\mathscr{M}}$ is a vector lattice.*

15 · 17 Proposition. *If (f_n) is a sequence of functions in $\tilde{\mathscr{M}}$ and $f(x) = \lim f_n(x)$ for a.a.x, then $f \in \tilde{\mathscr{M}}$.*

In general it is not true that $\tilde{\mathscr{M}}$ is an algebra. However, if the vector lattice $\mathscr{S}$ has the additional property introduced by the American mathematician Marshall H. Stone (1903–1989) and now called Stone's Axiom, then $\tilde{\mathscr{M}}$ is an algebra.

15 · 18 Stone's Axiom. *If $f \in \mathscr{S}$, then $1 \wedge f \in \mathscr{S}$.*

15 · 19 Proposition. *If $\mathscr{S}$ satisfies Stone's Axiom, then $1 \in \tilde{\mathscr{M}}$.*

15 · 20 Proposition. *If $\mathscr{S}$ satisfies Stone's Axiom, then $\tilde{\mathscr{M}}$ is an algebra.*

A subset E of X is called **measurable** if χ_E is a measurable function. Let $\mathscr{M}$ denote the set of measurable subsets of X.

15 · 21 Proposition. *$\mathscr{M}$ is a σ-ring.*

If $\mathscr{S}$ satisfies Stone's Axiom, then $\mathscr{M}$ is a σ-algebra.

The **measure** μ is a nonnegative extended real-valued function defined on the σ-ring $\mathscr{M}$ of measurable sets as follows:

$$\mu(E) = \begin{cases} \int \chi_E & \text{if } \chi_E \text{ is integrable} \\ \infty & \text{otherwise.} \end{cases}$$

Then μ is a countably additive function on $\mathscr{M}$ and has the properties 13 · 2–13 · 6, p. 145.

We now discuss briefly another approach to integration. We start with a nonempty set X, a σ-ring $\mathscr{M}$ of subsets of X, and a countably additive nonnegative extended real-valued function μ defined on $\mathscr{M}$ with the property that $\mu(\varnothing) = 0$. Then $(X, \mathscr{M}, \mu)$ is called a measure space, the subsets of X which arc in $\mathscr{M}$ are called measurable sets, and μ is called a measure. Having a measure space one can then define measurable functions, integration, and integrable functions. For an explication of this approach to integration many books are available, for example, [18], [4], or [39].

It is also true that if one has a measure space $(X, \mathscr{M}, \mu)$ then integration can be developed by the Daniell method. The set $\mathscr{S}$ of elementary functions can be taken to be the set of generalized step functions on X. A real-valued function f defined on X is called **a generalized** step function if it takes on a finite number of distinct nonzero values $c_1, \cdots, c_r$ and if each of the sets $E_k = \{x \in X : f(x) = c_k\}$ has finite measure. Thus,

$$f = \sum_{k=1}^{r} c_k \chi_{E_k}.$$

The elementary integral of the generalized step function f is taken to be

$$\int f = \sum_{k=1}^{r} c_k \mu(E_k). \tag{15 · 22}$$

It is not difficult to show that this set $\mathscr{S}$ and the integral on $\mathscr{S}$ possess the properties required for the Daniell development of integration. In general, the set of measurable sets obtained in this development will not be the same as the set $\mathscr{M}$ in the given measure space but it will contain $\mathscr{M}$.

PROBLEMS

1. Prove Proposition 15 · 1.

2. Prove Proposition 15 · 2.

3. Prove Proposition 15 · 3.

4. Prove Proposition 15 · 4.

5. Prove Propositions 15 · 19 and 15 · 20.

6. Let $(X, \mathscr{M}, \mu)$ be a measure space and let $\mathscr{S}$ be the set of generalized step functions on X. Show that $\mathscr{S}$ with the integral defined by 15 · 22 has the required properties for the Daniell development of integration.

7. Let X be the set of positive integers and $\mathscr{S}$ be the set of all real-valued functions defined on X which have only a finite number of nonzero values. If $f \in \mathscr{S}$, let $\int f = \sum_{n=1}^{\infty} f(n)$ (this is a finite sum). Show that $\mathscr{S}$ is a vector lattice and the integral is a linear functional on $\mathscr{S}$ which satisfies properties (1) and (2), p. 152. Show that a null set is empty, that $\mathscr{L}$ is the set of all sequences $f = (f(n))$ such that $\sum_{n=1}^{\infty} |f(n)|$ converges and $\int f = \sum_{n=1}^{\infty} |f(n)|$.

7 | *Normed Linear Spaces*

1 · INTRODUCTION

In this chapter we consider spaces which are both linear spaces and metric spaces. The Euclidean spaces $\mathbf{R}^n$ are familiar examples of such spaces. In $\mathbf{R}^n$ the metric is introduced via the length of a vector; that is, $d(x, y) = |x - y|$. In the linear spaces considered in this chapter we introduce a generalization of the length of a vector in $\mathbf{R}^n$, called a norm, and define a metric in terms of this norm.

A normed linear space is a linear space with a norm defined on it. Since such spaces are metric spaces, the theory developed for metric spaces applies to them. As one might expect, the introduction of algebraic structure allows a further development of the theory.

2 · THE HÖLDER AND MINKOWSKI INEQUALITIES

Before considering normed linear spaces we will prove some inequalities which will be useful in the construction of examples of these spaces. First, we extend a familiar inequality:

$$(1 + x)^n \geq 1 + nx \quad \text{if } x \geq -1 \text{ and } n \text{ is a positive integer.}$$

2 · 1 Lemma. *If x and p are real numbers such that $x \geq -1$ and $p \geq 1$, then*

$$(1 + x)^p \geq 1 + px.$$

PROOF. The result is obvious if $p = 1$. Assume $p > 1$. Let

$$f(x) = (1 + x)^p - 1 - px \quad \text{for } x \geq -1.$$

Then

$$f'(x) = p(1 + x)^{p-1} - p$$

and, thus, $f'(x) < 0$ for $-1 \leq x < 0$ and $f'(x) > 0$ for $x > 0$. This shows that f has a minimum value at $x = 0$ and, hence,

$$(1 + x)^p - 1 - px \geq 0 \quad \text{for } x \geq -1. \blacksquare$$

If we let $1 + x = u^{1/p}/v^{1/p}$, then we can rewrite the above inequality as follows:

$$\frac{u}{v} \geq 1 + p\frac{u^{1/p}}{v^{1/p}} - p \quad \text{for } u \geq 0, v > 0$$

and this in turn gives

$$(2\cdot 2)\qquad u^{1/p}v^{1-1/p} \le \frac{1}{p}u + \left(1 - \frac{1}{p}\right)v \quad \text{for } u \ge 0, v \ge 0.$$

Before proving Hölder's Inequality we introduce some notation. Let $\mathscr{L}_p$ be the set of complex-valued functions f defined on $\mathbf{R}^n$ such that f is measurable and $|f|^p$ is integrable. If E is a measurable set in $\mathbf{R}^n$, then $\mathscr{L}_p(E)$ denotes the set of functions f such that f is measurable over E and $|f|^p$ is integrable over E. Note that $\mathscr{L}_1(E) = \mathscr{L}(E)$.

2 · 3 Proposition (Hölder's Inequality). *Suppose that $p > 1$, $1/q = 1 - 1/p$, $f \in \mathscr{L}_p(E)$, and $g \in \mathscr{L}_q(E)$. Then $fg \in \mathscr{L}(E)$ and*

$$\int_E |fg| \le \left[\int_E |f|^p\right]^{1/p}\left[\int_E |g|^q\right]^{1/q}.$$

Proof. If $\int_E |f|^p = 0$, then $f = 0$ a.e. on E and, hence, $\int_E |fg| = 0$ so that equality holds $(0 = 0)$. If $\int_E |g|^q = 0$, then the result also holds. Suppose $\int_E |f|^p \ne 0$ and $\int_E |g|^q \ne 0$. In $2 \cdot 2$ replace u by $|f(x)|^p/\int_E |f|^p$ and v by $|g(x)|^q/\int_E |g|^q$. Then we obtain

$$\frac{|f(x)|}{[\int_E|f|^p]^{1/p}} \cdot \frac{|g(x)|}{[\int_E|g|^q]^{1/q}} \le \frac{|f(x)|^p}{p\int_E|f|^p} + \frac{|g(x)|^q}{q\int_E|g|^q}.$$

Thus, $fg \in \mathscr{L}(E)$ and integrating we have

$$\frac{\int_E|fg|}{[\int_E|f|^p]^{1/p}[\int_E|g|^q]^{1/q}} \le \frac{1}{p} + \frac{1}{q} = 1. \blacksquare$$

With the aid of Hölder's Inequality we now prove the Minkowski Inequality.

2 · 4 Proposition (Minkowski's Inequality). *If $p \ge 1$ and $f, g \in \mathscr{L}_p(E)$, then $f + g \in \mathscr{L}_p(E)$ and*

$$\left[\int_E |f + g|^p\right]^{1/p} \le \left[\int_E |f|^p\right]^{1/p} + \left[\int_E |g|^p\right]^{1/p}.$$

Proof. Since

$$|f + g|^p \le (|f| + |g|)^p \le [2(|f| \vee |g|)]^p \le 2^p(|f|^p + |g|^p),$$

$f + g \in \mathscr{L}_p(E)$. If $\int_E |f + g|^p = 0$, then Minkowski's Inequality obviously holds. Assume $\int_E |f + g|^p \ne 0$. The inequality follows directly from the Triangle Inequality in the case $p = 1$. For $p > 1$ we have

$$|f(x) + g(x)|^p \le |f(x)|\,|f(x) + g(x)|^{p-1} + |g(x)|\,|f(x) + g(x)|^{p-1}.$$

Since $(p - 1)q = p$, $|f + g|^{p-1} = |f + g|^{p/q} \in \mathscr{L}_q(E)$ and the above inequality and Hölder's Inequality imply

$$\int_E |f+g|^p \le \int_E |f|\,|f+g|^{p-1} + \int_E |g|\,|f+g|^{p-1}$$
$$\le \left[\int_E |f|^p\right]^{1/p}\left[\int_E |f+g|^{(p-1)q}\right]^{1/q}$$
$$+ \left[\int_E |g|^p\right]^{1/p}\left[\int_E |f+g|^{(p-1)q}\right]^{1/q}.$$

Dividing this inequality by $[\int_E |f+g|^p]^{1/q}$ we obtain Minkowski's Inequality. ∎

If x_k and y_k for $k = 1, \cdots, m$ are complex numbers, let

$$f(t) = |x_k| \quad \text{and} \quad g(t) = |y_k| \quad \text{for } t \in [k, k+1)$$

and $f(t) = 0 = g(t)$ for $t \notin [1, m+1)$. Then we obtain the summation form of these inequalities from the integral form.

2 · 5 Hölder's Inequality. *If $p > 1$ and $1/q = 1 - 1/p$, then*

$$\sum_{k=1}^{m} |x_k y_k| \le \left[\sum_{k=1}^{m} |x_k|^p\right]^{1/p}\left[\sum_{k=1}^{m} |y_k|^q\right]^{1/q}$$

for any complex numbers $x_1, \cdots, x_m, y_1, \cdots, y_m$.

2 · 6 Minkowski's Inequality. *If $p \ge 1$, then*

$$\left[\sum_{k=1}^{m} |x_k + y_k|^p\right]^{1/p} \le \left[\sum_{k=1}^{m} |x_k|^p\right]^{1/p} + \left[\sum_{k=1}^{m} |y_k|^p\right]^{1 p}$$

for any complex numbers $x_1, \cdots, x_m, y_1, \cdots, y_m$.

PROBLEMS

1. If $0 < r < s$ and $f \in \mathscr{L}_s([a, b])$ where $a, b \in \mathbf{R}$, show that $f \in \mathscr{L}_r([a, b])$ and

$$\left[\int_a^b |f|^r\right]^{1/r} \le (b-a)^{1/r - 1/s}\left[\int_a^b |f|^s\right]^{1/s}.$$

2. If f is a bounded function in $\mathscr{L}_p(E)$ and $r > p$, show that $f \in \mathscr{L}_r(E)$.

3. If (x_k) and (y_k) are sequences of complex numbers such that $\sum_{k=1}^{\infty} |x_k|^p$ and $\sum_{k=1}^{\infty} |y_k|^q$ converge where p and q are positive numbers such that $1/p + 1/q = 1$, show that

$$\sum_{k=1}^{\infty} |x_k y_k| \le \left[\sum_{k=1}^{\infty} |x_k|^p\right]^{1/p}\left[\sum_{k=1}^{\infty} |y_k|^q\right]^{1/q}.$$

4. If f is continuous on $[a, b]$ show that

$$\lim_{p\to\infty}\left[\int_a^b |f|^p\right]^{1/p} = \sup_{t\in[a,b]} |f(t)|.$$

5. For $x = (\alpha^1, \cdots, \alpha^n)$ and $y = (\beta^1, \cdots, \beta^n)$ in $V_n(\mathbf{R})$ and $p \geq 1$, define

$$d(x, y) = \left[\sum_{k=1}^{n} |\alpha^k - \beta^k|^p\right]^{1/p}.$$

Show that d is a metric on $V_n(\mathbf{R})$.

6. Show that, for $p \geq 1$, $\mathscr{L}_p$ is a linear space and if

$$d(f, g) = \left[\int |f - g|^p\right]^{1/p},$$

then d is a metric on $\mathscr{L}_p$ if $f = g$ in $\mathscr{L}_p$ means $f = g$ a.e..

3 · NORMED LINEAR SPACES

Let X be a linear space over the field $\mathbf{F}$ where $\mathbf{F}$ is either $\mathbf{R}$ or $\mathbf{C}$.

3 · 1 Definition. *A* ***norm*** *on X is a real-valued function, whose value at x is denoted by $\|x\|$, satisfying the following conditions for all $x, y \in X$ and $\alpha \in \mathbf{F}$:*

(1) $\|x\| > 0$ *if* $x \neq 0$
(2) $\|\alpha x\| = |\alpha|\,\|x\|$
(3) $\|x + y\| \leq \|x\| + \|y\|$.

A linear space X with a norm defined on it is called a ***normed linear space.***

From (2) we see that $\|0\| = \|0 \cdot x\| = 0\|x\| = 0$ and this along with (1) implies that $\|x\| = 0$ if and only if $x = 0$. Thus, a norm on a linear space has the basic properties of the length of a vector in a Euclidean space.

If we define

$$d(x, y) = \|x - y\|,$$

then it is easy to check that d is a metric on X (Problem 1). Noting that $d(x, 0) = \|x\|$, we now rephrase in normed linear space terminology some of the concepts considered in our discussion of metric spaces.

The set E in a normed linear space X is **bounded** if and only if there exists a positive number r such that $E \subset B(0; r)$; that is, if and only if there exists a positive number r such that $\|x\| < r$ for all $x \in E$.

The definition of continuity becomes: a function f from a normed linear space X into a normed linear space Y is **continuous at the point** $a \in X$ if for each $\varepsilon > 0$ there exists a $\delta > 0$ such that

$$\|f(x) - f(a)\| < \varepsilon \quad \text{whenever } \|x - a\| < \delta.$$

Notice that in this definition no notational distinction is made between the norms in the two spaces; the context makes clear which norm is intended.

In a normed linear space X the norm is a continuous function from X into $\mathbf{R}$, since for all $x, y \in X$

$$|\|x\| - \|y\|| \le \|x - y\| \qquad \text{(Problem 2).}$$

Rephrasing the definition of convergence in a normed linear space X, we have: a point b in X is the **limit** of the sequence (x_n) if for each $\varepsilon > 0$ there exists a positive integer n_0 such that

$$\|x_n - b\| < \varepsilon \quad \text{whenever } n \ge n_0.$$

If f is a function from a normed linear space X into a normed linear space Y, then we have the following result from Chapter 4 (6 · 2, p. 73 and Problem 11, p. 76): f is continuous at x if and only if $(f(x_n))$ converges to $f(x)$ whenever (x_n) converges to x.

If $(X_1, \|\cdot\|_1)$ and $(X_2, \|\cdot\|_2)$ are normed linear spaces over $\mathbf{F}$, we can make $X_1 \times X_2$ a normed linear space over $\mathbf{F}$ as follows. First, $X_1 \times X_2$ will be a linear space over $\mathbf{F}$, if, for $x = (x_1, x_2)$ and $y = (y_1, y_2)$, we define

$$x + y = (x_1 + y_1, x_2 + y_2) \quad \text{and} \quad \alpha x = (\alpha x_1, \alpha x_2).$$

For the norm we can take

$$(3 \cdot 2) \qquad \|x\| = \|x_1\|_1 + \|x_2\|_2$$

or

$$(3 \cdot 3) \qquad \|x\| = \max\{\|x_1\|_1, \|x_2\|_2\}.$$

These are the norms corresponding to the metrics introduced on $X_1 \times X_2$ in Chapter 4 and the topology corresponding to either of these norms is the product topology.

Let X be a normed linear space and let f denote the addition operation on X; that is, f is the function from $X \times X$ into X such that $f(x, y) = x + y$. Suppose that the sequence $((x_n, y_n))$ converges to (x, y) in $X \times X$. Since

$$\|f(x_n, y_n) - f(x, y)\| = \|(x_n + y_n) - (x + y)\| \le \|x_n - x\| + \|y_n - y\|,$$

using the norm 3 · 2 in $X \times X$ we see that f is continuous. Similarly, let g denote the scalar multiplication operation on X; that is, g is the function from $\mathbf{F} \times X$ into X such that $g(\alpha, x) = \alpha x$. Suppose that the sequence $((\alpha_n, x_n))$ converges to (α, x) in $\mathbf{F} \times X$. Since

$$\|g(\alpha_n, x_n) - g(\alpha, x)\| = \|\alpha_n x_n - \alpha x\| \le |\alpha_n| \, \|x_n - x\| + |\alpha_n - \alpha| \, \|x\|,$$

using the norm 3 · 3 in $\mathbf{F} \times X$ we see that g is continuous. Thus, in a normed linear space the operations of addition and scalar multiplication are continuous.

A **linear topological space (topological vector space)** is a generalization of a normed linear space. This is a linear space with a topology such that the operations of addition and scalar multiplication are continuous. In the above discussion we have shown that a normed linear space is a linear topological space.

In some linear spaces there occurs naturally a real-valued function that has the properties of a norm with the exception that a nonzero element may have value zero. A **semi-norm** on a linear space X is a real-valued function s such that

$$s(\alpha x) = |\alpha|\, s(x) \quad \text{and} \quad s(x + y) \leq s(x) + s(y).$$

It is easy to show that if s is a semi-norm then $s(0) = 0$ and $s(x) \geq 0$ for all $x \in X$. However, it is not true that $s(x) = 0$ implies $x = 0$.

Let $N = \{x \in X : s(x) = 0\}$. Then N is a subspace of X. If $x - y \in N$, then

$$s(x) \leq s(y) + s(x - y) = s(y)$$
$$s(y) \leq s(x) + s(y - x) = s(x)$$

and, therefore, $s(x) = s(y)$. Thus, on the quotient space X/N we can define

$$\|[x]\| = s(x)$$

and this will be a norm on the quotient space. Hence, from a linear space with a semi-norm we can obtain a normed linear space.

PROBLEMS

1. If X is a normed linear space and $d(x, y) = \|x - y\|$ for any $x, y \in X$, show that d is a metric on X.

2. If X is a normed linear space, show that for any $x, y \in X$:

$$|\, \|x\| - \|y\| \,| \leq \|x - y\|.$$

3. Show that $3 \cdot 2$ and $3 \cdot 3$ actually define norms on $X_1 \times X_2$.

4. If s is a semi-norm on the linear space X, show that:

a. $s(0) = 0$

b. $s(x) \geq 0$ for all $x \in X$

c. $s(x) - s(y) \leq s(x - y)$

d. $N = \{x \in X : s(x) = 0\}$ is a subspace of X

e. X/N with $\|[x]\| = s(x)$ is a normed linear space.

5. If X is a normed linear space and Y is a closed subspace of X, show that

$$\|[x]\| = \|x + Y\| = \inf\{\|x + y\| : y \in Y\}$$

defines a norm on the quotient space X/Y.

6. For a real-valued continuous function f on $[a, b]$, let

$$\|f\| = \sup_{t \in [a, b]} |f(t)|.$$

Show that this defines a norm on the space of continuous functions on $[a, b]$ and this normed linear space corresponds to the metric space $C[a, b]$.

4 · EXAMPLES OF NORMED LINEAR SPACES

We will obtain some examples of normed linear spaces by putting norms on the linear spaces of n-tuples of scalars, of sequences of scalars, and of scalar-valued functions. The field of scalars is either **R** or **C** and is denoted by **F** when it can be either.

4 · 1 The Space l_p^n. On the linear space $V_n(\mathbf{F})$ define, for any real number $p \geq 1$, a norm called the **p-norm** by the rule

$$\|x\|_p = \left[\sum_{i=1}^{n} |\alpha^i|^p\right]^{1/p} \quad \text{where } x = (\alpha^1, \cdots, \alpha^n).$$

That this defines a norm on $V_n(\mathbf{F})$ is easily verified (Problem 1). The resulting normed linear space is denoted by l_p^n. If we need to identify **F**, then we will call the space the real space l_p^n or the complex space l_p^n. In the real case if $p = 2$, then the p-norm is the Euclidean norm (length) and the space l_2^n is the Euclidean n-space $\mathbf{R}^n$. In the complex case if $p = 2$, the space l_2^n is called the **unitary** n-space and is denoted by $\mathbf{C}^n$.

4 · 2 The Space l_∞^n. On the linear space $V_n(\mathbf{F})$ define a norm, called the **infinity-norm**, by the rule

$$\|x\|_\infty = \max\{|\alpha^i| : i = 1, \cdots, n\} \quad \text{where } x = (\alpha^1, \cdots, \alpha^n).$$

Again, it is easy to verify that this defines a norm on $V_n(\mathbf{F})$ (Problem 1). The resulting normed linear space is denoted by l_∞^n. The notation used in this case is motivated by the fact that $\lim_{p\to\infty} \|x\|_p = \max\{|\alpha^i| : i = 1, \cdots, n\}$ (Problem 2).

4 · 3 The Space l_p. Let l_p be the subset of $V_\infty(\mathbf{F})$ consisting of all sequences $x = (\alpha^i)$ such that $\sum_{i=1}^\infty |\alpha^i|^p < \infty$ for the real number $p \geq 1$. This set is a subspace of $V_\infty(\mathbf{F})$. For, if $x, y \in l_p$ and $\alpha, \beta \in \mathbf{F}$, using the Minkowski Inequality we have for any positive integer n

$$\begin{aligned}\left[\sum_{i=1}^{n} |\alpha\alpha^i + \beta\beta^i|^p\right]^{1/p} &\leq |\alpha|\left[\sum_{i=1}^{n} |\alpha^i|^p\right]^{1/p} + |\beta|\left[\sum_{i=1}^{n} |\beta^i|^p\right]^{1/p} \\ &\leq |\alpha|\left[\sum_{i=1}^{\infty} |\alpha^i|^p\right]^{1/p} + |\beta|\left[\sum_{i=1}^{\infty} |\beta^i|^p\right]^{1/p}\end{aligned}$$

and, therefore,

$$(4 \cdot 4) \qquad \left[\sum_{i=1}^{\infty} |\alpha\alpha^i + \beta\beta^i|^p\right]^{1/p} \leq |\alpha|\left[\sum_{i=1}^{\infty} |\alpha^i|^p\right]^{1/p} + |\beta|\left[\sum_{i=1}^{\infty} |\beta^i|^p\right]^{1/p}.$$

This shows that $\alpha x + \beta y \in l_p$ and, hence, l_p is a subspace of $V_\infty(\mathbf{F})$. On the linear space l_p define the p-norm by the rule

$$\|x\|_p = \left[\sum_{i=1}^{\infty} |\alpha^i|^p\right]^{1/p} \quad \text{where } x = (\alpha^i).$$

The fact that this is a norm on l_p follows easily from the definition and 4 · 4 (Problem 3).

4 · 5 The Space l_∞. The set of all bounded sequences of scalars is also a subspace of $V_\infty(\mathbf{F})$ (Problem 4). By a bounded sequence we mean a sequence (α^i) such that the set $\{|\alpha^i| : i \in \mathbf{N}\}$ is bounded. If on the linear space of all bounded sequences of scalars we define the infinity-norm by the rule

$$\|x\|_\infty = \sup\{|\alpha^i| : i \in \mathbf{N}\} \quad \text{where } x = (\alpha^i),$$

we obtain the normed linear space l_∞.

4 · 6 The Space $\mathscr{L}_p(E)$. For any real number $p \geq 1$ let $\mathscr{L}_p(E)$ be the set of Lebesgue measurable scalar-valued functions f defined on a Lebesgue measurable set E in $\mathbf{R}^n$ such that $|f|^p$ is Lebesgue integrable. The Minkowski Inequality shows that if f and g are in $\mathscr{L}_p(E)$, then $f + g \in \mathscr{L}_p(E)$. Also, it is easy to see that if $f \in \mathscr{L}_p(E)$ and α is a scalar, then $\alpha f \in \mathscr{L}_p(E)$. Thus, $\mathscr{L}_p(E)$ is a linear space over $\mathbf{F}$.

On $\mathscr{L}_p(E)$ define a real-valued function s by the rule

$$s(f) = \left[\int_E |f|^p\right]^{1/p}.$$

Using the Minkowski Inequality we see that s is a semi-norm on $\mathscr{L}_p(E)$. It fails to be a norm because $s(f) = 0$ does not imply $f = 0$ but only that $f = 0$ a.e. on E. If we let $N = \{f : f = 0 \text{ a.e. on } E\}$, then the quotient space $\mathscr{L}_p(E)/N$ is a normed linear space with norm

$$\|[f]\| = \left[\int_E |f|^p\right]^{1/p}:$$

We will follow the common practice of not using equivalence class notation. Instead we will consider $\mathscr{L}_p(E)$ to be a normed linear space with equality meaning equality almost everywhere on E and with norm:

$$\|f\| = \left[\int_E |f|^p\right]^{1/p}.$$

4 · 7 The Space $\mathscr{L}_\infty(E)$. Let $\mathscr{L}_\infty(E)$ denote the set of Lebesgue measurable scalar-valued functions f defined on a Lebesgue measurable set E in $\mathbf{R}^n$ for which there exists a number M_f such that

$$|f(x)| \leq M_f \quad \text{for a.a.} x \in E.$$

Such functions are said to be **essentially bounded**. Clearly, $\mathscr{L}_\infty(E)$ is a linear space. If we define the function s on $\mathscr{L}_\infty(E)$ by the rule

$$s(f) = \inf\{M : |f| \leq M \text{ a.e. on } E\},$$

then s is a semi-norm on $\mathscr{L}_\infty(E)$. If $N = \{f : f = 0 \text{ a.e. on } E\}$, then $\mathscr{L}_\infty(E)/N$ is a normed linear space with norm

$$\|[f]\| = \inf\{M : |f| \le M \text{ a.e. on } E\}.$$

Again, we consider $\mathscr{L}_\infty(E)$ to be a normed linear space with norm

$$\|f\| = \inf\{M : |f| \le M \text{ a.e. on } E\}$$

and take equality in $\mathscr{L}_\infty(E)$ to mean equality almost everywhere on E.

4 · 8 The Space $C(X, \mathbf{R})$. Let X be a compact metric space and let $C(X, \mathbf{R})$ be the linear space of all continuous real-valued functions on X with norm defined by

$$\|f\|_\infty = \max\{|f(x)| : x \in X\}.$$

It is easy to check that $C(X, \mathbf{R})$ is a normed linear space. Note that the sequence (f_n) converges to f in $C(X, \mathbf{R})$ if and only if for each $\varepsilon > 0$ there exists an n_0 such that

$$\|f_n - f\| < \varepsilon \quad \text{whenever } n \ge n_0;$$

that is,

$$|f_n(x) - f(x)| < \varepsilon \quad \text{whenever } n \ge n_0 \text{ and } x \in X.$$

Thus, convergence of a sequence of functions in $C(X, \mathbf{R})$ is equivalent to the uniform convergence of the sequence on X.

If X is the interval $[a, b]$ in $\mathbf{R}$, then $C(X, \mathbf{R})$ is denoted by $C[a, b]$ and this corresponds to the metric space $C[a, b]$ considered in Chapter 4.

PROBLEMS

1. Verify that the p-norm ($p \ge 1$) and infinity-norm satisfy the conditions for a norm on $V_n(\mathbf{F})$.

2. In $V_n(\mathbf{F})$ show that $\lim_{p\to\infty} \|x\|_p = \|x\|_\infty$.

3. Verify that the p-norm satisfies the conditions for a norm on l_p.

4. Show that the set of all bounded sequences of scalars is a subspace of $V_\infty(\mathbf{F})$ and verify that the infinity-norm satisfies the conditions for a norm on this subspace.

5. If $1 \le p < q \le \infty$, show that $l_p \subset l_q$.

6. Let $(X, \|\cdot\|)$ and $(Y, \|\cdot\|')$ be normed linear spaces over the field $\mathbf{F}$. Show that

$$\|z\|_p = \{\|x\|^p + \|y\|'^p\}^{1/p}$$

defines a norm on $X \times Y$ for each $p \ge 1$ where $z = (x, y) \in X \times Y$. If $\|\cdot\|_\infty$ is the norm on $X \times Y$ defined by 3 · 3, then

$$\|z\|_\infty \le \|z\|_p \le \sqrt[p]{2}\,\|z\|_\infty.$$

Hence, by Problem 6, p. 71, the metrics defined on $X \times Y$ in terms of these norms are topologically equivalent and so define the same topology on $X \times Y$.

7. If E is a set of finite measure in $\mathbf{R}^n$ and $1 \leq p < q$, show that $\mathscr{L}_q(E) \subset \mathscr{L}_p(E)$.

8. Let $\mathbf{R}_n^m$ be the space of $m \times n$ matrices with entries in $\mathbf{R}$. If for $p \geq 1$ we define

$$\|[\tau_j^i]\|_p = \left[\sum_{i=1}^{m} \sum_{j=1}^{n} |\tau_j^i|^p\right]^{1/p},$$

verify that this is a norm on $\mathbf{R}_n^m$.

9. For $p \geq 1$ let $C_p[a, b]$ be the set of all continuous real-valued functions on the interval $[a, b]$ with

$$\|f\|_p = \left[\int_a^b |f|^p\right]^{1/p}.$$

Show that $C_p[a, b]$ is a normed linear space.

10. Let f be a continuous real-valued function on $[a, b]$. Take $x \in [a, b]$ and define $\varphi(f) = f(x)$. Show that φ is continuous on $C[a, b]$.

11. If $C(X, \mathbf{C})$ denotes the set of continuous complex-valued functions defined on a compact metric space X, show that $C(X, \mathbf{C})$ is a normed linear space if we define

$$\|f\| = \max\{|f(x)| : x \in X\}.$$

12. Show that l_p is a separable space.

13. Show that l_∞ is not a separable space.

5 · LINEAR TRANSFORMATIONS

In Chapter 3 we discussed linear transformations from a linear space to a linear space. We now consider linear transformations from a normed linear space to a normed linear space. In particular we will be interested in questions related to the continuity of such transformations.

5 · 1 Proposition. *Let T be a linear transformation from the normed linear space X into the normed linear space Y. If T is continuous at some point $x_0 \in X$, then it is continuous at every point of X and, in fact, it is uniformly continuous on X.*

Proof. If T is continuous at x_0, then corresponding to each $\varepsilon > 0$ there exists a $\delta > 0$ such that

$$\|T(x) - T(x_0)\| < \varepsilon \quad \text{whenever } \|x - x_0\| < \delta.$$

For any point $x_1 \in X$ if $\|x - x_1\| < \delta$ then $\|(x - x_1 + x_0) - x_0\| < \delta$ and, therefore,

$$\|T(x - x_1 + x_0) - T(x_0)\| < \varepsilon;$$

that is,

$$\|T(x) - T(x_1) + T(x_0) - T(x_0)\| = \|T(x) - T(x_1)\| < \varepsilon. \blacksquare$$

A linear transformation T from the normed linear space X into the normed linear space Y is said to be **bounded** if there exists a positive real number M such that $\|T(x)\| \le M\|x\|$ for all $x \in X$. We now show that for linear transformations the property of boundedness is equivalent to continuity.

5·2 Theorem. *Let T be a linear transformation from a normed linear space X into a normed linear space Y. Then T is continuous if and only if it is bounded.*

PROOF. Assume T is bounded; that is, there exists a real number M such that $\|T(x)\| \le M\|x\|$ for all $x \in X$. Then, for any $x, y \in X$

$$\|T(x) - T(y)\| = \|T(x - y)\| \le M\|x - y\|$$

which shows that T is continuous on X.

Now assume that T is continuous. The continuity of T at 0 implies that there exists a $\delta > 0$ such that

$$\|T(x)\| < 1 \quad \text{whenever } \|x\| < \delta.$$

For any nonzero $x \in X$, let $y = (\delta/(2\|x\|))x$. Then $\|y\| < \delta$ and therefore $\|T(y)\| < 1$; that is,

$$\left\|T\left(\frac{\delta}{2\|x\|}\,x\right)\right\| = \frac{\delta}{2\|x\|}\,\|T(x)\| < 1.$$

This shows that

$$\|T(x)\| \le \frac{2}{\delta}\,\|x\| \quad \text{for all nonzero } x \in X.$$

Clearly, this inequality holds also if $x = 0$. $\blacksquare$

Since the inverse of a linear transformation is also a linear transformation, Theorem 5·2 provides a characterization of linear homeomorphisms.

5·3 Corollary. *A linear transformation T from X into Y is a homeomorphism if and only if there exist positive numbers m and M such that*

$$m\|x\| \le \|T(x)\| \le M\|x\| \quad \textit{for all } x \in X. \tag{5·4}$$

PROOF. Assume T is a homeomorphism. Then T and T^{-1} are continuous linear transformations. Thus, there exist positive numbers M and m such that

$$\|T(x)\| \le M\|x\| \quad \text{for all } x \in X$$

and

$$\|T^{-1}(y)\| \le \frac{1}{m}\,\|y\| \quad \text{for all } y \in T(X).$$

If for any $x \in X$ we let $y = T(x)$, we obtain $5 \cdot 4$.

Now assume that the linear transformation T satisfies $5 \cdot 4$. By Theorem $5 \cdot 2$, T is continuous. Since $m\|x\| \le \|T(x)\|$, $T(x) = 0$ implies $x = 0$ and, hence, T is one-to-one. To show that the linear transformation T^{-1} is continuous, take any $y \in T(X)$ and let $x = T^{-1}(y)$. Then the inequality $m\|x\| \le \|T(x)\|$ becomes $\|T^{-1}(y)\| \le (1/m)\|y\|$ which shows that T^{-1} is continuous. ∎

The set of bounded (or, equivalently, continuous) linear transformations from a normed linear space X into a normed linear space Y is a subspace of $L(X, Y)$, the space of all linear transformations from X into Y. On this subspace, denoted by $BL(X, Y)$, define a norm as follows: for any $T \in BL(X, Y)$

$$(5 \cdot 5) \qquad \|T\| = \inf\{M : \|T(x)\| \le M\|x\| \text{ for all } x \in X\}.$$

Note that this implies that

$$\|T(x)\| \le \|T\| \, \|x\| \quad \text{for all } x \in X.$$

Also, if $X \ne \{0\}$, then $5 \cdot 5$ is equivalent to

$$\|T\| = \sup_{x \ne 0} \frac{\|T(x)\|}{\|x\|}.$$

The verification that this is a norm on $BL(X, Y)$ is straightforward and is left to the reader (Problem 3). Thus, $BL(X, Y)$ is a normed linear space.

If $X \ne \{0\}$, the following formulas for the norm also hold (Problem 4):

$$\|T\| = \sup_{\|x\| = 1} \|T(x)\| = \sup_{\|x\| \le 1} \|T(x)\|.$$

If X is a normed linear space over the field $\mathbf{F}$, then the normed linear space $BL(X, \mathbf{F})$ of bounded linear functionals is called the **dual** (or **conjugate**) **space** of X and is denoted by X^*.

In the case where X is l_p^n, the projection functions P_k, defined by $P_k(\alpha^1, \cdots, \alpha^n) = \alpha^k$, are bounded linear functionals and (P_k) is a basis for X^* (Problem 1). For a general normed linear space, projection functions (or, equivalently, coordinates) are not available. However, results for the space can often be obtained by consideration of bounded linear functionals on the space.

The Hahn-Banach Theorem which we now discuss ensures that there are nontrivial bounded linear functionals on normed linear spaces. This theorem states that a bounded linear functional defined on a subspace of a normed linear space can be extended to the whole space without changing the norm. We consider the real case first and this result is an immediate consequence of the following theorem, also called the Hahn-Banach Theorem.

5 · 6 Theorem (Hahn-Banach). *Let Z be a subspace of the linear space X over $\mathbf{R}$ and let p be a real-valued function on X such that for all $x, y \in X$*

$$p(x + y) \le p(x) + p(y) \quad \text{and} \quad p(\alpha x) = \alpha p(x) \quad \text{for } \alpha \ge 0.$$

If f is a linear functional on Z and $f(x) \le p(x)$ for all $x \in Z$, then there is a linear functional F on X such that $F(x) = f(x)$ for $x \in Z$ and $F(x) \le p(x)$ for all $x \in X$.

PROOF. Let $\mathscr{F}$ be the partially ordered set consisting of all linear extensions g of f such that $g \le p$ on D_g; $\mathscr{F}$ is partially ordered by extension. Note that $\mathscr{F}$ is nonempty since $f \in \mathscr{F}$. If $\{g_s : s \in S\}$ is a chain in $\mathscr{F}$, let g be the function with domain $\bigcup_{s \in S} D_{g_s}$ and $g(x) = g_s(x)$ for $x \in D_{g_s}$. Then g is a linear functional such that $g \le p$ on D_g. Thus, any chain in $\mathscr{F}$ has an upper bound and, hence, by Zorn's Lemma $\mathscr{F}$ has a maximal element, call it F. To complete the proof we need to show that the domain of F is X.

Suppose $D_F \neq X$ and take $x_0 \in X \sim D_F$. Then, for $y, z \in D_F$

$$\begin{aligned} F(z) - F(y) = F(z - y) \le p(z - y) &= p(z + x_0 - y - x_0) \\ &\le p(z + x_0) + p(-y - x_0). \end{aligned}$$

That is, for all $y, z \in D_F$

$$-p(-y - x_0) - F(y) \le p(z + x_0) - F(z).$$

Take β_0 such that

$$\sup_{y \in D_F} (-p(-y - x_0) - F(y)) \le \beta_0 \le \inf_{z \in D_F} (p(z + x_0) - F(z))$$

and define a linear functional on the subspace spanned by $D_F \cup \{x_0\}$ as follows:

$$g(x + \alpha x_0) = F(x) + \alpha\beta_0, \qquad x \in D_F, \quad \alpha \in \mathbf{R}.$$

Then, for $\alpha > 0$

$$g(x + \alpha x_0) \le F(x) + \alpha p\left(\frac{x}{\alpha} + x_0\right) - \alpha F\left(\frac{x}{\alpha}\right) = p(x + \alpha x_0)$$

and for $\alpha < 0$

$$g(x + \alpha x_0) \le F(x) - \alpha p\left(-\frac{x}{\alpha} - x_0\right) - \alpha F\left(\frac{x}{\alpha}\right) = p(x + \alpha x_0).$$

Thus, g is an extension of F which belongs to $\mathscr{F}$. Since F was maximal in $\mathscr{F}$ this contradiction shows that $D_F = X$. ▮

5 · 7 Corollary (Hahn-Banach). *If f is a bounded linear functional on the subspace Z of the real normed linear space X, then there is a bounded linear functional F on X such that $F = f$ on Z and $\|F\| = \|f\|$.*

PROOF. If we let $p(x) = \|f\| \, \|x\|$, then for all $x, y \in X$

$$p(x + y) \le p(x) + p(y) \quad \text{and} \quad p(\alpha x) = \alpha p(x) \quad \text{for } \alpha \ge 0.$$

Also, for all $x \in Z$

$$f(x) \le |f(x)| \le \|f\| \, \|x\| = p(x).$$

Thus, Theorem 5 · 6 implies that there is a linear functional F on X such that $F = f$ on Z and $F(x) \leq p(x)$ for all $x \in X$. Since

$$-F(x) = F(-x) \leq p(-x) = p(x),$$
$$|F(x)| \leq p(x) = \|f\| \, \|x\| \quad \text{for all } x \in X$$

and, hence, $\|F\| \leq \|f\|$.

Also, $F = f$ on Z implies

$$|f(x)| = |F(x)| \leq \|F\| \, \|x\| \quad \text{for all } x \in Z$$

and, therefore, $\|f\| \leq \|F\|$. Thus, $\|F\| = \|f\|$. ▮

We now obtain the complex form of the Hahn-Banach Theorem due to Bohnenblust and Sobczyk [6].

5 · 8 Theorem. *Let Z be a subspace of the complex linear space X and let p be a real-valued function on X such that for all $x, y \in X$ and all $\alpha \in \mathbf{C}$*

$$p(x + y) \leq p(x) + p(y) \quad \textit{and} \quad p(\alpha x) = |\alpha| p(x).$$

If f is a linear functional on Z and $|f(x)| \leq p(x)$ for all $x \in Z$, then there is a linear functional F on X such that $F = f$ on Z and $|F(x)| \leq p(x)$ for all $x \in X$.

PROOF. Let $f = f_1 + if_2$ where f_1 and f_2 are real-valued functions on Z. Then, f_1 and f_2 are real linear functionals on Z, i.e., for all $x, y \in Z$ and $\alpha \in \mathbf{R}$ $(j = 1, 2)$

$$f_j(x + y) = f_j(x) + f_j(y) \quad \text{and} \quad f_j(\alpha x) = \alpha f_j(x).$$

Also, for each $x \in Z$,

$$f_1(ix) + if_2(ix) = f(ix) = if(x) = if_1(x) - f_2(x).$$

Thus, $f_2(x) = -f_1(ix)$.

Since $f_1(x) \leq |f(x)| \leq p(x)$ for all $x \in Z$, by Theorem 5 · 6 there is real linear functional F_1 on X which is an extension of f_1 such that $F_1(x) \leq p(x)$ for all $x \in X$. Define

$$F(x) = F_1(x) - iF_1(ix), \quad x \in X.$$

Then, F is a complex linear functional, since it is a real linear functional and

$$F(ix) = F_1(ix) - iF_1(-x) = iF_1(x) + F_1(ix) = iF(x).$$

Also, F is an extensions of f. For, if $x \in Z$, then

$$F(x) = f_1(x) - if_1(ix) = f_1(x) + if_2(x) = f(x).$$

For each $x \in X$, let $F(x) = re^{i\theta}$, $r \geq 0$. Then

$$|F(x)| = r = e^{-i\theta}F(x) = F(e^{-i\theta}x) = F_1(e^{-i\theta}x) \leq p(e^{-i\theta}x) = p(x). \; ▮$$

From this theorem we can obtain the analogue of Corollary 5 · 7 for complex normed linear spaces.

PROBLEMS

1. Let P_k be a projection function from l_p^n into **F** defined by

$$P_k(x) = \alpha^k \quad \text{where } x = (\alpha^1, \cdots, \alpha^n).$$

Show that P_k is a bounded linear transformation and that $\|P_k\| = 1$. Also, show that (P_k) is a Hamel basis for $(l_p^n)^*$.

2. Let I be the linear transformation from l_∞^n into l_2^n such that $I(x) = x$. Show that I is a homeomorphism.

3. Verify that 5 · 5 defines a norm on $BL(X, Y)$.

4. If $X \neq \{0\}$ and $T \in BL(X, Y)$, prove

$$\|T\| = \sup_{\|x\|=1} \|T(x)\| = \sup_{\|x\|\leq 1} \|T(x)\|.$$

5. If T is the function on $\mathbf{R}^2$ defined by

$$T(\alpha^1, \alpha^2) = (\alpha^1 \cos\theta - \alpha^2 \sin\theta, \alpha^1 \sin\theta + \alpha^2 \cos\theta),$$

show that T is a bounded linear transformation and determine $\|T\|$.

6. If T is the linear transformation from $C[a, b]$ into **R** given by

$$T(f) = \int_a^b f,$$

determine $\|T\|$.

7. Let x_0 be a nonzero element in a real normed linear space X. Show that there is an $F \in X^*$ such that $F(x_0) = \|x_0\|$ and $\|F\| = 1$.

8. Let x be an element in a real normed linear space X such that, for all $f \in X^*$, $f(x) = 0$. Show that $x = 0$.

6 · ISOMORPHISMS

Two linear spaces X and Y over the field **F** are isomorphic if there is a one-to-one linear transformation (an isomorphism) T mapping X onto Y. If X and Y are normed linear spaces, then they will be essentially the same if they are isomorphic and if the isomorphism T preserves the norm: $\|T(x)\| = \|x\|$. We say that the normed linear spaces X and Y are **isometrically isomorphic** if there is an isomorphism T of X onto Y such that

$$\|T(x)\| = \|x\| \quad \text{for all } x \in X.$$

As the terminology suggests an isometric isomorphism T is an isometry:

$$d(T(x), T(y)) = \|T(x) - T(y)\| = \|T(x - y)\| = \|x - y\| = d(x, y).$$

We note also that in this definition we need only specify that the norm-preserving transformation T be linear. For, if $\|T(x)\| = \|x\|$, then $T(x) = 0$ implies $x = 0$ and, therefore, T is one-to-one.

If $\mathbf{R}_n{}^m$ is the space of $m \times n$ matrices with entries in $\mathbf{R}$ and $\|[\tau_j{}^i]\| = [\sum_{i=1}^m \sum_{j=1}^n (\tau_j{}^i)^2]^{1/2}$, then it is easy to show that the normed linear space $\mathbf{R}_n{}^m$ is isometrically isomorphic to $\mathbf{R}^{mn}$ (Problem 1).

Consideration of dual spaces provides other examples of isometric isomorphisms. For example, $(l_p{}^n)^*$ $(1 < p < \infty)$ is isometrically isomorphic to $l_q{}^n$ where $1/p + 1/q = 1$. If for $(l_p{}^n)^*$ we take the basis $(P_k)_{k=1}^n$ consisting of the projection functions (Problem 1, § 5) and for $l_q{}^n$ we take the natural basis $(e_k)_{k=1}^n$ where $e_k = (\delta_k{}^j)_{j=1}^n$, then we can define an isomorphism T by mapping P_k onto e_k; that is, for $f = \sum_{k=1}^n \beta^k P_k$ in $(l_p{}^n)^*$, define

$$T(f) = \sum_{k=1}^n \beta^k e_k = (\beta^k)_{k=1}^n.$$

Note that $\beta^k = f(e_k)$ where e_k is considered as an element of $l_p{}^n$.

It is easy to check that T is an isomorphism of $(l_p{}^n)^*$ onto $l_q{}^n$. The proof that T is an isometry can be deduced from the following more general result.

6 · 1 Proposition. *For* $1 < p < \infty$, $(l_p)^*$ *is isometrically isomorphic to* l_q *where* $1/p + 1/q = 1$.

PROOF. For $k = 1, 2, \cdots$, let $e_k = (\delta_k{}^j)_{j=1}^\infty$. Then, for any $x = (\alpha^k) \in lp$ and for any positive integer n

$$\|x - \sum_{k=1}^n \alpha^k e_k\| = \sum_{k=n+1}^\infty [|\alpha^k|^p]^{1/p}$$

and, hence, $x = \lim_{n\to\infty} \sum_{k=1}^n \alpha^k e_k$.

For $f \in (l_p)^*$, let $f(e_k) = \beta^k$ and define $T(f) = (\beta^k)_{k=1}^\infty$. Considered as a transformation from $(l_p)^*$ into $V_\infty(F)$, T is clearly linear and also is one-to-one. Suppose $f, g \in (l_p)^*$ and $T(f) = T(g)$. Then, $f(e_k) = g(e_k)$ for all positive integers k and, for any $x = (\alpha^k) \in l_p$,

$$f(x) = \lim_{n\to\infty} \sum_{k=1}^n \alpha^k f(e_k) = \lim_{n\to\infty} \sum_{k=1}^n \alpha^k g(e_k) = g(x)$$

and, hence, $f = g$.

We now show that T maps $(l_p)^*$ into l_q; that is, if $T(f) = (\beta^k)_{k=1}^\infty$, then $\sum_{k=1}^\infty |\beta^k|^q < \infty$. For $n = 1, 2, \ldots$, take $x_n = (\alpha^k)_{k=1}^\infty \in l_p$ where

$$\alpha^k = \begin{cases} |\beta^k|^{q-2}\overline{\beta^k} & \text{if } k = 1, \ldots, n \text{ and } \beta^k \neq 0 \\ 0 & \text{otherwise.} \end{cases}$$

Then

$$\|x_n\| = \left[\sum_{k=1}^n |\alpha^k|^p\right]^{1/p} = \left[\sum_{k=1}^n |\beta^k|^{pq-p}\right]^{1/p} = \left[\sum_{k=1}^n |\beta^k|^q\right]^{1/p}$$

and

$$|f(x_n)| = \left|f\left(\sum_{k=1}^n \alpha^k e_k\right)\right| = \left|\sum_{k=1}^n \alpha^k \beta^k\right| = \sum_{k=1}^n |\beta^k|^q.$$

Therefore,

$$\sum_{k=1}^{n} |\beta^k|^q = |f(x_n)| \le \|f\| \, \|x_n\| = \|f\| \left[\sum_{k=1}^{n} |\beta^k|^q \right]^{1/p};$$

that is, for $n = 1, 2, \cdots$

$$\left[\sum_{k=1}^{n} |\beta^k|^q \right]^{1/q} \le \|f\|.$$

Thus, $T(f) = (\beta^k) \in l_q$ and $\|T(f)\| \le \|f\|$.

To show that T maps $(l_p)^*$ onto l_q, take $y = (\beta^k) \in l_q$ and, for $x = (\alpha^k) \in l_p$, define

$$f(x) = \sum_{k=1}^{\infty} \alpha^k \beta^k.$$

By Hölder's Inequality,

$$|f(x)| \le \sum_{k=1}^{\infty} |\alpha^k \beta^k| \le \|x\| \, \|y\|.$$

Thus, the series defining $f(x)$ converges. Clearly, $f \in (l_p)^*$ and, since $f(e_k) = \sum_{j=1}^{\infty} \beta^j \delta_k{}^j = \beta^k$, $T(f) = y$. Also,

$$\|f\| \le \|y\| = \|T(f)\|.$$

Therefore, $\|T(f)\| = \|f\|$ and T is an isometric isomorphism of $(l_p)^*$ onto l_q. ∎

Since, as normed linear spaces, $(l_p)^*$ and l_q are the same we may write $(l_p)^* = l_q$ $(1 < p < \infty)$. It is also true that $(l_1)^* = l_\infty$ (Problem 2). However, $(l_\infty)^* \ne l_1$ (Problem 3). Similar results hold for the $\mathscr{L}_p$ spaces: for $1 < p < \infty$, $(\mathscr{L}_p)^* = \mathscr{L}_q$ where $1/p + 1/q = 1$ and $(\mathscr{L}_1)^* = \mathscr{L}_\infty$.

We now consider a less stringent notion of isomorphism for normed linear spaces. Since a normed linear space can be considered as a topological space, we can consider when two normed linear spaces are essentially the same both as linear spaces and as topological spaces. In this we understand that the spaces have the topology defined in terms of the metric determined by the norm. We say that the normed linear spaces X and Y over **F** are **homeomorphically isomorphic** if there is an isomorphism T from X onto Y which is also a homeomorphism. Thus, X and Y are homeomorphically isomorphic if and only if there is a linear transformation T from X onto Y such that for some positive numbers m and M

$$m\|x\| \le \|T(x)\| \le M\|x\| \quad \text{for all } x \in X.$$

Clearly, isometrically isomorphic normed linear spaces are homeomorphically isomorphic but the converse need not be true.

A case of particular interest is that of a linear space X with two different norms. Two norms on a linear space X are said to be **(topologically) equivalent** if they determine the same topology on X. Restating this definition we have: *two norms, denoted by* $\|\cdot\|$ *and* $\|\cdot\|'$, *are equivalent if the normed linear spaces* $(X, \|\cdot\|)$ *and* $(X, \|\cdot\|')$ *are homeomorphically isomorphic under the identity transformation,* $I(x) = x$. Thus, the two norms are equivalent if and only if there exist positive numbers m and M such that

$$m \|x\| \leq \|x\|' \leq M\|x\| \quad \text{for all } x \in X.$$

Since

$$\|x\|_\infty \leq \|x\|_p \leq n \|x\|_\infty \quad \text{for all } x \in V_n(\mathbf{F}),$$

the p-norms ($p \geq 1$) on $V_n(\mathbf{F})$ are all equivalent to the infinity-norm and, hence, are equivalent to each other. Thus, all the spaces l_p^n and l_∞^n are topologically indistinguishable. For example, for topological considerations in $\mathbf{R}^2$ if it is convenient we can take the open balls to be the interior of squares rather than the interior of circles; that is, we can use the infinity-norm rather than the Euclidean norm.

PROBLEMS

1. Verify that the normed linear space $\mathbf{R}_n^m$ of $m \times n$ real matrices with $\|[\tau_j^i]\| = [\sum_{i=1}^m \sum_{j=1}^n (\tau_j^i)^2]^{1/2}$ is isometrically isomorphic to $\mathbf{R}^{mn}$.

2. Prove that $(l_1)^*$ is isometrically isomorphic to l_∞.

3. Let c_0 be the subspace of l_∞ consisting of those sequences which converge to zero. Show that $(c_0)^*$ is isometrically isomorphic to l_1.

4. Show that the relation of being homeomorphically isomorphic is an equivalence relation.

5. For any positive integer define the continuous function f_n on $[0, 1]$ by the rule

$$f_n(t) = \begin{cases} 1 - nt, & 0 \leq t \leq 1/n \\ 0, & 1/n < t \leq 1. \end{cases}$$

a. Determine $\|f_n\|_\infty$ and $\|f_n\|_p$ for $p \geq 1$.

b. Show that the topology on $C[0, 1]$ is strictly stronger than the topology on $C_p[0, 1]$.

c. Show that if (g_n) converges to g in $C[0, 1]$, then (g_n) converges to g in $C_p[0, 1]$ for any $p \geq 1$.

7 · FINITE-DIMENSIONAL SPACES

In Chapter 3 we showed that two linear spaces over the same field are isomorphic if they have the same dimension. For finite-dimensional normed linear spaces we can extend this result. First we prove a preliminary proposition.

7 · 1 Proposition. *Any linear transformation T from the space l_2^n into a normed linear space Y is bounded.*

PROOF. Let $e_k = (\delta_k^1, \cdots, \delta_k^n)$. Then, if $x = (\alpha^1, \cdots, \alpha^n)$ is any point in l_2^n, we have $x = \sum_{k=1}^n \alpha^k e_k$ and

$$\|T(x)\| = \left\| \sum_{k=1}^n \alpha^k T(e_k) \right\| \leq \sum_{k=1}^n |\alpha^k| \, \|T(e_k)\| \leq \sum_{k=1}^n \|x\| \, \|T(e_k)\| = M\|x\|$$

where $M = \sum_{k-1}^n \|T(e_k)\|$. ∎

7 · 2 Theorem. *Two normed linear spaces over the field* **F** *which have the same finite dimension are homeomorphically isomorphic.*

PROOF. We prove this proposition by showing that any normed linear space Y of dimension n is homeomorphically isomorphic to l_2^n. Of course, we assume that the field $\mathbf{F}$ of scalars is the same for both spaces. Let $(y_1, \cdots, y_n)$ be a basis for Y. For each $x = (\alpha^1, \cdots, \alpha^n)$ in l_2^n define $T(x) = \sum_{k=1}^n \alpha^k y_k$. Then T is a one-to-one linear transformation from l_2^n onto Y. By Proposition $7 \cdot 1$ there exists a real number M such that $\|T(x)\| \leq M\|x\|$ for all $x \in l_2^n$.

We complete the proof by showing there exists a positive number m such that $m\|x\| \leq \|T(x)\|$ for all $x \in l_2^n$. Let $S = \{x : \|x\| = 1\} \subset l_2^n$ and let $f(x) = \|T(x)\|$ for $x \in S$. Since S is a closed and bounded set in the Euclidean space $\mathbf{R}^n$ (if $\mathbf{F} = \mathbf{R}$) or $\mathbf{R}^{2n}$ (if $\mathbf{F} = \mathbf{C}$), S is compact. Therefore, the non-negative continuous function f has a minimum value $m \geq 0$ on S. Since $\|T(x)\| \neq 0$ for $x \in S$, $m > 0$. If x is any nonzero point in l_2^n then $x/\|x\| \in S$ and, hence,

$$\|T(x)\| = \|x\| \left\| T\left(\frac{x}{\|x\|}\right) \right\| \geq m\|x\|. \blacksquare$$

Theorem $7 \cdot 2$ shows that any finite-dimensional normed linear space is the same as a Euclidean space from the point of view of algebraic and topological structure. Thus, for example, we can extend the Heine-Borel theorem to finite-dimensional normed linear spaces.

7 · 3 Proposition. *A closed and bounded set in a finite-dimensional normed linear space Y is compact.*

PROOF. Let T be a homeomorphic isomorphism from l_2^n onto Y. If E is a closed and bounded set in Y, then $T^{-1}(E)$ is a closed and bounded set in l_2^n. Thus, $T^{-1}(E)$ is a compact set in l_2^n and, hence, $T(T^{-1}(E)) = E$ is compact in Y. ∎

We now extend Proposition $7 \cdot 1$ and show that any linear transformation defined on a finite-dimensional normed linear space is bounded.

7 · 4 Proposition. *If T is a linear transformation from the finite-dimensional normed linear space X into the normed linear space Y, then T is bounded.*

PROOF. Let X have dimension n and let U be a homeomorphic isomorphism of l_2^n onto X. Then there exists a positive number m such that

$$\|U(z)\| \geq m\|z\| \quad \text{for all } z \in l_2^n.$$

Since $T \circ U$ is a linear transformation from l_2^n into Y, by Proposition $7 \cdot 1$ there exists a number M such that

$$\|T(U(z))\| \leq M\|z\| \quad \text{for all } z \in l_2^n.$$

For each $x \in X$ there exists a $z \in l_2^n$ such that $x = U(z)$. Then

$$\|T(x)\| = \|T(U(z))\| \leq M\|z\| \leq \frac{M}{m}\|U(z)\| = \frac{M}{m}\|x\|. \blacksquare$$

Thus, if X is a finite-dimensional normed linear space, $BL(X, Y) = L(X, Y)$. In particular, $BL(l_p{}^n, l_p{}^m) = L(l_p{}^n, l_p{}^m)$ where $p \in [1, \infty]$. Therefore, as a linear space $BL(l_p{}^n, l_p{}^m)$ is isomorphic to $\mathbf{F}_n{}^m$, the space of $m \times n$ matrices with entries in $\mathbf{F}$. To set up a particular isomorphism take the standard bases $(e_1, \cdots, e_n)$ where $e_k = (\delta_k{}^1, \cdots, \delta_k{}^n)$ and $(e'_1, \cdots, e'_m)$ where $e'_k = (\delta_k{}^1, \cdots, \delta_k{}^m)$ for the spaces $l_p{}^n$ and $l_p{}^m$, respectively. If

$$T(e_k) = \sum_{j=1}^{m} \tau_k{}^j e'_j,$$

let $\varphi(T) = [\tau_k{}^j]$. Then φ is an isomorphism of $BL(l_p{}^n, l_p{}^m)$ onto $\mathbf{F}_n{}^m$. If we define $\|[\tau_k{}^j]\|_p = \|\varphi^{-1}[\tau_k{}^j]\| = \|T\| = \sup_{\|x\|=1} \|T(x)\|_p$, then $\mathbf{F}_n{}^m$ will be a normed linear space which is isometrically isomorphic to $BL(l_p{}^n, l_p{}^m)$. The norms for $\mathbf{F}_n{}^m$ corresponding to $p = 1$ and $p = \infty$ are particularly convenient. They are given by the formulas:

$$\|[\tau_k{}^j]\|_1 = \max_k \sum_{j=1}^{m} |\tau_k{}^j|$$

and

$$\|[\tau_k{}^j]\|_\infty = \max_j \sum_{k=1}^{n} |\tau_k{}^j|.$$

We now derive the second of these two formulas; the other is left to the reader (Problem 3). Take $T \in BL(l_\infty{}^n, l_\infty{}^m)$. For any $x = (\alpha^1, \cdots, \alpha^n) \in l_\infty{}^n$ such that $\|x\| = 1$, we have

$$\begin{aligned}\|T(x)\| &= \left\| \sum_{k=1}^{n} \alpha^k T(e_k) \right\| = \left\| \sum_{j=1}^{m} \left(\sum_{k=1}^{n} \tau_k{}^j \alpha^k \right) e'_j \right\| \\ &= \max_j \left| \sum_{k=1}^{n} \tau_k{}^j \alpha^k \right| \le \max_j \sum_{k=1}^{n} |\tau_k{}^j|\,|\alpha^k| \le \max_j \sum_{k=1}^{n} |\tau_k{}^j|.\end{aligned}$$

If the maximum of $\sum_{k=1}^{n} |\tau_k{}^j|$ occurs for $j = i$, let $u = (\beta^1, \cdots, \beta^n)$ where

$$\beta^k = \begin{cases} |\tau_k{}^i|/\tau_k{}^i & \text{when } \tau_k{}^i \neq 0 \\ 1 & \text{when } \tau_k{}^i = 0. \end{cases}$$

Then $\|u\| = 1$ and

$$\left| \sum_{k=1}^{n} \tau_k{}^j \beta^k \right| \le \sum_{k=1}^{n} |\tau_k{}^j|\,|\beta^k| = \sum_{k=1}^{n} |\tau_k{}^j| \le \sum_{k=1}^{n} |\tau_k{}^i| \quad \text{for } j \neq i,$$

$$\left| \sum_{k=1}^{n} \tau_k{}^i \beta^k \right| = \sum_{k=1}^{n} |\tau_k{}^i|.$$

Therefore,

$$\|T(u)\| = \sum_{k=1}^{n} |\tau_k{}^i| = \max_j \sum_{k=1}^{n} |\tau_k{}^j|$$

and

$$\|T\| = \|[\tau_k{}^j]\|_\infty = \max_j \sum_{k=1}^{n} |\tau_k{}^j|.$$

PROBLEMS

1. If $\lim T_n = T$ in $BL(X, Y)$ show that for all $x \in X$

$$\lim T_n(x) = T(x) \quad \text{in } Y.$$

2. Show that a finite-dimensional normed linear space is separable.

3. Show that

$$\|[\tau_k{}^j]\|_1 = \max_k \sum_{j=1}^{m} |\tau_k{}^j|.$$

4. Let T be a linear transformation from $l_p{}^n$ into $l_p{}^m$ and let $[\tau_k{}^j]$ be the matrix associated with T using the natural bases for $l_p{}^n$ and $l_p{}^m$. Show that for $1 < p < \infty$

$$\|T\| \leq \left[\sum_{j=1}^{m} \left(\sum_{k=1}^{n} |\tau_k{}^j|^q\right)^{p/q}\right]^{1/p} \quad \text{where } \frac{1}{p} + \frac{1}{q} = 1.$$

5. Let Y be a finite-dimensional normed linear space and let $(x_1, \cdots, x_n)$ be a basis for Y. For $j = 1, \cdots, n$ let Y_j be the subspace $\{\alpha x_j : \alpha \in \mathbf{F}\}$ and let P_j be the projection function from Y into Y_j defined by

$$P_j(y) = \alpha^j x_j \quad \text{where } y = \sum_{j=1}^{n} \alpha^j x_j.$$

a. Show that $P_j \in BL(Y, Y_j)$.
b. If $y = \sum_{j=1}^{n} \alpha^j x_j$ and $y_k = \sum_{j=1}^{n} \alpha_k{}^j x_j$, $k = 1, 2, \cdots$, show that

$$\lim_{k\to\infty} y_k = y \quad \text{implies} \lim_{k\to\infty} \alpha_k{}^j = \alpha^j, \qquad j = 1, \ldots, n.$$

6. If Y is a finite-dimensional subspace of a normed linear space X, show that Y is closed.

8 · BANACH SPACES

8 · 1 Definition. *A complete normed linear space is called a* ***Banach space****.*

From Chapter 4 we know that the Euclidean spaces $\mathbf{R}^n$, the unitary spaces $\mathbf{C}^n$, and the space $C(X, \mathbf{R})$ of continuous real-valued functions on a compact metric space X are complete. Hence, they are Banach spaces. We also showed in Chapter 4 that completeness is not a topological property for metric spaces. However, for normed linear spaces we have the following useful result.

8 · 2 Proposition. *If X and Y are homeomorphically isomorphic normed linear spaces and X is a Banach space, then Y is a Banach space.*

PROOF. Let T be a linear homeomorphism from X onto Y. Take a Cauchy sequence (y_n) of points in Y and let $x_n = T^{-1}(y_n)$. Since there is a positive number m such that $m\|x\| \leq \|T(x)\|$ for all $x \in X$, (x_n) is a Cauchy sequence of points in X. The completeness of X implies that (x_n) converges to some point x in X. Since T is continuous, $(T(x_n))$ converges to $T(x)$ in Y. Thus, (y_n) converges in Y. ▮

The next two propositions follow easily.

8 · 3 Proposition. *Any finite-dimensional normed linear space is a Banach space.*

PROOF. If X is a normed linear space over $\mathbf{F}$ of finite dimension n, then X is homeomorphically isomorphic to the Banach space $l_2{}^n(\mathbf{R}^n$ or $\mathbf{C}^n)$. ▮

8 · 4 Proposition. *A finite-dimensional subspace E of a normed linear space X is closed in X.*

PROOF. Since E is a complete subspace of X it is closed in X. ▮

Proposition 8 · 3 implies that $l_p{}^n$ and $l_\infty{}^n$ are Banach spaces. The spaces l_p and l_∞ are also Banach spaces; this follows from the following result.

8 · 5 Proposition. *If Y is a Banach space, then BL(X, Y) is a Banach space.*

PROOF. Let (T_n) be a Cauchy sequence in $BL(X, Y)$. Since

$$\|T_n(x) - T_m(x)\| \le \|T_n - T_m\| \; \|x\| \quad \text{for each } x \in X,$$

$(T_n(x))$ is a Cauchy sequence in Y. The completeness of Y ensures that $(T_n(x))$ converges and we let $T(x) = \lim T_n(x)$ for each $x \in X$.

To show that $T \in BL(X, Y)$ take $x, y \in X$ and $\alpha, \beta \in \mathbf{F}$. Then,

$$\begin{aligned} T(\alpha x + \beta y) &= \lim T_n(\alpha x + \beta y) = \lim(\alpha T_n(x) + \beta T_n(y)) \\ &= \alpha \lim T_n(x) + \beta \lim T_n(y) = \alpha T(x) + \beta T(y). \end{aligned}$$

Thus, T is linear. Since (T_n) is a Cauchy sequence in $BL(X, Y)$, it is bounded; that is, there exists a number M such that $\|T_n\| \le M$ for all n. Then, for all $x \in X$,

$$\|T(x)\| = \|\lim T_n(x)\| = \lim \|T_n(x)\| \le M\|x\|.$$

Thus, T is bounded and, hence, $T \in BL(X, Y)$.

We complete the proof by showing that (T_n) converges to T in $BL(X, Y)$. Take $\varepsilon > 0$. There exists a positive integer n_0 such that

$$\|T_n - T_m\| < \varepsilon \quad \text{whenever } n, m \ge n_0.$$

Thus, for each $x \in X$ such that $\|x\| = 1$ we have

$$\|T_n(x) - T_m(x)\| < \varepsilon \quad \text{whenever } n, m \ge n_0$$

and taking the limit with respect to m

$$\|T_n(x) - T(x)\| \le \varepsilon \quad \text{whenever } n \ge n_0.$$

That is,

$$\|T_n - T\| \leq \varepsilon \quad \text{whenever } n \geq n_0$$

and, hence, $\lim T_n = T$. ∎

Since **F** is **R** or **C**, it is complete. Thus, for any normed linear space X, the dual space $X^* = BL(X, \mathbf{F})$ is a Banach space. Since by Proposition 6 · 1, for $1 < p < \infty$, l_p is isometrically isomorphic to the Banach space $(l_q)^*$, Proposition 8 · 2 implies that l_p is a Banach space. Also, since l_1 and l_∞ are isometrically isomorphic to $(c_0)^*$ and $(l_1)^*$, respectively (Problems 2 and 3, p. 174), l_1 and l_∞ are also Banach spaces. We now prove that the $\mathscr{L}_p$ spaces are Banach spaces.

8 · 6 Theorem (Riesz-Fischer). *The space $\mathscr{L}_p(E)$ $(p \geq 1)$ is a Banach space.*

Proof. To show completeness it is sufficient to show that any Cauchy sequence (f_k) in $\mathscr{L}_p(E)$ has a subsequence which converges in $\mathscr{L}_p(E)$. If (f_k) is a Cauchy sequence, there exists an increasing sequence (k_j) of positive integers such that

$$\|f_k - f_{k_j}\| < 1/2^j \quad \text{whenever } k \geq k_j.$$

In particular,

$$\|f_{k_{j+1}} - f_{k_j}\| < 1/2^j.$$

If we let

$$g_n = \sum_{j=1}^{n} |f_{k_{j+1}} - f_{k_j}|$$

then, for each positive integer n, $g_n \in \mathscr{L}_p(E)$ and

$$\left[\int_E g_n{}^p\right]^{1/p} = \left\| \sum_{j=1}^{n} |f_{k_{j+1}} - f_{k_j}| \right\| \leq \sum_{j=1}^{n} \|f_{k_{j+1}} - f_{k_j}\| < \sum_{j=1}^{n} \frac{1}{2^j} < 1.$$

Thus, $(g_n{}^p)$ is a nondecreasing sequence of functions in $\mathscr{L}(E)$ such that $(\int_E g_n{}^p)$ is bounded and, hence, by the Monotone Convergence Theorem $(g_n{}^p(x))$ converges for a.a.$x \in E$. Therefore, $(g_n(x))$ converges for a.a.$x \in E$, or equivalently, $\sum_{j=1}^{\infty} |f_{k_{j+1}}(x) - f_{k_j}(x)|$ converges for a.a.$x \in E$. This implies that $\sum_{j=1}^{\infty} [f_{k_{j+1}}(x) - f_{k_j}(x)]$ converges for a.a.$x \in E$. Since

$$\sum_{j=1}^{n} [f_{k_{j+1}}(x) - f_{k_j}(x)] = f_{k_{n+1}}(x) - f_{k_1}(x),$$

$(f_{k_j}(x))$ converges for a.a.$x \in E$. Let $f(x) = \lim f_{k_j}(x)$ for a.a.$x \in E$.

For fixed j, since

$$\|f_{k_i} - f_{k_j}\| < 1/2^j \quad \text{whenever } i \geq j,$$

the sequence $(|f_{k_i} - f_{k_j}|^p)$ satisfies the conditions of Fatou's Lemma. That is, for all i, $|f_{k_i} - f_{k_j}|^p \geq 0$ and

$$\underline{\lim}_i \int_E |f_{k_i} - f_{k_j}|^p < \infty.$$

Thus, since $\underline{\lim}_i |f_{k_i}(x) - f_{k_j}(x)|^p = |f(x) - f_{k_j}(x)|^p$ for a.a.$x \in E$, $|f - f_{k_j}|^p \in \mathscr{L}(E)$. Therefore, $f - f_{k_j} \in \mathscr{L}_p(E)$ and, hence, $f \in \mathscr{L}_p(E)$. Also,

$$\|f - f_{k_j}\| \leq 1/2^j$$

and, therefore, (f_{k_j}) converges to f in $\mathscr{L}_p(E)$. ∎

The space $\mathscr{L}_\infty(E)$ is also a Banach space (Problem 1).

Any normed linear space can be densely embedded in a Banach space. That is, if X is a normed linear space, then its completion $\tilde{X}$ as a metric space is a Banach space containing an isometrically isomorphic image of X which is dense in $\tilde{X}$(Problem 7).

PROBLEMS

1. Show that $\mathscr{L}_\infty(E)$ is a Banach space.

2. Let c denote the set of convergent sequences of scalars. Show that c is a closed subspace of l_∞ and hence is a Banach space.

3. Let

$$f_n(t) = \begin{cases} 0, & 0 \leq t \leq \dfrac{1}{2} - \dfrac{1}{n} \\[2ex] 1 + nt - \dfrac{n}{2}, & \dfrac{1}{2} - \dfrac{1}{n} < t < \dfrac{1}{2} \\[2ex] 1, & \dfrac{1}{2} \leq t \leq 1 \end{cases}$$

for $n = 2, 3, 4, \cdots$.

a. Show that (f_n) is a Cauchy sequence in $C_p[0, 1]$ for any $p \geq 1$.

b. Show that (f_n) does not converge in $C_p[0, 1]$ and, hence, $C_p[0, 1]$ is not a Banach space.

4. If X is a normed linear space and (x_n) is a Cauchy sequence in X, show that $(\|x_n\|)$ converges.

HH **5.** If X is a normed linear space, prove that X is a Banach space if and only if the set $\{x : \|x\| = 1\}$ is complete.

6. Let X and Y be normed linear spaces and let $X \times Y$ be the normed linear space with norm

$$\|(x, y)\| = \|x\| + \|y\|.$$

Show that $X \times Y$ is a Banach space if and only if X and Y are Banach spaces.

7. Let X be a normed linear space and let $\tilde{X}$ be the completion of X as a metric space described in Chapter 4. Prove the following:

a. If (x_n) and (y_n) are Cauchy sequences of elements of X, then $(x_n + y_n)$ and (αx_n) are Cauchy sequences.

b. Addition and scalar multiplication can be defined in $\tilde{X}$ in the following way:

$$\tilde{x} + \tilde{y} = [(x_n + y_n)] \quad \text{and} \quad \alpha\tilde{x} = [(\alpha x_n)]$$

where $\tilde{x} = [(x_n)]$ and $\tilde{y} = [(y_n)]$.

c. $\tilde{X}$ is a linear space with the operations defined in part b.

d. If we define $\|\tilde{x}\| = \tilde{d}(\tilde{x}, 0)$, then $\tilde{X}$ is a normed linear space.

e. If $f(x) = [(\bar{x})]$, where $(\bar{x})$ denotes the sequence all of whose terms are x, then f is an isometric isomorphism of X onto a dense subset of $\tilde{X}$.

9 · SERIES

Since we can add and take limits in a normed linear space, we can define series in such a space. If (a_k) is a sequence of points in a normed linear space X and $s_n = \sum_{k=1}^{n} a_k$, then the sequence (s_n) is called a **series** and the terms of (a_k) are called the **terms** of the series. Using the standard notation $\sum a_k$ for the series (s_n), we say that $\sum a_k$ has **sum** a or **converges** to a if the sequence (s_n) converges to a. We denote the sum of the series $\sum a_k$ by $\sum_{k=1}^{\infty} a_k$. If a series does not converge, then we say that it diverges.

9 · 1 Proposition. *If (a_k) does not converge to zero, then $\sum a_k$ diverges.*

PROOF. If $\sum a_k$ converges to some point a, then

$$\lim a_k = \lim (s_k - s_{k-1}) = a - a = 0. \blacksquare$$

A series $\sum a_k$ is said to be **absolutely convergent** if the series $\sum \|a_k\|$ of nonnegative real numbers is convergent. In a Banach space we will show that the convergence of the series $\sum \|a_k\|$ of nonnegative real numbers implies the convergence of the series $\sum a_k$ of points.

9 · 2 Proposition. *If $\sum a_k$ is an absolutely convergent series of points in a Banach space X, then $\sum a_k$ is convergent and $\|\sum_{k=1}^{\infty} a_k\| \le \sum_{k=1}^{\infty} \|a_k\|$.*

PROOF. The convergence of $\sum \|a_k\|$ implies that the corresponding sequence of partial sums is a Cauchy sequence. Thus, for each $\varepsilon > 0$ there exists a positive integer n_0 such that

$$\|a_{n+1}\| + \cdots + \|a_{n+p}\| < \varepsilon \quad \text{whenever } n \ge n_0 \text{ and } p \ge 1.$$

By the Triangle Inequality it follows that

$$\|a_{n+1} + \cdots + a_{n+p}\| < \varepsilon \quad \text{whenever } n \ge n_0 \text{ and } p \ge 1.$$

That is, the sequence of partial sums of $\sum a_k$ is a Cauchy sequence in X. Since X is a Banach space, $\sum a_k$ converges. Also, for any positive integer n

$$\left\| \sum_{k=1}^{n} a_k \right\| \leq \sum_{k=1}^{n} \|a_k\| \leq \sum_{k=1}^{\infty} \|a_k\|$$

and, therefore,

$$\left\| \sum_{k=1}^{\infty} a_k \right\| \leq \sum_{k=1}^{\infty} \|a_k\|. \blacksquare$$

We can also prove a converse to this theorem and thus obtain a characterization of Banach spaces.

9 · 3 Proposition. *If in a normed linear space X every absolutely convergent series is convergent, then X is a Banach space.*

Proof. Let (x_n) be a Cauchy sequence in X. Choose a subsequence (x_{n_k}) of (x_n) such that

$$\|x_{n_k} - x_{n_{k-1}}\| < 1/2^k.$$

Let $a_1 = x_{n_1}$ and $a_k = x_{n_k} - x_{n_{k-1}}$, $k > 1$. Then $x_{n_k} = \sum_{i=1}^{k} a_i$. Since $\|a_i\| < 1/2^i$ for $i > 1$, $\sum \|a_i\|$ converges. Therefore, $\sum a_i$ converges; that is, the subsequence (x_{n_k}) converges. Thus, (x_n) converges and X is complete. $\blacksquare$

Next we prove a result concerning rearrangement of series. A series $\sum b_k$ is a **rearrangement** of $\sum a_k$ if there is a one-to-one mapping f from the positive integers **N** onto **N** such that $b_k = a_{f(k)}$. A series $\sum a_k$ is said to be **unconditionally convergent** if it converges to a sum a and every rearrangement of $\sum a_k$ has sum a. We now show that an absolutely convergent series is unconditionally convergent.

9 · 4 Proposition. *If $\sum a_k$ is an absolutely convergent series in a Banach space X and its sum is a, then any rearrangement of $\sum a_k$ converges to a.*

Proof. Let $\sum b_k$ be a rearrangement of $\sum a_k$ with $b_k = a_{f(k)}$ and let $s_n = \sum_{k=1}^{n} a_k$ and $t_n = \sum_{k=1}^{n} b_k$. Take $\varepsilon > 0$. Since $\sum a_k$ is absolutely convergent, there exists a positive integer n_0 such that

$$\sum_{k=n_0+1}^{n_0+p} \|a_k\| < \varepsilon \quad \text{for all } p \geq 1.$$

Take m_0 such that $\{k : k \leq n_0\} \subset \{f(k) : k \leq m_0\}$. Then $m_0 \geq n_0$. For each $n, m \geq m_0$, $s_n - t_m$ is a sum of terms a_k with $k > n_0$ and, therefore, for some $p \geq 1$

$$\|s_n - t_m\| \leq \sum_{k=n_0+1}^{n_0+p} \|a_k\| < \varepsilon.$$

Taking the limit with respect to n, we have

$$\|a - t_m\| \leq \varepsilon \quad \text{for all } m \geq m_0.$$

Thus, $\lim t_m = a$. ∎

PROBLEMS

1. If $\sum a_k$ and $\sum b_k$ are convergent series in a normed linear space, show that

$$\sum_{k=1}^{\infty} (a_k + b_k) = \sum_{k=1}^{\infty} a_k + \sum_{k=1}^{\infty} b_k \quad \text{and} \quad \sum_{k=1}^{\infty} \alpha a_k = \alpha \sum_{k=1}^{\infty} a_k.$$

2. If (f_k) is a sequence of functions in $C(X, \mathbf{R})$ (4 · 8, p. 165) such that $\sum f_k$ converges to f in $C(X, \mathbf{R})$, show that

$$\sum_{k=1}^{\infty} f_k(x) = f(x) \quad \text{for all } x \in X.$$

3. If (f_k) is a sequence of continuous real-valued functions defined on a compact metric space X such that $\sum f_k$ converges uniformly to f on X, show that f is continuous.

4. If $a_k(t) = t^k(t-1)$ and $a(t) = -t$ for $t \in [0, 1]$, show that

a. $\sum a_k$ converges to a in $C_2[0, 1]$ (c.f. Problem 9, p. 166)

b. $\sum \|a_k\|_2$ converges

c. $\sum a_k(t)$ converges for each $t \in [0, 1]$

d. $\sum a_k$ diverges in $C[0, 1]$.

5. Use Proposition 9 · 3 to show that if X is a Banach space and Y is a closed subspace of X, then X/Y is a Banach space (c.f. Problem 5, p. 162).

10 · THE SPACE OF BOUNDED FUNCTIONS

Let E be any set and Y be a normed linear space with scalar field $\mathbf{F}$. The set of all functions from E into Y, with the usual pointwise definitions of addition and multiplication by a scalar:

$$(f + g)(t) = f(t) + g(t) \quad \text{and} \quad (\alpha f)(t) = \alpha f(t),$$

is a linear space over $\mathbf{F}$. An element f of this space is said to be **bounded** if $\sup \{\|f(t)\| : t \in E\}$ is finite. Since

$$\|f(t) + g(t)\| \leq \|f(t)\| + \|g(t)\| \quad \text{and} \quad \|\alpha f(t)\| = |\alpha| \, \|f(t)\|,$$

the set of bounded functions from E into Y is a subspace. It is easily shown that

$$\|f\| = \sup \{\|f(t)\| : t \in E\}$$

defines a norm on this subspace (Problem 1). The resulting normed linear space is denoted by $B(E, Y)$.

If (f_n) is a sequence in $B(E, Y)$ which converges to f in the norm of $B(E, Y)$, then (f_n) converges to f uniformly on E. In contrast if, for each $t \in E$, $(f_n(t))$ converges to $f(t)$ in Y, then we say that (f_n) converges to f pointwise on E. Uniform convergence on E implies pointwise convergence on E but the converse does not hold.

We now show that completeness in Y implies completeness in $B(E, Y)$.

10 · 1 Proposition. *If Y is a Banach space, then $B(E, Y)$ is a Banach space.*

PROOF. Let (f_n) be a Cauchy sequence in $B(E, Y)$. Then, for each $\varepsilon > 0$ there exists a positive integer n_0 such that

$$\|f_n - f_m\| < \varepsilon \quad \text{whenever } n, m \geq n_0$$

and, therefore, for all $t \in E$

$$\|f_n(t) - f_m(t)\| < \varepsilon \quad \text{whenever } n, m \geq n_0 .$$

Thus, $(f_n(t))$ is a Cauchy sequence in the complete space Y and, hence, converges to some point $f(t) \in Y$. Also, we have for all $t \in E$

$$\lim_{m \to \infty} \|f_n(t) - f_m(t)\| = \|f_n(t) - f(t)\| \leq \varepsilon \quad \text{whenever } n \geq n_0 .$$

This implies that

$$\|f(t)\| \leq \|f_{n_0}(t)\| + \varepsilon \leq \|f_{n_0}\| + \varepsilon \quad \text{for all } t \in E$$

and, hence, $f \in B(E, Y)$. Also

$$\|f_n - f\| \leq \varepsilon \quad \text{whenever } n \geq n_0;$$

that is, (f_n) converges to f in $B(E, Y)$. ∎

Let the set E be replaced by a metric space X. Then the set of continuous bounded functions from X into Y is a subspace, denoted by $C(X, Y)$, of the normed linear space $B(X, Y)$. If the space X is compact then a continuous function on X must be bounded; but this is not true for a general metric space X.

We now show that the subspace $C(X, Y)$ is closed in $B(X, Y)$; that is, the limit of a uniformly convergent sequence of continuous bounded functions is a continuous bounded function.

10 · 2 Proposition. *The subspace $C(X, Y)$ is closed in $B(X, Y)$.*

PROOF. If f is in the closure of $C(X, Y)$, then there exists a sequence (f_n) of functions in $C(X, Y)$ which converges to f in $B(X, Y)$. Thus, for each $\varepsilon > 0$ there exists a positive integer n_0 such that $\|f_{n_0} - f\| < \varepsilon/3$. Since f_{n_0} is continuous, for each $x_0 \in X$ there exists a $\delta > 0$ such that

$$\|f_{n_0}(x) - f_{n_0}(x_0)\| < \varepsilon/3 \quad \text{whenever } d(x, x_0) < \delta.$$

Thus, whenever $d(x, x_0) < \delta$

$$\|f(x) - f(x_0)\| \le \|f(x) - f_{n_0}(x)\| + \|f_{n_0}(x) - f_{n_0}(x_0)\| + \|f_{n_0}(x_0) - f(x_0)\| < \varepsilon\,;$$

that is, f is continuous at x_0. ∎

Since a closed subspace of a Banach space is a Banach space, we have the following result: *If Y is a Banach space, then $C(X, Y)$ is a Banach space.* Note that if X is compact and Y is $\mathbf{R}$, then $C(X, Y)$ becomes the space $C(X, \mathbf{R})$ considered previously.

If in $B(E, \mathbf{F})$, the space of bounded scalar-valued functions from a set E, we define multiplication pointwise:

$$(fg)(t) = f(t)g(t),$$

then $B(E, \mathbf{F})$ is a commutative algebra with unit. Also,

$$\|fg\| = \sup_{t \in E}|f(t)g(t)| \le \sup_{t \in E}|f(t)| \cdot \sup_{t \in E}|g(t)| = \|f\|\,\|g\|.$$

An algebra A with unit e is called a **normed algebra** if it is a normed linear space and the norm has the additional properties:

$$\|xy\| \le \|x\|\,\|y\| \quad \text{for all } x, y \in A$$

and

$$\|e\| = 1.$$

A complete normed algebra is called a **Banach algebra.** The algebra $B(E, \mathbf{F})$, where $\mathbf{F} = \mathbf{R}$ or $\mathbf{C}$, is an example of a Banach algebra. Also, if X is a compact metric space and $C(X, \mathbf{F})$ is the space of continuous scalar-valued functions on X, then $C(X, \mathbf{F})$ is a Banach algebra (Problem 4).

Another important example of a Banach algebra is the space $BL(Y, Y)$ of bounded linear transformations of a Banach space Y into itself (Problem 6). In this algebra multiplication is composition of transformations.

PROBLEMS

1. Verify that $\|f\| = \sup\{\|f(t)\| : t \in E\}$ defines a norm on $B(E, Y)$.

2. Let (f_k) be a sequence of functions from a set E into a Banach space Y such that, for each k, $\|f_k(x)\| \le M_k$ for all $x \in E$. If $\sum M_k$ converges, show that $\sum f_k$ converges in $B(E, Y)$.

3. Let $D^n[a, b]$ be the set of real-valued functions which are continuous together with their first n derivatives on $[a, b]$. Let

$$\|f\| = \max_{0 \le k \le n} \sup_{t \in [a, b]} |f^{(k)}(t)|$$

where $f^{(0)} = f$ and, for $k \ge 1$, $f^{(k)}$ is the kth derivative of f. Show that $D^n[a, b]$ is a normed linear space.

4. If X is a compact metric space, show that $C(X, \mathbf{F})$ is a Banach algebra.

5. If $\mathscr{A}$ is a subalgebra of $C(X, \mathbf{F})$, show that the closure $\overline{\mathscr{A}}$ of $\mathscr{A}$ in $C(X, \mathbf{F})$ is also a subalgebra of $C(X, \mathbf{F})$.

6. Let $BL(Y, Y)$ be the normed linear space of bounded linear transformations from the normed linear space Y into Y and let $ST = S \circ T$.

a. Show that $BL(Y, Y)$ is a normed algebra.

b. If Y is a Banach space, show that $BL(Y, Y)$ is a Banach algebra.

c. If $T \in BL(Y, Y)$ where Y is a Banach space and $\|T\| < 1$, show that $\sum T^k$ converges in $BL(Y, Y)$.

8 | *Approximation*

1 · INTRODUCTION

In this chapter we will consider a number of topics involving approximation in which it is advantageous to work in normed linear spaces. First is the uniform approximation of continuous functions by polynomials—the Weierstrass Approximation Theorem. Next, we consider approximation of integrable functions by continuous functions.

We also discuss differentiable functions. These are functions from a normed linear space into a normed linear space which can be locally approximated by affine transformations. The material on differentiable functions is an extension to a more general situation of familiar material of calculus.

2 · THE STONE-WEIERSTRASS THEOREM

Let $(x_1, \cdots, x_n)$ be a linearly independent family of elements in a linear space X. Consider the problem of obtaining the best approximation to an element y in X by a linear combination of the elements $x_1, \cdots, x_n$. This best approximation is measured in terms of some norm on X and the solution depends on the norm used. We shall show that in any normed linear space X a solution exists; that is, there exist values for the scalars $a_1, \cdots, a_n$ such that $\|y - \sum_{k=1}^{n} a_k x_k\|$ attains its minimum value. In Chapter 11 we will consider this problem in an inner product space and obtain formulas for the values of the a_k which give the minimum.

If we let E be the subspace of X spanned by $(x_1, \cdots, x_n)$, then the result can be stated as follows.

2·1 Proposition. *If E is a finite-dimensional subspace of a normed linear space X and y is an element in X, then $\{\|y - x\| : x \in E\}$ has a minimum.*

PROOF. Let $f(x) = \|y - x\|$ for $x \in E$. Then f is a continuous real-valued function on the normed linear space E. Take any point x_0 in E and let $\|y - x_0\| = r$. Then, if f has a minimum value on

$$B[y; r] = \{x \in E : \|y - x\| \leq r\},$$

this will also be the minimum value of f on E. Since $B[y; r]$ is a closed and bounded set in the finite-dimensional space E, it is compact and, hence, the continuous function f has a minimum value on $B[y; r]$. ∎

As a particular application of this theorem, consider the space $C[a, b]$ of continuous real-valued functions defined on the interval $[a, b]$ with

$$\|f\| = \max\{|f(t)| : t \in [a, b]\}.$$

The family $(1, I, \cdots, I^n)$ where $I^k(t) = t^k$, is linearly independent in $C[a, b]$ and the space spanned by this family is the set of all polynomials of degree $\leq n$. Proposition $2 \cdot 1$ implies that for each $f \in C[a, b]$ there exists a polynomial p_n of degree $\leq n$ such that

$$\|f - p_n\| \leq \|f - q_n\|$$

for all polynomials q_n of degree $\leq n$; i.e., p_n is the best uniform approximation to f among all polynomials of degree $\leq n$. The approximation theorem due to the German mathematician Karl Weierstrass (1815–1897) states that this approximation can be made arbitrarily good by taking n sufficiently large; that is, the set of polynomials is dense in $C[a, b]$.

We shall prove a generalization of the Weierstrass Approximation Theorem due to M. H. Stone and called the Stone-Weierstrass Theorem [7]. In this generalization the space $C[a, b]$ is replaced by the space $C(X, \mathbf{R})$ of continuous real-valued functions on a compact metric space X and the set of polynomials is replaced by a suitable subset of $C(X, \mathbf{R})$. One property of this subset is that it should separate points of X. A subset E of $C(X, \mathbf{R})$ is said to **separate points** of X if for each pair of distinct points $x, y \in X$, there exists a function $f \in E$ such that $f(x) \neq f(y)$.

We now prove some preliminary results. The first is due to the Italian mathematician Ulisse Dini (1845–1918).

2·2 Proposition (Dini). *If X is a compact metric space and (f_n) is a monotonic sequence in $C(X, \mathbf{R})$ which converges pointwise to $f \in C(X, \mathbf{R})$, then (f_n) converges to f in $C(X, \mathbf{R})$.*

PROOF. Suppose (f_n) is nondecreasing; that is, $f_n(x) \leq f_{n+1}(x)$ for all $x \in X$. Take $\varepsilon > 0$. For each $x \in X$ there exists a positive integer $n(x)$ such that

$$0 \leq f(x) - f_m(x) < \varepsilon/3 \quad \text{whenever } m \geq n(x).$$

Since f and $f_{n(x)}$ are continuous at x, there exists an open neighborhood U_x of x such that

$$|f(y) - f(x)| < \varepsilon/3 \quad \text{and} \quad |f_{n(x)}(y) - f_{n(x)}(x)| < \varepsilon/3 \quad \text{whenever } y \in U_x.$$

Thus, if $y \in U_x$,

$$\begin{aligned} 0 &\leq f(y) - f_{n(x)}(y) \\ &\leq |f(y) - f(x)| + |f(x) - f_{n(x)}(x)| + |f_{n(x)}(x) - f_{n(x)}(y)| < \varepsilon. \end{aligned}$$

Since X is compact there exists a finite number of points $x_1, \cdots, x_k \in X$ such that $\bigcup_{i=1}^k U_{x_i}$ covers X. Let

$$n_0 = \max\{n(x_i) : i = 1, \cdots, k\}.$$

If $n \geq n_0$ and $y \in X$, then $y \in U_{x_i}$ for some $i = 1, \cdots, k$ and

$$0 \leq f(y) - f_n(y) \leq f(y) - f_{n_0}(y) \leq f(y) - f_{n(x_i)}(y) < \varepsilon. \quad \blacksquare$$

2·3 Proposition. *There exists a sequence (p_n) of real-valued polynomials which converges to the square root function $I^{1/2}$ in $C[0, 1]$.*

PROOF. Define the sequence (p_n) inductively as follows:

$$p_1 = 0,$$
$$p_{n+1}(t) = p_n(t) + \tfrac{1}{2}[t - p_n{}^2(t)] \quad \text{for } n \geq 1.$$

We first show by induction that $0 \leq p_n(t) \leq \sqrt{t}$ for all $t \in [0, 1]$. Obviously, $0 \leq p_1(t) \leq \sqrt{t}$ for all $t \in [0, 1]$. Also, if $0 \leq p_m(t) \leq \sqrt{t}$ for all $t \in [0, 1]$, then

$$\begin{aligned} 0 \leq p_{m+1}(t) &= p_m(t) + \tfrac{1}{2}[t - p_m{}^2(t)] \\ &= p_m(t) + \tfrac{1}{2}[\sqrt{t} - p_m(t)][\sqrt{t} + p_m(t)] \\ &\leq p_m(t) + \sqrt{t} - p_m(t) \\ &= \sqrt{t} \quad \text{for all } t \in [0, 1]. \end{aligned}$$

Since $p_n{}^2(t) \leq t$, it is clear from the definition of p_{n+1} that $p_n(t) \leq p_{n+1}(t)$. Thus, for each $t \in [0, 1]$, $(p_n(t))$ is a bounded nondecreasing sequence and, hence, converges. It is easy to see that it converges to $\sqrt{t}$. Then, by Proposition 2 · 2, (p_n) converges to $I^{1/2}$ in $C[0,1]$. ∎

2 · 4 Lemma. *If X is a compact metric space and $\mathscr{A}$ is a closed subalgebra of $C(X, \mathbf{R})$ which contains the constant functions, then $\mathscr{A}$ is a lattice.*

PROOF. First, we show that if $f \in \mathscr{A}$, then $|f| \in \mathscr{A}$. If $f = 0$, then $|f| = f = 0 \in \mathscr{A}$. Assume now that $f \neq 0$ and let $\alpha = \|f\|$. Using Proposition 2 · 3 we see that there exists a sequence (p_n) of polynomials on $[0, 1]$ such that $(p_n \circ (f^2/\alpha^2))$ converges to $\sqrt{f^2/\alpha^2} = |f|/\alpha$ in $C(X, \mathbf{R})$. Since $\mathscr{A}$ is an algebra containing the constant functions, $p_n \circ (f^2/\alpha^2)$ is in $\mathscr{A}$ and, since $\mathscr{A}$ is closed, $|f|/\alpha$ is in $\mathscr{A}$. Then, $|f| \in \mathscr{A}$.

If f and g are in $\mathscr{A}$, then, since

$$f \wedge g = \tfrac{1}{2}(f + g - |f - g|)$$
$$f \vee g = \tfrac{1}{2}(f + g + |f - g|),$$

$f \wedge g$ and $f \vee g$ are in $\mathscr{A}$; that is, $\mathscr{A}$ is a lattice. ∎

2·5 Theorem (Stone-Weierstrass). *Let X be a compact metric space. If $\mathscr{A}$ is a subalgebra of $C(X, \mathbf{R})$ which contains the constant functions and separates points of X, then $\mathscr{A}$ is dense in $C(X, \mathbf{R})$.*

PROOF. Take $f \in C(X, \mathbf{R})$. Our purpose is to show that $f \in \bar{\mathscr{A}}$. For each pair of distinct points x, y and each pair of real numbers α, β there is a function h in $\mathscr{A}$ such that $h(x) = \alpha$ and $h(y) = \beta$. For, since $\mathscr{A}$ separates points of X, there is a function $g \in \mathscr{A}$ such that $g(x) \neq g(y)$. Then, let

$$h = \alpha + (\beta - \alpha)\frac{g - g(x)}{g(y) - g(x)}.$$

Thus, for each pair of distinct points x, y in X there is a function h_{xy} in $\mathscr{A}$ such that

$$h_{xy}(x) = f(x) \quad \text{and} \quad h_{xy}(y) = f(y).$$

Take $\varepsilon > 0$. For fixed x and for each $y \neq x$ let

$$U(y) = \{z : h_{xy}(z) < f(z) + \varepsilon\}.$$

Since f and h_{xy} are continuous, $U(y)$ is open. Also, x and y are in $U(y)$. Since X is compact, the open cover $\{U(y) : y \in X, y \neq x\}$ of X has a finite subcover $\{U(y_i) : i = 1, \cdots, m\}$. If

$$h_x = h_{xy_1} \wedge \cdots \wedge h_{xy_m}$$

then $h_x \in \bar{\mathscr{A}}$ by Lemma 2 · 4. Also, $h_x(x) = f(x)$ and

$$h_x(z) < f(z) + \varepsilon \quad \text{for all } z \in X.$$

For each $x \in X$ let

$$V(x) = \{z : h_x(z) > f(z) - \varepsilon\}.$$

Then $x \in V(x)$ and $V(x)$ is open since h_x and f are continuous. The open cover $\{V(x) : x \in X\}$ of X has a finite subcover $\{V(x_i) : i = 1, \cdots, n\}$. If

$$h = h_{x_1} \vee \cdots \vee h_{x_n},$$

then $h \in \bar{\mathscr{A}}$ and

$$f(z) - \varepsilon < h(z) < f(z) + \varepsilon \quad \text{for all } z \in X.$$

Thus,

$$\|f - h\| = \sup\{|f(z) - h(z)| : z \in X\} \leq \varepsilon$$

and, hence, $f \in \bar{\mathscr{A}}$. ▮

It is now easy to obtain the Weierstrass Approximation Theorem.

2 · 6 Corollary. *If the real-valued function f is continuous on $[a, b]$, then corresponding to each $\varepsilon > 0$ there exists a polynomial p such that*

$$|f(t) - p(t)| \leq \varepsilon \quad \textit{for all } t \in [a, b].$$

PROOF. The restriction to $[a, b]$ of the set of all polynomials from **R** into **R** is a subalgebra $\mathscr{A}$ of $C[a, b]$. $\mathscr{A}$ contains the constant functions on $[a, b]$ and separates points of $[a, b]$ since for two distinct points c and d in $[a, b]$ the polynomial x defined by $x(t) = t$ will have different values at c and d. Therefore, by the Stone-Weierstrass Theorem $\mathscr{A}$ is dense in $C[a, b]$. ▮

The extension of the Weierstrass Theorem to functions defined on $\mathbf{R}^n$ is also a direct consequence of the Stone-Weierstrass Theorem.

2·7 Corollary. *If f is a real-valued continuous function on a compact subset E of $\mathbf{R}^n$, then there exists a sequence of polynomials which converges uniformly to f on E.*

There is also a form of the Stone-Weierstrass Theorem which applies to complex-valued functions. In this we put an additional requirement on the subalgebra $\mathscr{A}$; if a function f belongs to $\mathscr{A}$, then its complex conjugate $\bar{f}$ must also belong to $\mathscr{A}$.

2·8 Theorem (Complex Stone-Weierstrass Theorem). *Let X be a compact metric space. If $\mathscr{A}$ is a subalgebra of $C(X, \mathbf{C})$ which contains the constant functions, separates points of X, and contains $\bar{f}$ whenever it contains f, then $\mathscr{A}$ is dense in $C(X, \mathbf{C})$.*

PROOF. If $f \in \mathscr{A}$, then $\text{Rl}\, f = \frac{1}{2}(f + \bar{f})$ and $\text{Im}\, f = -\frac{1}{2}i(f - \bar{f})$ belong to $\mathscr{A}$. Hence, if $\mathscr{A}'$ is the subalgebra of $C(X, \mathbf{R})$ containing all real-valued functions in $\mathscr{A}$, then $\mathscr{A}'$ separates points of X and contains the real-valued constant functions on X. Therefore, $\mathscr{A}'$ is dense in $C(X, \mathbf{R})$.

If $f \in C(X, \mathbf{C})$, then $\text{Rl}\, f$ and $\text{Im}\, f$ are in $C(X, \mathbf{R})$ and hence in $\bar{\mathscr{A}}'$. Then $\text{Rl}\, f$ and $\text{Im}\, f$ are in $\bar{\mathscr{A}}$ and, since $f = \text{Rl}\, f + i\text{Im}\, f$ and $\bar{\mathscr{A}}$ is an algebra, $f \in \bar{\mathscr{A}}$. ∎

2·9 Corollary (Weierstrass). *If the complex-valued function f is continuous on $\mathbf{R}$ and has period 2π, then corresponding to each $\varepsilon > 0$ there exists a trigonometric polynomial T, $T(t) = \sum_{k=-n}^{n} c_k e^{ikt}$, such that*

$$|f(t) - T(t)| \le \varepsilon \quad \textit{for all } t \in \mathbf{R}.$$

PROOF. Let $g(t) = e^{it}$, $t \in [0, 2\pi)$. Then g is a one-to-one mapping of $[0, 2\pi)$ onto $E = \{e^{it} : t \in [0, 2\pi)\}$. The set E is the unit circle in $\mathbf{C}$ (or, equivalently $\mathbf{R}^2$) and is a compact set.

Let $\mathscr{A}$ be the set of functions P of the form

$$P(z) = \sum_{k=-n}^{n} c_k z^k, \qquad z \in E, \quad c_k \in \mathbf{C}, \quad n = 0, 1, 2, \cdots.$$

Then, $\mathscr{A}$ is a subalgebra of $C(E, \mathbf{C})$ which contains the constant functions on E, separates points of E, and contains the complex conjugate of each of its elements. Thus, by Theorem 2 · 8, $\mathscr{A}$ is dense in $C(E, \mathbf{C})$.

Since $f \circ g^{-1} \in C(E, \mathbf{C})$, corresponding to each $\varepsilon > 0$ there exists a trigonometric polynomial $P \circ g$, $P(e^{it}) = \sum_{k=-n}^{n} c_k e^{ikt}$, such that

$$|f \circ g^{-1}(e^{it}) - P(e^{it})| \le \varepsilon \quad \text{for all } e^{it} \in E.$$

Thus, letting $T(t) = P(e^{it})$,

$$|f(t) - T(t)| \le \varepsilon \quad \text{for all } t \in [0, 2\pi).$$

For each $t \in \mathbf{R}$ there exists an integer m and $t' \in [0, 2\pi)$ such that

$$t = t' + 2m\pi.$$

Then, since f and T have period 2π,

$$|f(t) - T(t)| = |f(t') - T(t')| \le \varepsilon \quad \text{for all } t \in \mathbf{R}. \blacksquare$$

We now use the Weierstrass Approximation Theorem to obtain a result on the extension of continuous functions which is a special case of a theorem due to the German mathematician Heinrich Tietze (1880–1954). If a real-valued function f is continuous on an open set in $\mathbf{R}^n$, it may not be possible to find a continuous extension of f with domain $\mathbf{R}^n$. For example, if $f(x) = 1/x$ for $x > 0$, then it is not possible to find a function g with domain $\mathbf{R}$ which is a continuous extension of f. We will show, however, that if a real-valued function is continuous on a closed set in $\mathbf{R}^n$, then there exists an extension which is continuous on $\mathbf{R}^n$.

2 · 10 Lemma. *If f is continuous on a nonempty compact set E in $\mathbf{R}^n$ then f can be extended to a function which is continuous on $\mathbf{R}^n$.*

PROOF. The Weierstrass Theorem 2 · 7 implies that for each $k = 0, 1, 2, \cdots$ there is a polynomial p_k such that

$$|f(x) - p_k(x)| < 2^{-k-2} \quad \text{for all } x \in E.$$

Let $q_0 = p_0$ and $q_k = p_k - p_{k-1}$ for $k = 1, 2, \cdots$. Then $p_k = \sum_{j=0}^{k} q_j$ and $\sum q_k$ converges uniformly to f on E. Since f is continuous on the compact set E, f is bounded on E and

$$|p_0(x)| < 2^{-2} + |f(x)| \le 2^{-2} + M \quad \text{for all } x \in E$$

where $M = \sup\{|f(x)| : x \in E\}$. Also for $k = 1, 2, \cdots$

$$|q_k(x)| < 2^{-k} \quad \text{for all } x \in E.$$

If we let

$$h_0 = (-2^{-2} - M) \vee (q_0 \wedge (2^{-2} + M))$$

and

$$h_k = (-2^{-k}) \vee (q_k \wedge 2^{-k}), \qquad k = 1, 2, \cdots,$$

then $h_k(x) = q_k(x)$ for $x \in E$, h_k is continuous on $\mathbf{R}^n$, and for $k = 1, 2, \cdots$, $|h_k(x)| \le 2^{-k}$ for all $x \in \mathbf{R}^n$. Thus $\sum h_k$ converges uniformly on $\mathbf{R}^n$. If $\sum_{k=0}^{\infty} h_k = h$, then h is continuous in $\mathbf{R}^n$ and $h(x) = f(x)$ for $x \in E$. $\blacksquare$

2 · 11 Theorem (Tietze). *If f is continuous on a nonempty closed set $E \subset \mathbf{R}^n$, then there is an extension of f which is continuous on $\mathbf{R}^n$.*

PROOF. If E is bounded, then E is compact and the result follows from Lemma 2 · 10. Assume E is not bounded. For some closed ball $B[0; m]$, $E \cap B[0; m]$ is a nonempty compact set.

Let f_m be the restriction of f to $E \cap B[0; m]$. By Lemma $2 \cdot 10$, f_m has an extension h_m which is continuous on $\mathbf{R}^n$. Let

$$g_m(x) = \begin{cases} h_m(x), & x \in B[0; m] \\ f(x), & x \in E \cap B[0; m+1]. \end{cases}$$

Then, g_m is continuous on the compact set $B[0; m] \cup (E \cap B[0; m+1])$ and so has an extension h_{m+1} which is continuous on $\mathbf{R}^n$. Let

$$g_{m+1}(x) = \begin{cases} h_{m+1}(x), & x \in B[0; m+1] \\ f(x), & x \in E \cap B[0; m+2]. \end{cases}$$

Continuing in this way, we obtain a sequence (g_k) such that

$$\mathrm{D}_{g_m} \subset \mathrm{D}_{g_{m+1}} \subset \mathrm{D}_{g_{m+2}} \subset \cdots.$$

Define the function g on $\mathbf{R}^n$ by

$$g(x) = g_k(x) \quad \text{if } x \in B[0; k].$$

Then g is a continuous extension of f. ▮

PROBLEMS

1. Let $y = (1/k) \in l_2$ and let E be the subspace of l_2 spanned by $(u_1, \cdots, u_n)$ where $u_k = (\delta_k{}^j)$. Determine $\min\{\|y - x\| : x \in E\}$ and find a point in E at which this minimum is attained. Do the same if the space l_2 is replaced by l_∞.

2. Let E be a compact set in $\mathbf{R}^n$ and let P_k be the kth projection function from E into $\mathbf{R} : P_k(\alpha^1, \cdots, \alpha^n) = \alpha^k$. Show that if $\mathscr{A}$ is a subset of $C(E, \mathbf{R})$ which contains the projective functions P_k, $k = 1, \cdots, n$, then $\mathscr{A}$ separates points of E.

3. Starting with the Binomial Formula

$$(a + b)^n = \sum_{k=0}^{n} \binom{n}{k} a^k b^{n-k} \quad \text{where } \binom{n}{k} = \frac{n!}{k!(n-k)!}$$

show that:

a. $\displaystyle\sum_{k=0}^{n} \binom{n}{k} x^k (1 - x)^{n-k} = 1$ for all $x \in \mathbf{R}$.

b. $\displaystyle\sum_{k=0}^{n} k \binom{n}{k} x^k (1 - x)^{n-k} = nx$ for all $x \in \mathbf{R}$.

Suggestion. Differentiate the Binomial Formula with respect to a.

c. $\displaystyle\sum_{k=0}^{n} k^2 \binom{n}{k} x^k (1 - x)^{n-k} = nx(1 - x + nx)$ for all $x \in \mathbf{R}$.

d. $\displaystyle\sum_{k=0}^{n} \binom{n}{k} (k - nx)^2 x^k (1 - x)^{n-k} \le \frac{n}{4}$ for all $x \in \mathbf{R}$.

4. Let f be defined on $[0, 1]$. The polynomial

$$B_n(x) = \sum_{k=0}^{n} f\left(\frac{k}{n}\right)\binom{n}{k}x^k(1-x)^{n-k}$$

is called the **Bernstein polynomial** of degree n associated with f. Prove: If f is continuous on $[0, 1]$, then for each $\varepsilon > 0$ there exists a number n_0 such that

$$|B_n(x) - f(x)| < \varepsilon \quad \text{for all } n > n_0 \text{ and all } x \in [0, 1].$$

Suggestion.

$$|B_n(x) - f(x)| \le \sum_{k=0}^{n} \left|f\left(\frac{k}{n}\right) - f(x)\right| \binom{n}{k}x^k(1-x)^{n-k}.$$

Since f is uniformly continuous on $[0, 1]$ corresponding to $\varepsilon > 0$, there exists a $\delta > 0$ such that $x', x'' \in [0, 1]$ and $|x' - x''| < \delta$ imply that $|f(x') - f(x'')| < \varepsilon$. Separate the terms in the above sum into those for which $|k/n - x| < \delta$ and those for which $|k/n - x| \ge \delta$. In the latter case $(k - nx)^2/(n^2\delta^2) \ge 1$ and $|f(k/n) - f(x)| \le 2M$ where $M = \max\{|f(x)| : x \in [0, 1]\}$.

5. Use Problem 4 to prove the Weierstrass Approximation Theorem.

6. If f is continuous on $[a, b]$ and

$$\int_a^b f(x)x^n\,dx = 0, \qquad n = 0, 1, 2, \cdots,$$

show that $f = 0$ on $[a, b]$.

7. If f is a real-valued continuous function with period 2π, show that for each $\varepsilon > 0$ there exists a real-valued trigonometric polynomial T such that

$$|f(t) - T(t)| \le \varepsilon \quad \text{for all } t \in \mathbf{R}.$$

8. If f is a complex-valued continuous function on $\mathbf{R}$ with period $b - a$, show that for each $\varepsilon > 0$ there exists a trigonometric polynomial T, $T(x) = \sum_{k=-n}^{n} c_k e^{i2\pi kx/(b-a)}$, such that

$$|f(x) - T(x)| \le \varepsilon \quad \text{for all } x \in \mathbf{R}.$$

9. If f is a measurable function on $\mathbf{R}^n$, show that for each $\delta > 0$ there exists a continuous function g on $\mathbf{R}^n$ such that $\mu\{x : f(x) \ne g(x)\} < \delta$.

Suggestion. Use Lusin's Theorem (14 · 7, p. 151) and the Tietze Extension Theorem.

3 · THE SPACE $\mathscr{L}_p[a, b]$

We will prove by a sequence of steps that the set of continuous complex-valued functions on $[a, b]$ is dense in the complex space $\mathscr{L}_p[a, b]$ where $[a, b]$ is an interval in $\mathbf{R}$.

3·1 Lemma. *If I is an open interval in $[a, b]$, then there exists a sequence (g_n) of continuous functions such that*

$$\lim\|\chi_I - g_n\| = 0.$$

PROOF. Let $h(x) = d(x, I)$, the distance from x to I; then h is continuous and $h(x) = 0$ for $x \in \bar{I}$ and $h(x) > 0$ for $x \notin \bar{I}$. Let

$$g_n(x) = \frac{1}{1 + nh(x)}.$$

Then $(|\chi_I - g_n|^p)$ is a sequence of measurable functions such that $|\chi_I - g_n|^p \leq 1$ for all n and $\lim|\chi_I(x) - g_n(x)|^p = 0$ for a.a.$x \in [a, b]$. Thus, by the Lebesgue Dominated Convergence Theorem (p. 136)

$$\lim \int_a^b |\chi_I - g_n|^p = 0. \blacksquare$$

3·2 Lemma. *If f is a step function on $[a, b]$, there exists a sequence (g_n) of continuous functions such that*

$$\lim\|f - g_n\| = 0.$$

PROOF. Let $\{I_k : k = 1, \cdots, r\}$ be a partition of $[a, b]$ such that $f(x) = c_k$ for $x \in I_k$. Then

$$f = \sum_{k=1}^{r} c_k \chi_{I_k} \quad \text{a.e. on } [a, b].$$

Let (g_n^k) be a sequence of continuous functions such that

$$\lim_{n\to\infty} \|\chi_{I_k} - g_n^k\| = 0, \qquad k = 1, \cdots, r,$$

and let $g_n = \sum_{k=1}^r c_k g_n^k$. Since

$$\|f - g_n\| = \left\|\sum_{k=1}^{r} c_k \chi_{I_k} - \sum_{k=1}^{r} c_k g_n^k\right\| \leq \sum_{k=1}^{r} |c_k| \, \|\chi_{I_k} - g_n^k\|,$$

$\lim\|f - g_n\| = 0.$ ∎

3·3 Lemma. *If f is a nonnegative function in $\mathscr{P}$, then for each $\varepsilon > 0$ there exists a step function h such that*

$$\int_a^b |f - h| < \varepsilon.$$

PROOF. Since $f \in \mathscr{P}$ and $f \geq 0$, there is a nondecreasing sequence (f_n) of nonnegative step functions such that $\lim f_n(x) = f(x)$ for a.a.$x \in [a, b]$ and

$$\int_a^b f = \lim \int_a^b f_n.$$

Since

$$\int_a^b |f - f_n| = \int_a^b (f - f_n) = \int_a^b f - \int_a^b f_n,$$

we may take $h = f_n$ for some sufficiently large n. ∎

3·4 Lemma. *If f is a nonnegative integrable function on $[a, b]$, then for each $\varepsilon > 0$ there exists a step function h such that*

$$\int_a^b |f - h| < \varepsilon.$$

PROOF. There exist nonnegative functions f_1 and f_2 in $\mathscr{G}$ such that $f = f_1 - f_2$ a.e. on $[a, b]$. Then by Lemma 3 · 3 for each $\varepsilon > 0$ there exist step functions h_1 and h_2 such that

$$\int_a^b |f_1 - h_1| < \frac{\varepsilon}{2} \quad \text{and} \quad \int_a^b |f_2 - h_2| < \frac{\varepsilon}{2}.$$

Thus, for $h = h_1 - h_2$ we have

$$\int_a^b |f - h| \le \int_a^b |f_1 - h_1| + \int_a^b |f_2 - h_2| < \varepsilon. \ ∎$$

3·5 Lemma. *If f is a nonnegative function in $\mathscr{L}_p[a, b]$, then for each $\varepsilon > 0$ there exists a step function h and a continuous function g such that*

$$\|f - h\| < \varepsilon \quad \textit{and} \quad \|f - g\| < \varepsilon.$$

PROOF. For each positive integer, n, let

$$f_n = f \wedge n \quad \text{on } [a, b].$$

Since f is measurable and the constant function n is integrable over $[a, b]$, (f_n) is a nondecreasing sequence of nonnegative integrable functions such that

$$\lim f_n(x) = f(x) \quad \text{for } x \in [a, b].$$

Then $(|f - f_n|^p)$ is a sequence of measurable functions such that $|f - f_n|^p \le |f|^p$ for all n and

$$\lim |f(x) - f_n(x)|^p = 0 \quad \text{for all } x \in [a, b].$$

Hence, by the Lebesgue Dominated Convergence Theorem

$$\lim \int_a^b |f - f_n|^p = 0.$$

Thus, for each $\varepsilon > 0$ there exists a nonnegative integrable function f_n such that

$$\|f - f_n\| < \varepsilon/3.$$

By Lemma 3 · 4 there exists a step function h_n such that

$$\int_a^b |f_n - h_n| < \left(\frac{\varepsilon}{3n}\right)^p.$$

If $h = h_n^+ \wedge n$, then h is a nonnegative step function bounded by n and h^p is integrable. Also

$$|f_n - h| = |f \wedge n - h_n^+ \wedge n| \le n$$

and

$$|f_n - h| \le |f_n - h_n|.$$

Therefore,

$$\|f_n - h\|^p = \int_a^b (|f_n - h|^{p-1}\, |f_n - h|) \le n^{p-1}\left(\frac{\varepsilon}{3n}\right)^p < \left(\frac{\varepsilon}{3}\right)^p.$$

Thus

$$\|f - h\| \le \|f - f_n\| + \|f_n - h\| < \varepsilon.$$

By Lemma 3 · 2 there exists a continuous function g such that

$$\|h - g\| < \varepsilon/3.$$

Then

$$\|f - g\| \le \|f - f_n\| + \|f_n - h\| + \|h - g\| < \varepsilon. \blacksquare$$

Letting $f = f^+ - f^-$, we easily obtain:

3 · 6 Theorem. *If f is a real-valued function in $\mathscr{L}_p[a, b]$, then for each $\varepsilon > 0$ there exists a step function h and a continuous function g such that*

$$\|f - h\| < \varepsilon \quad \textit{and} \ \|f - g\| < \varepsilon.$$

Then, letting $f = f_1 + if_2$, we can prove:

3 · 7 Theorem. *If f is a complex-valued function in $\mathscr{L}_p[a, b]$, then for each $\varepsilon > 0$ there exists a continuous function g such that*

$$\|f - g\| < \varepsilon.$$

Thus, we have:

3 · 8 Theorem. *The set of continuous functions is dense in $\mathscr{L}_p[a, b]$.*

Using this theorem we can prove (Problem 3):

3 · 9 Theorem. *$\mathscr{L}_p[a, b]$ is separable.*

PROBLEMS

1. Prove Theorem 3 · 6.
2. Prove Theorem 3 · 7.
3. Prove the following:
   a. The set of polynomials with rational coefficients is dense in the real space $\mathscr{L}_p[a, b]$.
   b. The complex space $\mathscr{L}_p[a, b]$ is separable.

4 · SYSTEMS OF LINEAR EQUATIONS

In §3 of Chapter 5 we considered the problem of solving in the real field a system of n equations in n unknowns:

$$(4 \cdot 1) \qquad \sum_{j=1}^{n} a_j^{\ i} x^j = y^i, \qquad i = 1, \ldots, n$$

where $a_j^{\ i}$ and y^i are real numbers. If we let $A = [a_j^{\ i}]$, $x = [x^j]$, and $y = [y^i]$ where A is an $n \times n$ matrix and x and y are $n \times 1$ matrices, then this system of linear equations becomes the matrix equation

$$(4 \cdot 2) \qquad Ax = y.$$

We can now discuss this problem in a more general setting.

The problem of solving 4 · 2 was converted into a fixed-point problem by defining, for fixed y, the transformation S_y from $V_n(\mathbf{R})$ into itself by the formula

$$S_y(x) = (I - A)x + y$$

where I is the $n \times n$ identity matrix. Clearly, x is a solution of $Ax = y$ if and only if x is a fixed-point of S_y. If we let $T(x) = (I - A)x$, we have

$$S_y(x) = T(x) + y.$$

Thus, S_y is the linear transformation T followed by a translation through y. A transformation which is a linear transformation followed by a translation is called an **affine transformation.**

In order to apply the Banach Fixed-Point Theorem (p. 94), S_y must be a contraction of a complete metric space into itself. We consider S_y to be a transformation on $l_p^{\ n}$ into $l_p^{\ n}$ where $p \in [1, \infty]$. We previously restricted consideration to the cases $p = 1$ and $p = \infty$. The linear transformation $T \in BL(l_p^{\ n}, l_p^{\ n})$ and if x and x' are points in $l_p^{\ n}$ then

$$\|S_y(x) - S_y(x')\| = \|T(x - x')\| \leq \|T\| \, \|x - x'\| = \|I - A\|_p \|x - x'\|.$$

Thus, if $\|I - A\|_p < 1$, S_y is a contraction of $l_p^{\ n}$ into $l_p^{\ n}$ and, hence, has a unique fixed-point $\bar{x}$; that is, $Ax = y$ has a unique solution $\bar{x}$. Also, if x_0 is any point in $l_p^{\ n}$ and $x_k = S_y(x_{k-1})$, then (x_k) converges to $\bar{x}$ in $l_p^{\ n}$; in fact,

$$\|\bar{x} - x_k\| \leq \frac{\lambda^k}{1 - \lambda} \|x_1 - x_0\| \quad \text{where } \lambda = \|I - A\|_p.$$

In particular, if $\|I - A\|_1 = \max_j \sum_{i=1}^n |\delta_j^{\,i} - a_j^{\,i}| < 1$ or $\|I - A\|_\infty = \max_i \sum_{j=1}^n |\delta_j^{\,i} - a_j^{\,i}| < 1$, then $4 \cdot 1$ has a unique solution which can be approximated by iteration.

If we choose $x_0 = 0$, then $x_1 = y$, $x_2 = (I - A)y + y$, and in general

$$x_k = \sum_{j=0}^{k-1} (I - A)^j y,$$

where $(I - A)^0 = I$. Since the sequence (x_k) converges to $\bar{x}$ in $l_p^{\,n}$, we have

$$\bar{x} = \sum_{j=0}^{\infty} (I - A)^j y$$

where the series converges in $l_p^{\,n}$. The observation that if the matrix A has an inverse A^{-1}, then $x = A^{-1}y$ is a solution of $Ax = y$ suggests that under appropriate conditions $A^{-1} = \sum_{j=0}^{\infty} (I - A)^j$.

We now prove a general result for Banach algebras which will have this suggested result for matrices as a consequence.

4 · 3 Proposition. *If $\mathscr{A}$ is a Banach algebra with unit e and x is an element of $\mathscr{A}$ such that $\|e - x\| < 1$, then x is invertible in $\mathscr{A}$ and*

$$x^{-1} = \sum_{k=0}^{\infty} (e - x)^k \quad \textit{and} \quad \|x^{-1}\| \leq \frac{1}{1 - \|e - x\|}.$$

PROOF. In this statement we understand that $(e - x)^0 = e$. Since $\|(e - x)^k\| \leq \|e - x\|^k$ and $\|e - x\| < 1$, $\sum_{k=0}^{\infty} (e - x)^k$ is absolutely convergent in $\mathscr{A}$ and

$$\left\| \sum_{k=0}^{\infty} (e - x)^k \right\| \leq \sum_{k=0}^{\infty} \|e - x\|^k = \frac{1}{1 - \|e - x\|}.$$

Also, we have

$$x\left(\sum_{k=0}^{\infty} (e - x)^k \right) = (e - (e - x)) \sum_{k=0}^{\infty} (e - x)^k$$

$$= \sum_{k=0}^{\infty} (e - x)^k - \sum_{k=0}^{\infty} (e - x)^{k+1} = e$$

and, similarly,

$$\left(\sum_{k=0}^{\infty} (e - x)^k \right) x = e.$$

Therefore,

$$x^{-1} = \sum_{k=0}^{\infty} (e - x)^k. \ \blacksquare$$

If we change notation in $4 \cdot 3$, we obtain a familiar looking result: If y is an element of a Banach algebra such that $\|y\| < 1$, then

$$\sum_{k=0}^{\infty} y^k = (e - y)^{-1}.$$

4 · 4 Corollary. *If A is a matrix in $\mathbf{F}_n^n$ and, for some $p \in [1, \infty]$, $\|I - A\|_p < 1$, then A has an inverse and*

$$A^{-1} = \sum_{k=0}^{\infty} (I - A)^k$$

where convergence is in the p-norm of $\mathbf{F}_n^n$.

PROOF. This follows immediately from 4 · 3 since $\mathbf{F}_n^n$ with the p-norm is a Banach algebra. ▌

The following result which will be used later in this chapter is also a corollary of Proposition 4 · 3.

4 · 5 Corollary. *Let X and Y be Banach spaces and let $\mathscr{U}$ be the set of transformations in the Banach space $BL(X, Y)$ which have bounded inverses. If $T \in \mathscr{U}$ and $\|T^{-1}\| = \alpha$, then the open ball $B(T; 1/\alpha)$ is contained in $\mathscr{U}$ and, hence, $\mathscr{U}$ is open in $BL(X, Y)$. Also, the mapping f from $\mathscr{U}$ into $BL(Y, X)$ defined by $f(T) = T^{-1}$ is continuous.*

PROOF. Let $T \in \mathscr{U}$ and $\|T^{-1}\| = \alpha$. If $S \in B(T; 1/\alpha)$, then $T^{-1}S$ is an element of the Banach algebra $BL(X, X)$ and

$$\|I - T^{-1}S\| = \|T^{-1}T - T^{-1}S\| \le \|T^{-1}\| \, \|T - S\| < 1.$$

By Proposition 4 · 3, $T^{-1}S$ is invertible in $BL(X, X)$ and, hence, $S = T(T^{-1}S) \in \mathscr{U}$. Thus, $B(T; 1/\alpha) \subset \mathscr{U}$ and $\mathscr{U}$ is open in $BL(X, Y)$.

Also, from Proposition 4 · 3

$$\|(T^{-1}S)^{-1}\| = \|S^{-1}T\| \le \frac{1}{1 - \|I - T^{-1}S\|} \le \frac{1}{1 - \alpha\|T - S\|}.$$

Then,

$$\begin{aligned}\|f(S) - f(T)\| &= \|S^{-1} - T^{-1}\| = \|S^{-1}T(I - T^{-1}S)T^{-1}\| \\ &\le \|S^{-1}T\| \, \|I - T^{-1}S\| \, \|T^{-1}\| \le \frac{\alpha^2\|T - S\|}{1 - \alpha\|T - S\|}.\end{aligned}$$

This shows that f is continuous at T. ▌

If A is a matrix in $\mathbf{F}_n^n$ such that, for some $p \in [1, \infty]$, $\|I - A\|_p < 1$, then Corollary 4 · 4 gives a formula for A^{-1}: $A^{-1} = \sum_{k=0}^{\infty} (I - A)^k$. However, in general, this series converges too slowly for computational purposes. We now give a rapidly converging iterative method for computing the inverse of an invertible matrix A. This method requires an initial approximation to A^{-1} and such an approximation can be obtained from some method such as Gauss elimination. In general, Gauss elimination gives only an approximation to A^{-1} due to roundoff errors in the computations and the method described below then provides a way of improving the approximation.

Let A be an invertible matrix. Suppose D_0 is an initial approximation to A^{-1} such that $\|I - AD_0\|_p < 1$ for some $p \in [1, \infty]$. If we let $E_0 = I - AD_0$, then we have

$$A^{-1} = D_0(I - E_0)^{-1} = D_0 \sum_{k=0}^{\infty} E_0{}^k.$$

Take the first two terms of this series as an approximation D_1 to A^{-1}. Then

$$D_1 = D_0(I + E_0) = D_0 + D_0 E_0.$$

In general, let

$$D_n = D_{n-1} + D_{n-1}E_{n-1} \quad \text{where } E_{n-1} = I - AD_{n-1} \quad \text{for } n \geq 1.$$

Then

$$\begin{aligned} E_n &= I - AD_n = I - AD_{n-1} - AD_{n-1}E_{n-1} \\ &= I - AD_{n-1} - AD_{n-1}(I - AD_{n-1}) = (I - AD_{n-1})^2 = E_{n-1}^2 \end{aligned}$$

and, hence, $E_n = E_0^{2^n}$. Since $D_n = A^{-1}(I - E_n)$,

$$\begin{aligned} \|A^{-1} - D_n\|_p &= \|A^{-1}E_n\|_p = \|D_0(I - E_0)^{-1}E_0^{2^n}\|_p \\ &\leq \|D_0\|_p \frac{\|E_0\|_p^{2^n}}{1 - \|E_0\|_p}. \end{aligned}$$

Thus, we have proved the following result.

4 · 6 Proposition. *Let D_0 be an approximation to the inverse of an invertible matrix A for which $\|I - AD_0\|_p = \lambda < 1$. The sequence (D_n) defined recursively by the formula*

$$D_n = D_{n-1}(2I - AD_{n-1}) \quad \textit{for } n \geq 1$$

converges to A^{-1} and

$$\|A^{-1} - D_n\|_p \leq \|D_0\|_p \frac{\lambda^{2^n}}{1 - \lambda}.$$

4 · 7 Example. Let

$$A = \begin{pmatrix} 1.00 & 0.54 & 0.42 & 0.44 \\ 0.54 & 1.00 & 0.31 & 0.24 \\ 0.42 & 0.31 & 1.00 & 0.68 \\ 0.44 & 0.24 & 0.68 & 1.00 \end{pmatrix}$$

and let the initial approximation to A^{-1} be

$$D_0 = \begin{pmatrix} 1.6648 & -0.7406 & -0.1718 & -0.4379 \\ -0.7406 & 1.4377 & -0.2262 & 0.1346 \\ -0.1718 & -0.2262 & 1.9605 & -1.2032 \\ -0.4379 & 0.1346 & -1.2032 & 1.9786 \end{pmatrix}$$

Find the elements of A^{-1} with error less than 10^{-6}.

SOLUTION. Although A and D_0 are symmetric, E_0 is not symmetric. We have

$$E_0 = I - AD_0 = 10^{-6}\begin{pmatrix} -44 & 22 & -54 & -24 \\ -38 & 42 & -15 & -6 \\ -58 & 37 & -46 & -56 \\ -44 & 32 & -60 & -52 \end{pmatrix}\begin{matrix}(144)\\(101)\\(197)\\(188)\end{matrix}$$
$$\qquad (184)\ (133)\ (175)\ (138)$$

The terms in parentheses on the right and below the matrix are the sums of the absolute values of the elements in that row or column, respectively. We have $\|E_0\|_1 = 0.000184$ and $\|E_0\|_\infty = 0.000197$. Since $\|E_0\|_1 < \|E_0\|_\infty$, we use the 1-norm for our error estimates. We also note that since D_0 is symmetric, $\|D_0\|_1 = \|D_0\|_\infty = 3.7543$. Thus, by Proposition 4 · 6 we have

$$\|D_1 - A^{-1}\|_1 \le 3.7543\,\frac{(0.000184)^2}{1 - 0.000184} < 0.000\,000\,13.$$

Thus D_1 approximates A^{-1} to about one unit in the seventh decimal place for each of its elements. Computing D_1 we obtain (D_1 is symmetric)

$$D_1 = D_0 + D_0 E_0 =$$
$$\begin{pmatrix} 1.6647841 & -0.7406148 & -0.1718446 & -0.4379031 \\ & 1.4377400 & -0.2261792 & 0.1346148 \\ & & 1.9604947 & -1.2032417 \\ & & & 1.9785742 \end{pmatrix}$$

We find that

$$E_1 = I - AD_1 = 10^{-9}\begin{pmatrix} -12 & -44 & -26 & -24 \\ -48 & -8 & -48 & -32 \\ -16 & -34 & 40 & 4 \\ 18 & 17 & 26 & 68 \end{pmatrix}. \blacksquare$$

PROBLEM

Let

$$A = \begin{pmatrix} 1.00 & -0.64 & 0.10 & -0.31 \\ -0.21 & 1.00 & -0.47 & -0.14 \\ 0.55 & -0.99 & 1.00 & -0.23 \\ -0.37 & 0.03 & 0.68 & 1.00 \end{pmatrix}$$

and let D_0 be an approximation to A^{-1} where

$$D_0 = \begin{pmatrix} 1.427 & 0.763 & -0.136 & 0.518 \\ 0.231 & 1.582 & 0.451 & 0.397 \\ -0.377 & 1.039 & 1.303 & 0.328 \\ 0.777 & -0.472 & -0.950 & 0.956 \end{pmatrix}.$$

a. Compute D_1. Assuming no mistakes in arithmetic, how many decimal digits are certain and what is the maximum error in the first uncertain digit?

b. If D_2 were computed from this data, to what accuracy would its terms approximate those of A^{-1}?

5 · DIFFERENTIABLE FUNCTIONS

In this section we consider a class of mappings from one normed linear space into another which can be locally approximated by affine transformations.

5 · 1 Definition. *Let X and Y be normed linear spaces over* **F** (**R** *or* **C**) *and let a be a point of X. A function f which maps a neighborhood N of a into Y is* ***differentiable*** *at a if there is a bounded linear transformation T_a from X into Y such that*

$$\lim_{h\to 0} \frac{\|f(a+h) - f(a) - T_a(h)\|}{\|h\|} = 0. \tag{5 · 2}$$

The bounded linear transformation T_a is called the ***differential*** *of f at a and is denoted by df_a. The set of mappings from X into Y which are differentiable at a will be denoted by $D_a(X, Y)$.*

If, for $a + h \in N$, we let $r_a(h) = f(a+h) - f(a) - T_a(h)$ then 5 · 2 can be written as

$$f(a+h) = f(a) + T_a(h) + r_a(h) \quad \text{where} \lim_{h\to 0} \frac{\|r_a(h)\|}{\|h\|} = 0. \tag{5 · 3}$$

Rewriting this as

$$\begin{aligned} f(x) &= f(a) + T_a(x-a) + r_a(x-a) \\ &= T_a(x) + f(a) - T_a(a) + r_a(x-a), \end{aligned}$$

we see that, in a neighborhood of a, f is approximated by the affine transformation composed of the bounded linear transformation T_a followed by the translation through $f(a) - T_a(a)$ with the error being given by r_a.

Since T_a is continuous it is clear that: *If $f \in D_a(X, Y)$, then f is continuous at a.*

We now show that the differential of f is uniquely determined by 5 · 3. Suppose there exist T_a and T_a' in $BL(X, Y)$ such that, for a neighborhood N of a and for $a + h \in N$,

$$f(a+h) = f(a) + T_a(h) + r_a(h) \quad \text{where} \lim_{h\to 0} \frac{\|r_a(h)\|}{\|h\|} = 0$$

and

$$f(a+h) = f(a) + T_a'(h) + s_a(h) \quad \text{where} \lim_{h\to 0} \frac{\|s_a(h)\|}{\|h\|} = 0.$$

Then,

$$T_a(h) - T_a'(h) = s_a(h) - r_a(h)$$

and, hence,

$$\lim_{h \to 0} \frac{\|T_a(h) - T_a'(h)\|}{\|h\|} = 0.$$

Thus, for each $\varepsilon > 0$ there exists a $\delta > 0$ such that

$$\|(T_a - T_a')(h)\| \le \varepsilon \|h\| \quad \text{whenever } h \in B(0; \delta).$$

For any $x \in X$, let $x = \alpha h$ where $h \in B(0; \delta)$. Then

$$\|(T_a - T_a')(x)\| = |\alpha| \, \|(T_a - T_a')(h)\| \le |\alpha| \varepsilon \|h\| = \varepsilon \|x\|.$$

Thus, $(T_a - T_a')(x) = 0$; that is, $T_a = T_a'$.

5 · 4 Proposition. *If c is a constant function from X into Y, then, for all $a \in X$, $c \in D_a(X, Y)$ and*

$$dc_a = 0.$$

Proof. For all $a + h \in X$,

$$c(a + h) = c(a).$$

Thus, 5 · 3 holds with $T_a = 0$ and $r_a = 0$. ∎

We now show that $BL(X, Y) \subset D_a(X, Y)$ for any $a \in X$.

5 · 5 Proposition. *If $T \in BL(X, Y)$, then, for all $a \in X$, $T \in D_a(X, Y)$ and*

$$(dT)_a = T.$$

Proof. For any $a + h \in X$,

$$T(a + h) = T(a) + T(h).$$

Thus, 5 · 3 holds with $T_a = T$ and $r_a = 0$. ∎

With the usual definitions of addition of functions and scalar multiplication, $D_a(X, Y)$ is a linear space.

5 · 6 Proposition. *If $f, g \in D_a(X, Y)$ and $\alpha \in \mathbf{F}$, then $f + g$ and αf are in $D_a(X, Y)$ and*

$$d(f + g)_a = df_a + dg_a \quad \textit{and} \quad d(\alpha f)_a = \alpha df_a.$$

The proof of this is left to the reader (Problem 3).

The following two propositions give some additional rules for the differential of combinations of functions.

5·7 Proposition. *If $\varphi \in D_a(X, \mathbf{F})$ and $f \in D_a(X, Y)$, then $\varphi f \in D_a(X, Y)$ and*

$$d(\varphi f)_a = \varphi(a)df_a + d\varphi_a f(a)$$

PROOF. First a word about $d\varphi_a f(a)$. At any $h \in X$, $(d\varphi_a f(a))(h) = d\varphi_a(h)f(a)$ where $d\varphi_a(h) \in \mathbf{F}$ and $f(a) \in Y$. Also, $d\varphi_a f(a) \in BL(X, Y)$, since

$$\begin{aligned}(d\varphi_a f(a))(\alpha x + \beta y) &= d\varphi_a(\alpha x + \beta y)f(a) = (\alpha d\varphi_a(x) + \beta d\varphi_a(y))f(a)\\ &= \alpha(d\varphi_a f(a))(x) + \beta(d\varphi_a f(a))(y).\end{aligned}$$

Suppose for each $a + h$ in a neighborhood N of a,

$$\varphi(a + h) = \varphi(a) + d\varphi_a(h) + \rho_a(h) \quad \text{where } \lim_{h\to 0} \frac{|\rho_a(h)|}{\|h\|} = 0$$

and

$$f(a + h) = f(a) + df_a(h) + r_a(h) \quad \text{where } \lim_{h\to 0} \frac{\|r_a(h)\|}{\|h\|} = 0.$$

Then

$$\begin{aligned}(\varphi f)(a + h) &= (\varphi(a) + d\varphi_a(h) + \rho_a(h))(f(a) + df_a(h) + r_a(h))\\ &= (\varphi f)(a) + (\varphi(a)df_a + d\varphi_a f(a))(h) + s_a(h)\end{aligned}$$

where

$$\begin{aligned}s_a(h) = \varphi(a)r_a(h) + d\varphi_a(h)df_a(h) + d\varphi_a(h)r_a(h) + \rho_a(h)f(a)\\ + \rho_a(h)df_a(h) + \rho_a(h)r_a(h).\end{aligned}$$

Since $\varphi(a)df_a + d\varphi_a f(a) \in BL(X, Y)$ and

$$\lim_{h\to 0} \frac{\|s_a(h)\|}{\|h\|} = 0,$$

$\varphi f \in D_a(X, Y)$ and

$$d(\varphi f)_a = \varphi(a)df_a + d\varphi_a f(a). \blacksquare$$

5·8 Proposition. (Chain Rule). *If $f \in D_a(X, Y)$ and $g \in D_{f(a)}(Y, Z)$, then $g \circ f \in D_a(X, Z)$ and*

$$d(g \circ f)_a = dg_{f(a)} \circ df_a.$$

PROOF. For a neighborhood M of $f(a)$ and for $f(a) + k \in M$,

$$g(f(a) + k) = g(f(a)) + dg_{f(a)}(k) + s_{f(a)}(k)$$

where

$$\lim_{k\to 0} \frac{\|s_{f(a)}(k)\|}{\|k\|} = 0.$$

Also, there is a neighborhood N of a such that $f(N) \subset M$ and for $a + h \in N$

$$f(a + h) = f(a) + df_a(h) + r_a(h) \quad \text{where } \lim_{h\to 0} \frac{\|r_a(h)\|}{\|h\|} = 0.$$

Then, for $a + h \in N$,

$$\begin{aligned}(g \circ f)(a + h) &= g(f(a + h)) = g(f(a) + df_a(h) + r_a(h)) \\ &= g(f(a)) + dg_{f(a)}(df_a(h) + r_a(h)) \\ &\qquad + s_{f(a)}(df_a(h) + r_a(h)) \\ &= (g \circ f)(a) + (dg_{f(a)} \circ df_a)(h) + t_a(h)\end{aligned}$$

where

$$t_a(h) = dg_{f(a)}(r_a(h)) + s_{f(a)}(df_a(h) + r_a(h)).$$

If

$$\lim_{h \to 0} \frac{\|t_a(h)\|}{\|h\|} = 0,$$

then since $dg_{f(a)} \circ df_a \in BL(X, Z)$ the theorem is proved. Take $\varepsilon > 0$. There exists an $\eta > 0$ such that for $\|k\| < \eta$,

$$\|s_{f(a)}(k)\| < \varepsilon \|k\|.$$

Also, there exists a $\delta > 0$ such that for $\|h\| < \delta$

$$\|r_a(h)\| < \varepsilon \|h\| \quad \text{and} \quad \|df_a(h) + r_a(h)\| < \eta.$$

Then, for $\|h\| < \delta$,

$$\begin{aligned}\|t_a(h)\| &\leq \|dg_{f(a)}(r_a(h))\| + \|s_{f(a)}(df_a(h) + r_a(h))\| \\ &\leq \|dg_{f(a)}\| \, \|r_a(h)\| + \varepsilon \|df_a(h) + r_a(h)\| \\ &< (\varepsilon \|dg_{f(a)}\| + \varepsilon \|df_a\| + \varepsilon^2) \|h\|.\end{aligned}$$

Thus,

$$\lim_{h \to 0} \frac{\|t_a(h)\|}{\|h\|} = 0. \blacksquare$$

PROBLEMS

1. If f is the function from $\mathbf{R}^2$ into $\mathbf{R}^3$ defined by

$$f(x, y) = (x^2 y, xy + y^2, x^3 - 2),$$

show that f is differentiable at (1, 2) and determine $df_{(1,2)}$.

2. If b and c are points in Y and T is a transformation from $\mathbf{R}$ into Y defined by

$$T(\tau) = b + \tau c,$$

show that for each $\alpha \in \mathbf{R}$

$$dT_\alpha(\eta) = \eta c.$$

3. Prove Proposition $5 \cdot 6$.

4. Let g and f be functions from the normed linear space X into the normed algebra Y which are differentiable at a. If gf is defined by $(gf)(x) = g(x)f(x)$, $x \in X$, show that $gf \in D_a(X, Y)$ and

$$d(gf)_a = g(a)df_a + dg_a f(a).$$

5. If f is a mapping of a normed linear space X into $\mathbf{R}$ which is differentiable at a and $f(a) \neq 0$, show that $1/f$ is differentiable at a and

$$d\left(\frac{1}{f}\right)_a = -\frac{df_a}{f^2(a)}.$$

6. Let $f \in D_a(X, Z)$, $g \in D_b(Y, Z)$, and S be the mapping from $X \times Y$ into Z defined by

$$S = f \circ P_1 + g \circ P_2$$

where P_1 and P_2 are the projection functions from $X \times Y$ into X and Y, respectively. Show that $S \in D_{(a,b)}(X \times Y, Z)$ and

$$dS_{(a,\ b)} = df_a \circ P_1 + dg_b \circ P_2.$$

7. Let $f \in D_a(X, Y)$ and $g \in D_b(Y, X)$ where $b = f(a)$. If $f \circ g = I_Y$ and $g \circ f = I_X$, show that

$$dg_b = (df_a)^{-1}.$$

6 · DIRECTIONAL DERIVATIVES AND PARTIAL DERIVATIVES

Let X and Y be normed linear spaces over $\mathbf{F}$.

6 · 1 Definition. *If f is a function from X into Y and u is a unit vector in X, the **derivative of f in the direction** u is denoted by $D_u f$ and its value at x is*

$$D_u f(x) = \lim_{\tau \to 0} \frac{f(x + \tau u) - f(x)}{\tau}$$

if this limit exists.

The following theorem gives a relation between directional derivatives and differentials.

6 · 2 Theorem. *If f is differentiable at a, then all the directional derivatives of f exist at a and*

$$D_u f(a) = df_a(u).$$

PROOF. Since f is differentiable at a there is a neighborhood N of a such that for all $a + h \in N$

$$f(a + h) = f(a) + df_a(h) + r_a(h) \quad \text{where} \lim_{h \to 0} \frac{\|r_a(h)\|}{\|h\|} = 0.$$

Thus, if u is a unit vector in X and τ is sufficiently small,

$$f(a+\tau u) = f(a) + df_a(\tau u) + r_a(\tau u).$$

Also,

$$0 = \lim_{\tau\to 0} \frac{\|r_a(\tau u)\|}{\|\tau u\|} = \lim_{\tau\to 0} \left\| \frac{r_a(\tau u)}{\tau} \right\|$$

and, hence,

$$\lim_{\tau\to 0} \frac{r_a(\tau u)}{\tau} = 0.$$

Therefore,

$$\lim_{\tau\to 0} \frac{f(a+\tau u) - f(a)}{\tau} = \lim_{\tau\to 0} \left[df_a(u) + \frac{r_a(\tau u)}{\tau} \right] = df_a(u);$$

that is,

$$D_u f(a) = df_a(u). \blacksquare$$

It is not true that if all directional derivatives exist at a point, then the function is differentiable at that point. For example, let f be the function from $\mathbf{R}^2$ to $\mathbf{R}$ defined by

$$f(0) = 0 \quad \text{and} \quad f(x) = \frac{\alpha^3}{\alpha^2 + \beta^2} \quad \text{where } x = (\alpha, \beta) \neq 0.$$

Note that $f(\tau x) = \tau f(x)$. Then, for any unit vector u in $\mathbf{R}^2$,

$$\frac{f(0+\tau u) - f(0)}{\tau} = f(u)$$

and, hence, the derivative of f in the direction u exists at 0 and

$$D_u f(0) = f(u).$$

If f is differentiable at 0, then, for any unit vector u,

$$df_0(u) = D_u f(0) = f(u).$$

Furthermore, if for any $x \in \mathbf{R}^2$ we write $x = \tau u$ where u is a unit vector, then we have

$$df_0(x) = df_0(\tau u) = \tau df_0(u) = \tau f(u) = f(\tau u) = f(x).$$

This shows that if f is differentiable at 0, then $f = df_0$ and, hence, f is linear. But f is not linear since $f(1, 0) = 1$, $f(0, 1) = 0$, and $f(1, 1) = 1/2$.

For a function f from $\mathbf{R}$ into Y, the derivative of f in the direction 1 is called the **derivative** of f and is denoted by f'. Thus,

$$f'(x) = D_1 f(x) = \lim_{t\to 0} \frac{f(x+t) - f(x)}{t}.$$

If the function f is differentiable at a, then using Theorem 6 · 2 we have

$$df_a(h) = hdf_a(1) = hf'(a). \tag{6 · 3}$$

Conversely, if f is defined in a neighborhood of a and the derivative of f exists at a, then f is differentiable at a and $df_a(h) = hf'(a)$ (Problem 1).

We now obtain a Mean Value Theorem for functions from **R** into Y.

6 · 4 Theorem. *If f is differentiable on $[a, b]$ and $\|f'(t)\| \le m$ for all $t \in (a, b)$, then*

$$\|f(b) - f(a)\| \le m(b - a).$$

PROOF. Take $\varepsilon > 0$ and let

$$\varphi(t) = \|f(t) - f(a)\| - (m + \varepsilon)(t - a), \quad t \in [a, b].$$

Then φ is continuous on $[a, b]$ since the differentiability of f implies continuity of f. Let

$$S = \{t \in [a, b] : \varphi(t) \le \varepsilon\}.$$

Since $\varphi(a) = 0$ and φ is continuous, S contains some point $t > a$. Also, S is closed. If $c = \sup S$, then $c \in S \cap (a, b]$. We shall show that $c = b$. Suppose $c < b$. Then $\|f'(c)\| \le m$ and, hence, there exists a point $d \in (c, b)$ such that

$$\|f(d) - f(c)\| \le (m + \varepsilon)(d - c).$$

Then,

$$\begin{aligned}\|f(d) - f(a)\| &\le \|f(d) - f(c)\| + \|f(c) - f(a)\| \\ &\le (m + \varepsilon)(d - c) + (m + \varepsilon)(c - a) + \varepsilon \\ &= (m + \varepsilon)(d - a) + \varepsilon;\end{aligned}$$

that is, $d \in S$, But since $d > c = \sup S$, this is a contradiction and, hence, $c = b \in S$. Thus,

$$\|f(b) - f(a)\| \le m(b - a) + \varepsilon(1 + b - a).$$

Therefore, since $\varepsilon > 0$ is arbitrary,

$$\|f(b) - f(a)\| \le m(b - a). \blacksquare$$

Consider now a function f from $\mathbf{R}^n$ into Y. Let $(u_i)_{i=1}^n$ be the standard basis for $\mathbf{R}^n$; that is, $u_i = (\delta_i^k)_{k=1}^n$. This standard basis gives preferred unit vectors, and directional derivatives in these directions are called partial derivatives.

6 · 5 Definition. *The derivative of f in the direction u_i is called the* ith ***partial derivative*** *of f and is denoted by $D_i f$ rather than $D_{u_i} f$.*

Thus, if $a = (\alpha^1, \cdots, \alpha^n)$,

$$D_i f(a) = \lim_{\tau \to 0} \frac{1}{\tau} [f(a + \tau u_i) - f(a)]$$

$$= \lim_{\tau \to 0} \frac{1}{\tau} [f(\alpha^1, \cdots, \alpha^i + \tau, \cdots, \alpha^n) - f(\alpha^1, \cdots, \alpha^i, \cdots, \alpha^n)]$$

From the preceding discussion of the general case we know that if f is differentiable at a point a then all the partial derivatives exist at a and

$$D_i f(a) = df_a(u_i).$$

Also, we know that the existence of all the partial derivatives at a point does not ensure differentiability at that point. We now show that if all the partial derivatives exist and are continuous in a neighborhood of a point then the function is differentiable at the point. In fact, we will show that the differential is continuous at the given point. Note that if f is a function from X into Y and if A is the set of points in X at which f is differentiable, then the differential df of f is the function from A into $BL(X, Y)$ whose value at $a \in A$ is df_a. If df is continuous at a, we say that f is **continuously differentiable** at a.

6 · 6 Definition. *If all the partial derivatives of a function f from $\mathbf{R}^n$ into Y are continuous on a set E, then we say that f belongs to class C^1 on E and write $f \in C^1(E)$.*

6 · 7 Theorem. *Let f be a function from $\mathbf{R}^n$ into Y and let E be an open set in $\mathbf{R}^n$. If $f \in C^1(E)$, then, for each $a \in E$ and $h = (\eta^1, \cdots, \eta^n) \in \mathbf{R}^n$,*

$$df_a(h) = \sum_{j=1}^{n} \eta^j D_j f(a)$$

and, hence, f is continuously differentiable on E.

PROOF. Take a point $a \in E$ and take $\varepsilon > 0$. Let $B(a; \delta)$ be an open ball such that $B(a; \delta) \subset E$ and for all $x \in B(a; \delta)$

$$\| D_j f(x) - D_j f(a) \| < \varepsilon, \qquad j = 1, \cdots, n.$$

Take $a + h \in B(a; \delta)$ where $h = (\eta^1, \cdots, \eta^n) = \sum_{i=1}^{n} \eta^i u_i$. For $j = 1, \cdots, n$ let $x_j = a + \sum_{i=1}^{j} \eta^i u_i$ and let $x_0 = a$. Then

$$f(a + h) - f(a) - \sum_{j=1}^{n} \eta^j D_j f(a)$$

$$= \sum_{j=1}^{n} [f(x_{j-1} + \eta^j u_j) - f(x_{j-1}) - \eta^j D_j f(a)].$$

If we let

$$g_j(\tau) = f(x_{j-1} + \tau \eta^j u_j) - \tau \eta^j D_j f(a), \quad \tau \in [0, 1],$$

then

$$g_j'(\tau) = \eta^j[D_j f(x_{j-1} + \tau\eta^j u_j) - D_j f(a)].$$

Hence, for $\tau \in [0, 1]$,

$$\|g_j'(\tau)\| < \varepsilon\,|\eta^j|$$

and by the Mean Value Theorem 6 · 4

$$\|g_j(1) - g_j(0)\| \le \varepsilon\,|\eta^j|;$$

that is,

$$\|f(x_{j-1} + \eta^j u_j) - f(x_{j-1}) - \eta^j D_j f(a)\| \le \varepsilon\,|\eta^j|.$$

Therefore,

$$\|f(a+h) - f(a) - \sum_{j=1}^{n} \eta^j D_j f(a)\| \le \varepsilon \sum_{j=1}^{n} |\eta^j| \le \varepsilon n\,\|h\|.$$

Since the transformation T from $\mathbf{R}^n$ into Y defined by $T(h) = \sum_{j=1}^{n} \eta^j D_j f(a)$ is linear, this shows that f is differentiable at a and

$$df_a(h) = \sum_{j=1}^{n} \eta^j D_j f(a). \;\blacksquare$$

Consider a function f from $\mathbf{R}^n$ into $\mathbf{R}^m$ and assume that f is differentiable at a. Then, df_a is a linear transformation from $\mathbf{R}^n$ into $\mathbf{R}^m$ and, hence, there is a matrix associated with df_a with respect to the standard bases for $\mathbf{R}^n$ and $\mathbf{R}^m$. We now obtain that matrix. Let P_i be the ith projection function from $\mathbf{R}^m$ into $\mathbf{R}$ and let $f^i = P_i \circ f$. Then f^i is a function from $\mathbf{R}^n$ into $\mathbf{R}$. Since P_i is linear, f^i is differentiable at a and

$$df_a^{\,i} = d(P_i \circ f)_a = P_i \circ df_a, \qquad i = 1, \cdots, m.$$

Then the jth column of the matrix associated with df_a is

$$df_a(u_j) = (df_a^{\,i}(u_j))_{i=1}^{m} = (D_j f^i(a))_{i=1}^{m}, \qquad j = 1, \cdots, n.$$

Thus the matrix associated with df_a is the $m \times n$ matrix $(D_j f^i(a))$ where $D_j f^i(a)$ is the element in the ith row and jth column. The element $D_j f^i(a)$ is often denoted by $\partial f^i/\partial x_j$. The matrix $(D_j f^i(a))$ is called the **Jacobian matrix** of f at a.

Since

$$(6 \cdot 8) \qquad df_a(h) = (D_j f^i(a))h$$

where the right hand side is the product of an $m \times n$ matrix and the $n \times 1$ matrix h, the similarity of 6 · 8 to 6 · 3 suggests that we may call the Jacobian matrix of f the derivative of f and denote it by f' or Df.

It is clear that if the Jacobian matrix is continuous in a neighborhood of a, that is, if all the partial derivatives $D_j f^i$ are continuous in a neighborhood of a, then f is differentiable at a (Problem 2).

PROBLEMS

1. If f is a function from $\mathbf{R}$ into a normed linear space Y such that f is defined in a neighborhood of a point a and $f'(a)$ exists, show that f is differentiable at a and $df_a(h) = hf'(a)$.

2. If f is a function from $\mathbf{R}^n$ into $\mathbf{R}^m$ such that, for all $i = 1, \cdots, m$ and all $j = 1, \cdots, n$, $D_j f^i$ is continuous in a neighborhood N of a, show that $D_j f$ is continuous on N and therefore f is continuously differentiable on N.

3. If f is the function from $\mathbf{R}^2$ into $\mathbf{R}^3$ defined by

$$f(x, y) = (x^2 y, xy + y^2, x^3 - 2),$$

determine the partial derivatives of f. Show that f is differentiable at all points in $\mathbf{R}^2$. Determine the Jacobian matrix of f and use it to find $df_{(1,2)}$.

4. If $f \in D_a(X, \mathbf{R})$ and f has a relative maximum at a, show that $df_a = 0$.

5. Let f be a function from $\mathbf{R}^n$ into a normed linear space Y and let $a = (\alpha^1, \cdots, \alpha^n)$ be a point in $\mathbf{R}^n$. The jth **partial differential** of f at a, denoted by $d_j f_a$, is defined to be an element of $BL(\mathbf{R}, Y)$ such that

$$\lim_{\eta \to 0} \frac{\|f(a + \eta u_j) - f(a) - d_j f_a(\eta)\|}{|\eta|} = 0.$$

Show that:

a. $d_j f_a$ exists if and only if $D_j f(a)$ exists and

$$d_j f_a(\eta) = \eta D_j f(a).$$

b. If all of the partial differentials $d_j f$ are continuous in a neighborhood of a, then f is differentiable at a and

$$df_a = \sum_{j=1}^{n} d_j f_a \circ P_j$$

where P_j is the jth projection function from $\mathbf{R}^n$ into $\mathbf{R}$.

7 · THE IMPLICIT FUNCTION THEOREM

First, we define a convex set. A set E in the linear space X is said to be **convex** if, for each pair of points x and y in E, the line segment $[x, y]$ joining x and y lies in E; i.e., if $x, y \in E$, then $[x, y] = \{\tau x + (1 - \tau)y : \tau \in [0, 1]\} \subset E$. Clearly, any subspace of a linear space is convex.

We now give a mean value theorem which is more general than Theorem 6 · 4 and which will be needed later in this section. In this theorem X and Y are normed linear spaces.

7 · 1 Theorem. (Mean Value Theorem) *If f is a function from X into Y which is differentiable on a convex set $C \subset X$ and $\|df_x\| \leq \delta$ for all $x \in C$, then for any points a and $a + h$ in C*

$$\|f(a + h) - f(a)\| \leq \delta \|h\|.$$

PROOF. Take a and $a + h$ in C and let

$$g(\tau) = f(a + \tau h), \quad \tau \in [0, 1].$$

Since C is convex, $a + \tau h \in C$ for $\tau \in [0, 1]$. By the Chain Rule

$$g'(\tau) = dg_\tau(1) = df_{a+\tau h}(h).$$

Thus, for $\tau \in [0, 1]$,

$$\|g'(\tau)\| = \|df_{a+\tau h}(h)\| \leq \|df_{a+\tau h}\| \, \|h\| \leq \delta \|h\|.$$

Then, by Theorem 6 · 4,

$$\|g(1) - g(0)\| \leq \delta \|h\|;$$

that is,

$$\|f(a + h) - f(a)\| \leq \delta \|h\|. \blacksquare$$

The Implicit Function Theorem is concerned with the problem of solving an equation $f(x, y) = 0$ for one of the variables in terms of the other. In §4 we considered a special case of this problem where f is the function from $\mathbf{R}^n \times \mathbf{R}^n$ into $\mathbf{R}^n$ defined by $f(x, y) = Ax - y$. If for fixed $y \in \mathbf{R}^n$ we let $f(\cdot, y)$ be the function from $\mathbf{R}^n$ into $\mathbf{R}^n$ whose value at x is $f(x, y)$, then $f(\cdot, y)$ is an affine transformation. Also, we note that in this case, for any point (a, b) in $\mathbf{R}^n \times \mathbf{R}^n$, $(df(\cdot, b))_a(h) = Ah$. Thus, the requirement that A be invertible is the same as requiring that this differential be invertible.

Since a differentiable function is locally approximated by an affine transformation, we might expect that, in the general case of the equation $f(x, y) = 0$, the imposition of differentiability conditions will yield a local result. Before stating this result we introduce some terminology.

7 · 2 Definition. *Let X, Y, and Z be normed linear spaces and let (a, b) be a point in $X \times Y$. If $f(\cdot, b)$ maps a neighborhood of a into Z, the **first partial differential** of f at (a, b), denoted by $d_1 f_{(a,b)}$, is defined to be an element of $BL(X, Z)$ such that*

$$\lim_{h \to 0} \frac{\|f(a + h, b) - f(a, b) - d_1 f_{(a,b)}(h)\|}{\|h\|} = 0.$$

The second partial differential is defined in a similar way and is denoted by $d_2 f$. It is clear that $d_1 f_{(a,b)} = (df(\cdot, b))_a$ and $d_2 f_{(a,b)} = (df(a, \cdot))_b$. Also, if f is differentiable at (a, b), then the partial differentials exist at (a, b) and

$$df_{(a,b)}(h, 0) = d_1 f_{(a,b)}(h) \quad \text{and} \quad df_{(a,b)}(0, k) = d_2 f_{(a,b)}(k).$$

Then

$$df_{(a,b)}(h, k) = d_1 f_{(a,b)}(h) + d_2 f_{(a,b)}(k);$$

that is, if P_1 and P_2 are the projection functions from $X \times Y$ into X and Y, respectively, then

$$df_{(a,b)} = d_1 f_{(a,b)} \circ P_1 + d_2 f_{(a,b)} \circ P_2. \tag{7 · 3}$$

7 · 4 Theorem. (Implicit Function Theorem). *Let X, Y, and Z be Banach spaces, (a, b) be a point in $X \times Y$, and $N \times M$ be a neighborhood of (a, b). Suppose that f is a continuously differentiable function from $N \times M$ into Z such that $d_1 f_{(a,b)}$ has a bounded inverse and $f(a, b) = 0$. Then, there are open balls $B(b; \rho) \subset M$ and $B(a; \varepsilon) \subset N$ and a unique continuously differentiable function g from $B(b; \rho)$ into $B(a; \varepsilon)$ such that $g(b) = a$ and $f(g(y), y) = 0$ for all $y \in B(b; \rho)$.*

PROOF. Let $d_1 f_{(a,b)} = A$. Since $d_1 f$ is continuous on $N \times M$ there exist open balls $B(a; \varepsilon) \subset N$ and $B(b; \delta) \subset M$ such that for all $(x, y) \in B(a; \varepsilon) \times B(b; \delta)$

$$\|d_1 f_{(x,y)} - A\| \le \tfrac{1}{2} \|A^{-1}\|^{-1}. \tag{7 · 5}$$

Take $y \in B(b; \delta)$ and let $S(\cdot, y)$ be the mapping from X into X defined by

$$S(\cdot, y) = I - A^{-1} \circ f(\cdot, y)$$

where I is the identity transformation from X into X. It is clear that x satisfies $f(x, y) = 0$ if and only if x is a fixed-point of $S(\cdot, y)$. For any $x \in B(a; \varepsilon)$

$$(dS(\cdot, y))_x = I - A^{-1} \circ d_1 f_{(x,y)}$$

and, therefore,

$$\|(dS(\cdot, y))_x\| = \|A^{-1} \circ A - A^{-1} \circ d_1 f_{(x,y)}\| \le \|A^{-1}\| \, \|A - d_1 f_{(x,y)}\| \le \tfrac{1}{2}.$$

Applying the Mean Value Theorem to $S(\cdot, y)$, we have, for x_1 and x_2 in $B(a; \varepsilon)$,

$$\|S(x_1, y) - S(x_2, y)\| \le \tfrac{1}{2}\|x_1 - x_2\|. \tag{7 · 6}$$

Thus, $S(\cdot, y)$ is a contraction on $B(a; \varepsilon)$ and if $S(\cdot, y)$ has a fixed-point in $B(a; \varepsilon)$, it is unique.

Since f is continuous at (a, b) and $f(a, b) = 0$ there exists an open ball $B(b; \rho) \subset B(b; \delta)$ such that for all $y \in B(b; \rho)$

$$\|f(a, y)\| < \tfrac{1}{2}\|A^{-1}\|^{-1} \varepsilon.$$

Now, take $y \in B(b; \rho)$ and let $\|f(a, y)\| = \frac{1}{2} \|A^{-1}\|^{-1}\eta$ where $\eta < \varepsilon$. Then, for x in the closed ball $B[a; \eta]$,

$$\begin{aligned} \|S(x, y) - a\| &\le \|S(x, y) - S(a, y)\| + \|S(a, y) - a\| \\ &\le \tfrac{1}{2}\|x - a\| + \|A^{-1}\| \, \|f(a, y)\| \le \eta. \end{aligned}$$

Thus, $S(\cdot, y)$ is a contraction of the complete metric space $B[a; \eta]$ into itself and, hence, by the Banach Fixed-Point Theorem $S(\cdot, y)$ has a unique fixed-point x_y in $B[a; \eta]$. That is, for each $y \in B(b; \rho)$ there is a unique point x_y in $B(a; \varepsilon)$ such that $S(x_y, y) = x_y$ or, equivalently, such that $f(x_y, y) = 0$. Letting $g(y) = x_y$ we obtain a function g from $B(b; \rho)$ into $B(a; \varepsilon)$ such that for all $y \in B(b; \rho)$

$$f(g(y), y) = 0.$$

Also, from the definition of $S(\cdot, b)$ it is clear that $S(a, b) = a$ and, therefore, $g(b) = a$.

It remains to be shown that g is continuously differentiable on $B(b; \rho)$. Take y and $y + k$ in $B(b; \rho)$ and let $x = g(y)$ and $x + h = g(y + k)$. Then, x and $x + h$ are in $B(a; \varepsilon)$. Since f is differentiable at (x, y),

$$(7 \cdot 7) \qquad \begin{aligned} 0 &= f(x + h, y + k) - f(x, y) = df_{(x, y)}(h, k) + r_{(x, y)}(h, k) \\ &= d_1 f_{(x, y)}(h) + d_2 f_{(x, y)}(k) + r_{(x, y)}(h, k), \end{aligned}$$

where

$$\lim_{(h, k) \to 0} (\|(h, k)\|^{-1} \|r_{(x, y)}(h, k)\|) = 0.$$

Since $(x, y) \in B(a; \varepsilon) \times B(b; \delta)$, $7 \cdot 5$ states that

$$\|d_1 f_{(x, y)} - A\| \leq \tfrac{1}{2}\|A^{-1}\|^{-1}$$

and, therefore, Corollary $4 \cdot 5$ (p. 200) implies that $d_1 f_{(x, y)}$ has a bounded inverse transformation. Letting $(d_1 f_{(x, y)})^{-1} = T$ and applying this linear transformation to $7 \cdot 7$, we obtain

$$0 = h + T(d_2 f_{(x, y)}(k)) + T(r_{(x, y)}(h, k));$$

that is,

$$g(y + k) - g(y) = -T(d_2 f_{(x, y)}(k)) - T(r_{(x, y)}(h, k)).$$

Since $-T \circ d_2 f_{(x, y)} \in BL(Y, X)$, if $\lim_{k \to 0}(\|k\|^{-1}\|T(r_{(x, y)}(h, k))\|) = 0$, then g is differentiable at y and $dg_y = -T \circ d_2 f_{(x, y)}$. Using $7 \cdot 6$ and the differentiability of f at (x, y), we have

$$\begin{aligned} \|h\| &= \|S(x + h, y + k) - S(x, y)\| \\ &\leq \|S(x + h, y + k) - S(x, y + k)\| + \|S(x, y + k) - S(x, y)\| \\ &\leq \tfrac{1}{2}\|h\| + \|A^{-1}(f(x, y + k))\| \\ &\leq \tfrac{1}{2}\|h\| + \|A^{-1}\|(\|df_{(x, y)}\| \, \|k\| + \|r_{(x, y)}(0, k)\|). \end{aligned}$$

Thus, for $\|k\|$ sufficiently small,

$$\|g(y + k) - g(y)\| = \|h\| \leq \mu\|k\|$$

where $\mu = 2\|A^{-1}\|(\|df_{(x, y)}\| + 1)$. This shows that $\lim_{k \to 0} h = 0$; that is, g is continuous at y. Also, for $\|k\|$ sufficiently small,

$$\|(h, k)\| \leq \|h\| + \|k\| \leq (\mu + 1)\|k\|.$$

Then,

$$\lim_{k \to 0} \frac{\|T(r_{(x, y)}(h, k))\|}{\|k\|} \leq \lim_{k \to 0} \left(\|T\| \, \frac{\|(h, k)\|}{\|k\|} \, \frac{\|r_{(x, y)}(h, k)\|}{\|(h, k)\|} \right) = 0.$$

Thus, g is differentiable on $B(b; \rho)$ and for $y \in B(b; \rho)$

$$dg_y = -(d_1 f_{(g(y), y)})^{-1} \circ d_2 f_{(g(y), y)}.$$

Since g is continuous at y and $d_1 f$ and $d_2 f$ are continuous at $(g(y), y)$ and the operation of taking the inverse is continuous at $d_1 f_{(g(y), y)}$(Corollary 4 · 5), then dg is continuous at y. ∎

As a special case of the Implicit Function Theorem consider the problem of solving the following system of n equations in $n + m$ real variables:

$$f^i(x_1, \ldots, x_n, y_1, \ldots, y_m) = 0$$

for $x_1, \ldots, x_n$ in terms of $y_1, \ldots, y_m$. If we let f be the function from $\mathbf{R}^n \times \mathbf{R}^m = \mathbf{R}^{n+m}$ into $\mathbf{R}^n$ whose ith component function is f^i, we can write this system of equations as

$$f(x, y) = 0.$$

If (a, b) is a solution of this equation, the Implicit Function Theorem states that for y near b the equation can be solved for x in terms of y provided f has continuous partial derivatives in a neighborhood of (a, b) and $d_1 f_{(a,b)}$ has an inverse (in this case boundedness is automatic). With the standard basis for $\mathbf{R}^{n+m}$ the matrix associated with $d_1 f_{(a,b)}$ is the square matrix $(D_j f^i(a, b))$, $i, j = 1, \ldots, n$. If the determinant of this matrix is nonzero, then the matrix has an inverse and, hence, $d_1 f_{(a,b)}$ has a (bounded) inverse.

The following theorem, the Inverse Function Theorem, is a special case of the Implicit Function Theorem.

7 · 8 Theorem. (The Inverse Function Theorem). *Let X and Y be Banach spaces, a be a point of X, and N be a neighborhood of a. Suppose that f is a continuously differentiable function from N into Y such that df_a has a bounded inverse and $f(a) = b$. Then, there is an open ball $B(b; \rho)$ and a unique continuously differentiable function g from $B(b; \rho)$ into N such that $g(b) = a$ and $f(g(y)) = y$ for all $y \in B(b; \rho)$.*

PROOF. Let F be the function from $N \times Y$ into Y such that

$$F(x, y) = f(x) - y;$$

that is, if P_1 and P_2 are the projection functions from $X \times Y$ into X and Y, respectively, then

$$F = f \circ P_1 - P_2.$$

Clearly, $F(a, b) = 0$. Since for $(x, y) \in N \times Y$

$$dF_{(x,y)} = df_x \circ P_1 - P_2,$$

F is continuously differentiable on $N \times Y$. Also, since $d_1 F_{(a,b)} = df_a$, then $d_1 F_{(a,b)}$ has a bounded inverse. Thus, F satisfies the conditions of the Implicit Function Theorem and, hence, there is an open ball $B(b; \rho)$ and a unique continuously differentiable function g from $B(b; \rho)$ into N such that $g(b) = a$ and $F(g(y), y) = 0$ for all $y \in B(b; \rho)$. Since $F(g(y), y) = 0$ implies $f(g(y)) = y$, the proof is complete. ∎

PROBLEMS

1. If f is a function from X into Y which is differentiable on a convex set $C \subset X$ and $df_x = 0$ for all $x \in C$, show that f is constant on C.

2. Suppose that the conditions for the Implicit Function Theorem are satisfied for the system

$$f^1(x_1, x_2, y_1, y_2, y_3) = 0$$
$$f^2(x_1, x_2, y_1, y_2, y_3) = 0$$

and that $x_1 = g^1(y_1, y_2, y_3)$ and $x_2 = g^2(y_1, y_2, y_3)$. If $h^i(y_1, y_2, y_3) = f^i(g^1(y_1, y_2, y_3), g^2(y_1, y_2, y_3), y_1, y_2, y_3)$, $i = 1, 2$, write out $D_1 h^i$ and determine $D_1 g^i$, $i = 1, 2$.

3. Show that the equations

$$xyu - yv^2 + 2x^3 = 0, \qquad 4u^2 + 2v^2 - x^3 y = 0$$

can be solved for u and v in terms of x and y near the point $(u, v, x, y) = (0, 1, 1, 2)$ and determine $\partial u/\partial x$, $\partial u/\partial y$, $\partial v/\partial x$, and $\partial v/\partial y$.

4. Show that the equations

$$x^2u^2 + xzv + y^2 = 0, \qquad yzu + xyv^2 - 3x = 0$$

can be solved for u and v in terms of x, y, and z near $(u, v, x, y, z) = (0, 1, 3, 3, -3)$ and determine $\partial u/\partial x$, $\partial u/\partial y$, $\partial u/\partial z$, $\partial v/\partial x$, $\partial v/\partial y$, $\partial v/\partial z$.

5. If $f(x, y) = (e^x \cos y, e^x \sin y)$, show that f has a local inverse around any point in $\mathbf{R}^2$.

9 | *The Fundamental Theorems of Calculus*

1 · INTRODUCTION

In this chapter we will limit our attention to real-valued functions of a real variable. For such functions there are relationships between the derivative and the Lebesgue integral which are analogous to the Fundamental Theorems of Calculus which relate the derivative and Riemann integral. Recall that the First Fundamental Theorem of Calculus states that if f is continuous on an interval I, $a \in I$, and $F(x) = \int_a^x f$ for $x \in I$, then $F'(x) = f(x)$. The Second Fundamental Theorem states that if f' is continuous on an interval I and $[a, b] \subset I$, then $\int_a^b f' = f(b) - f(a)$. We will obtain the analogous theorems for the Lebesgue integral in this chapter.

In the course of obtaining these Fundamental Theorems for the Lebesgue integral a number of concepts are introduced which lead us in a natural way into a discussion of the Stieltjes integral which we consider in Chapter 10.

2 · SEMICONTINUITY

Let f be a real-valued function defined on a subset E of a metric space X. Then f is continuous at a point $a \in E$ if for each $\varepsilon > 0$ there exists a $\delta > 0$ such that

$$f(a) - \varepsilon < f(x) < f(a) + \varepsilon \quad \text{for all } x \in B(a; \delta) \cap E$$

where $B(a; \delta)$ is the open ball in X with center a and radius δ. The function f will be called **semicontinuous** at a if one of the above inequalities holds.

Suppose for each $\varepsilon > 0$ there exists a $\delta > 0$ such that

$$(2 \cdot 1) \qquad f(x) < f(a) + \varepsilon \quad \text{for all } x \in B(a; \delta) \cap E.$$

Then f is said to be **upper semicontinuous** at a. Similarly, f is said to be **lower semicontinuous** at a if for each $\varepsilon > 0$ there exists a $\delta > 0$ such that

$$f(a) - \varepsilon < f(x) \quad \text{for all } x \in B(a; \delta) \cap E.$$

Inequality $2 \cdot 1$ implies that

$$\sup\{f(x) : x \in B(a; \delta) \cap E\} \le f(a) + \varepsilon$$

and, hence,

$$\inf_{\delta > 0} \sup\{f(x) : x \in B(a; \delta) \cap E\} \le f(a).$$

In fact, since $a \in B(a; \delta) \cap E$ for all $\delta > 0$, 2 · 1 implies that

$$f(a) = \inf_{\delta>0} \sup\{f(x): x \in B(a; \delta) \cap E\}.$$

Since $\sup\{f(x) : x \in B(a; \delta) \cap E\}$ is nonincreasing as δ decreases toward 0,

$$\inf_{\delta>0} \sup\{f(x): x \in B(a; \delta) \cap E\} = \lim_{\delta\to 0^+} \sup\{f(x): x \in B(a; \delta) \cap E\}.$$

This quantity will be called the limit superior of f at a and will be denoted by $\overline{\mathrm{L}}f(a)$. Thus, the function f from E into $\mathbf{R}$ will be upper semicontinuous at $a \in E$ if

$$f(a) = \overline{\mathrm{L}}f(a).$$

We now give some general definitions that are suggested by the above discussion. For any function f from E into $\mathbf{R}$ (not necessarily upper semicontinuous) and any $a \in E$, $\inf_{\delta>0} \sup\{f(x) : x \in B(a; \delta) \cap E\}$ exists in the extended real number system $\overline{\mathbf{R}}$. As a matter of fact, this quantity exists in $\overline{\mathbf{R}}$ if f is a function from E into $\overline{\mathbf{R}}$ and $a \in \bar{E}$.

2 · 2 Definition. *Let f be a function from the subset E of the metric space X into $\overline{\mathbf{R}}$ and let $a \in \bar{E}$. Then the* ***nondeleted limit superior*** *of f at a, denoted by $\overline{\mathrm{L}}f(a)$ or $\overline{\mathrm{Lim}}_{x\to a} f(x)$, is defined by*

$$\overline{\mathrm{L}}f(a) = \inf_{\delta>0} \sup\{f(x): x \in B(a; \delta) \cap E\}.$$

Note that Definition 2 · 2 defines a function $\overline{\mathrm{L}}f$ from $\bar{E}$ into $\overline{\mathbf{R}}$. The qualifying "nondeleted" refers to the fact that we are using nondeleted neighborhoods $B(a; \delta)$ of a. The **deleted limit superior** of f at a, denoted by $\overline{\lim}_{x\to a} f(x)$, has the same definition except the point a is deleted from $B(a; \delta)$. Similarly, the **nondeleted limit inferior** of f at $a \in \bar{E}$, denoted by $\underline{\mathrm{L}}f(a)$ or $\underline{\mathrm{Lim}}_{x\to a} f(x)$, is the extended real number

$$\underline{\mathrm{L}}f(a) = \sup_{\delta>0} \inf\{f(x): x \in B(a; \delta) \cap E\}.$$

For $a \in E$, since $\inf\{f(x) : x \in B(a; \delta) \cap E\} \le f(a) \le \sup\{f(x) : x \in B(a; \delta) \cap E\}$ for any $\delta > 0$, it is clear that

$$\underline{\mathrm{L}}f(a) \le f(a) \le \overline{\mathrm{L}}f(a). \tag{2 · 3}$$

2 · 4 Definition. *The function f from E into $\overline{\mathbf{R}}$ is* ***upper semicontinuous*** *at a point $a \in E$ if and only if*

$$f(a) = \overline{\mathrm{L}}f(a).$$

Similarly, f is ***lower semicontinuous*** *at a point $a \in E$ if and only if*

$$\underline{\mathrm{L}}f(a) = f(a).$$

If $f(a) = \infty$, then equality must hold on the right side of $2 \cdot 3$ and f is upper semicontinuous at a. Similarly, if $f(a) = -\infty$, then f is lower semicontinuous at a. Since $\underline{L}f(a) = -\overline{L}(-f)(a)$, f is lower semicontinuous at a if and only if $-f$ is upper semicontinuous at a. Also, f is continuous at a point $a \in E$ if and only if it is both upper and lower semicontinuous at a (Problem 1).

We now show that $\overline{L}f$ is upper semicontinuous. As usual, we say that a function is upper semicontinuous if it is upper semicontinuous at all points in its domain.

2 · 5 Proposition. *If f maps E into $\overline{\mathbf{R}}$, then the function $\overline{L}f$ from $\overline{E}$ into $\overline{\mathbf{R}}$ is upper semicontinuous.*

Proof. For $a \in \overline{E}$ let $g(a) = \overline{L}f(a)$. If $g(a) = \infty$, then $g = \overline{L}f$ is upper semicontinuous at a.

Assume now that $g(a) \in \mathbf{R}$ and take $\varepsilon > 0$. Since $g(a) = \overline{L}f(a) < g(a) + \varepsilon$, there exists a $\delta_1 > 0$ such that

$$\sup\{f(x) : x \in B(a; \delta_1) \cap E\} < g(a) + \varepsilon.$$

For each $y \in B(a; \delta_1) \cap \overline{E}$ there exists an open ball $B(y; \eta_1) \subset B(a; \delta_1)$ and, hence,

$$g(y) = \inf_{\eta > 0} \sup\{f(x) : x \in B(y; \eta) \cap E\} < g(a) + \varepsilon.$$

Therefore,

$$\sup\{g(y) : y \in B(a; \delta_1) \cap \overline{E}\} \leq g(a) + \varepsilon$$

and this implies that

$$\overline{L}g(a) \leq g(a) + \varepsilon.$$

Since the above inequality holds for all $\varepsilon > 0$, $\overline{L}g(a) \leq g(a)$ and, therefore, $\overline{L}g(a) = g(a)$. ∎

Let (x_n) be a sequence of points in E which converges to a point $a \in E$. We know that if f is continuous at a, then the sequence $(f(x_n))$ converges to $f(a)$. If f is only upper semicontinuous at a, the sequence $(f(x_n))$ may fail to converge and the best we can say is that $\overline{\lim}\, f(x_n) \leq f(a)$ where $\overline{\lim}\, f(x_n)$ is the upper limit of the sequence $(f(x_n))$. This can be inferred from the following result.

2 · 6 Proposition. *If f maps E into $\mathbf{R}$ and (x_n) is a sequence of points in E which converges to $a \in \overline{E}$, then*

$$\overline{\lim}\, f(x_n) \leq \overline{L}f(a).$$

Proof. Since $\lim x_n = a$, for each $\delta > 0$ there exists an n_0 such that $x_n \in B(a; \delta)$ for all $n \geq n_0$ and, hence,

$$f(x_n) \leq \sup\{f(x) : x \in B(a; \delta) \cap E\} \quad \text{for all } n \geq n_0.$$

Then, for each $\delta > 0$

$$\overline{\lim} f(x_n) \leq \sup\{f(x) : x \in B(a; \delta) \cap E\}$$

and, therefore,

$$\overline{\lim} f(x_n) \leq \inf_{\delta>0} \sup\{f(x) : x \in B(a; \delta) \cap E\} = \overline{L}f(a). \blacksquare$$

Another important property of real-valued continuous functions is that they have a maximum and minimum value on a compact set. We now state the corresponding result for upper semicontinuous functions.

2 · 7 Theorem. *If f is an upper semicontinuous function on the compact subset E of the metric space X, then f has a maximum value on E.*

PROOF. If $b = \sup\{f(x) : x \in E\}$, then there is a sequence (x_n) of points in E such that $\lim f(x_n) = b$. Since E is compact, (x_n) has a subsequence (x_{n_k}) which converges in E. Let $\lim x_{n_k} = a$. Then, by Proposition 2 · 6,

$$b = \lim f(x_{n_k}) \leq \overline{L}f(a) = f(a) \leq b$$

and, therefore,

$$b = f(a). \blacksquare$$

Consider now a function g which maps a finite open interval (a, b) of the real line into $\mathbf{R}$. For such functions we give some technical results which will be needed in the next section.

2 · 8 Proposition. *Let g be a function from (a, b) into $\mathbf{R}$, $G = \overline{L}g$, and assume that the set*

$$E = \{x \in (a, b) : \text{for some } y \in (x, b),\ G(x) < g(y)\}$$

is nonempty. Then E is open and, hence, is the union of a countable number of disjoint open intervals (a_k, b_k). Also, $g(x) \leq G(b_k)$ for all $x \in (a_k, b_k)$.

PROOF. Take $x_0 \in E$. Then there exists a $y_0 \in (x_0, b)$ such that $G(x_0) < g(y_0)$. Since G is upper semicontinuous, $\overline{L}G(x_0) = G(x_0) < g(y_0)$. This implies that there exists an open interval $(x_0 - \delta, x_0 + \delta) \subset (a, y_0)$ such that $G(x) < g(y_0)$ for all $x \in (x_0 - \delta, x_0 + \delta)$. Thus, $(x_0 - \delta, x_0 + \delta) \subset E$ and, therefore, E is open. Then, E is the union of a countable disjoint collection of open intervals (a_k, b_k).

Now take $x \in (a_k, b_k)$. Let x_1 be a point in $[x, b]$ where G assumes its maximum value on $[x, b]$. We note that $x_1 \notin E$; for, if $x_1 \in E$, then there is a point $y_1 \in (x_1, b)$ such that $G(x_1) < g(y_1) \leq G(y_1)$ which is impossible. Then, since $[x, b_k) \subset E$, we conclude that $x_1 \in [b_k, b]$. Suppose now that

$$G(b_k) < G(x_1) = \inf_{\delta>0} \sup\{g(y) : y \in (x_1 - \delta, x_1 + \delta) \cap (a, b)\}.$$

Then, $b_k < x_1$ and there exists a $y \in (b_k, b)$ such that $G(b_k) < g(y)$. This implies that $b_k \in E$ but such is not the case. Therefore, $G(b_k) = G(x_1)$ and

$$g(x) \leq G(x) \leq G(x_1) = G(b_k). \quad \blacksquare$$

2·9 Proposition. *Let g be a function from (a, b) into $\mathbf{R}$, $G = \underline{L}g$, and assume that the set*

$$E = \{x \in (a, b) : \textit{for some } y \in (a, x),\ G(x) < g(y)\}$$

is nonempty. Then E is open and, hence, is the union of a countable number of disjoint open intervals (a_k, b_k). Also, $g(x) \leq G(a_k)$ for all $x \in (a_k, b_k)$.

The proof of this proposition is left to the reader (Problem 10). The proof can be obtained by simple modifications of the proof of Proposition 2·8 or by applying Proposition 2·8 to the function h on $(-b, -a)$ where $h(x) = g(-x)$.

For functions f from an arbitrary open interval I on the real line into $\mathbf{R}$ it is often useful to consider left and right hand limits. Recall that if $c \in I$, then $\lim_{x \to c^+} f(x) = d$ if for each $\varepsilon > 0$ there exists a $\delta > 0$ such that $|f(x) - d| < \varepsilon$ whenever $x \in (c, c + \delta) \cap I$. This right hand limit is also denoted by $f(c+)$. The left hand limit, $\lim_{x \to c^-} f(x)$ or $f(c-)$, is defined similarly. Also, for functions from $\mathbf{R}$ into $\mathbf{R}$, we can introduce (deleted) left and right hand limits inferior and superior. For example, the (deleted) left hand limit superior of f at c is:

$$\overline{\lim_{x \to c^-}} f(x) = \inf_{\delta > 0} \sup\{f(x) : x \in (c - \delta, c) \cap I\}.$$

The reader should have no difficulty supplying the definitions of the other limits (Problem 9).

PROBLEMS

1. If f maps the subset E of the metric space X into $\mathbf{R}$ and $a \in E$, show that f is continuous at a if and only if it is both upper and lower semicontinuous at a.

2. Show that the greatest integer function is upper semicontinuous.

3. If f is the function from $\mathbf{R}$ into $\mathbf{R}$ such that $f(x) = 1$ for x rational and $f(x) = 0$ for x irrational, show that f is upper semicontinuous at the rationals and lower semicontinuous at the irrationals.

4. Show that

$$\overline{\lim_{x \to a}} (f + g)(x) \leq \overline{\lim_{x \to a}} f(x) + \overline{\lim_{x \to a}} g(x).$$

5. If f and g are upper semicontinuous at a, show that $f + g$ is upper semicontinuous at a. Also if $c > 0$, then cf is upper semicontinuous at a.

6. If f and g are upper semicontinuous at a, show that $f \wedge g$ is upper semicontinuous at a.

7. If f is a mapping of the metric space X into $\mathbf{R}$ show that f is upper semicontinuous if and only if, for each $c \in \mathbf{R}$, $f^{-1}(-\infty, c)$ is open.

8. Let f be a function from the subset E of the metric space X into $\mathbf{R}$ and let $a \in \bar{E}$. Show that

a. $\underline{\lim}_{x \to a} f(x) \leq \overline{\lim}_{x \to a} f(x)$
b. f has a limit at a if and only if $\underline{\lim}_{x \to a} f(x)$ and $\overline{\lim}_{x \to a} f(x)$ are finite and equal.

9. Let f be a function from an open interval $I \subset \mathbf{R}$ into $\mathbf{R}$ and let $c \in I$.
a. Define $\overline{\lim}_{x \to c^+} f(x)$, $\underline{\lim}_{x \to c^-} f(x)$, and $\underline{\lim}_{x \to c^+} f(x)$.
b. Show that f has a limit at c if and only if $\overline{\lim}_{x \to c^-} f(x)$, $\overline{\lim}_{x \to c^+} f(x)$, $\underline{\lim}_{x \to c^-} f(x)$, and $\underline{\lim}_{x \to c^+} f(x)$ are all finite and equal.

10. Prove Proposition 2 · 9.

3 · DIFFERENTIABILITY OF A MONOTONIC FUNCTION

Let f be a nondecreasing function from an open interval I of $\mathbf{R}$ into $\mathbf{R}$; that is, $x < y$ in I implies $f(x) \leq f(y)$. It is easy to see that a nondecreasing function has a left and right hand limit at every point of I. It follows from this that the set of discontinuities of f is countable and, hence, of measure zero (Problem 1).

Our objective in this section is to prove the following more difficult result:

3 · 1 Theorem. *A nondecreasing function f on an open I is differentiable almost everywhere.*

Lebesgue established a result of this sort in 1904 in the first edition of his book on integration [26] using the additional hypothesis of continuity. It appeared at the end of the book as the final result of the whole theory. However, the concept of the integral does not appear in the statement of the theorem and is not needed in the proof.

We observe that we can make some simplifying assumptions. If $I = (a, b)$ where a and b are extended real numbers, take a sequence of finite open intervals (a_n, b_n) where (a_n) is a decreasing sequence in (a, b) with $\lim a_n = a$ and (b_n) is an increasing sequence in (a, b) with $\lim b_n = b$. If $x \in (a_n, b_n)$, then $f(a_n) \leq f(x) \leq f(b_n)$. Thus, f is bounded on (a_n, b_n). Since Theorem 3 · 1 will hold on I if it holds on each (a_n, b_n), it is sufficient to prove the following:

3 · 2 Theorem. *A bounded nondecreasing function f on a finite open interval (a, b) is differentiable almost everywhere.*

In the proof of this theorem we use **Dini Derivatives** which are defined as follows:

upper right derivative $$D^+f(x) = \overline{\lim_{h\to 0^+}} \frac{f(x+h)-f(x)}{h}$$

lower right derivative $$D_+f(x) = \underline{\lim_{h\to 0^+}} \frac{f(x+h)-f(x)}{h}$$

upper left derivative $$D^-f(x) = \overline{\lim_{h\to 0^-}} \frac{f(x+h)-f(x)}{h}$$

lower left derivative $$D_-f(x) = \underline{\lim_{h\to 0^-}} \frac{f(x+h)-f(x)}{h}$$

From these definitions it is clear that for a function f satisfying the condition of Theorem 3 · 2 all of the Dini derivatives of f at a point $x \in (a, b)$ exist in the extended real number system and that

$$(3 \cdot 3) \qquad 0 \le D_+f(x) \le D^+f(x) \quad \text{and} \quad 0 \le D_-f(x) \le D^-f(x).$$

Also, f has a derivative at x if and only if all of the Dini derivatives are finite and equal.

The proof of Theorem 3 · 2 is accomplished by showing that at almost all points in (a, b) the Dini derivatives are all finite and equal. Since this proof is rather lengthy we split it into a number of lemmas.

3 · 4 Lemma. *If f is a bounded nondecreasing function on (a, b), then $D^+f(x) < \infty$ for a.a. $x \in (a, b)$.*

Proof. Let $A = \{x \in (a, b) : D^+f(x) = \infty\}$. Then, for each positive real number r,

$$A \subset A_r = \{x \in (a, b) : D^+f(x) > r\}.$$

Observe that if $D^+f(x) > r$, then, from some $y \in (x, b)$,

$$\frac{f(y) - f(x)}{y - x} > r$$

or, equivalently,

$$f(x) - rx < f(y) - ry.$$

Let $g(x) = f(x) - rx$. We then have

$$A \subset A_r \subset B_r = \{x \in (a, b) : \text{for some } y \in (x, b),\ g(x) < g(y)\}.$$

Now let $G = \overline{L}g$. Since f and, hence, g is continuous on (a, b) except for a set S of measure zero, then $G = g$ on $(a, b) \sim S$ and, therefore,

$$(A \sim S) \subset E_r = \{x \in (a, b) : \text{for some } y \in (x, b),\ G(x) < g(y)\}.$$

If E_r is empty, then $A \sim S$ is empty and, hence, A is of measure zero. Assume E_r is nonempty. Then, we know from Proposition 2 · 8 that E_r is the union of a countable disjoint collection of open intervals (a_k, b_k) and

$$g(x) \leq G(b_k) \quad \text{for all } x \in (a_k, b_k).$$

Then, for each k, $g(a_k+) \leq G(b_k)$ or, equivalently,

$$f(a_k+) - ra_k \leq G(b_k).$$

Take $x \in (a, b)$. For $y \in (x - \delta, x] \subset (a, b)$, $g(y) \leq f(x) - r(x - \delta)$ and, for $y \in (x, x + \delta) \subset (a, b)$, $g(y) \leq f(x + \delta) - rx$. Then

$$G(x) \leq f(x+) - rx.$$

Thus, if $b_k \neq b$, then

$$f(a_k+) - ra_k \leq G(b_k) \leq f(b_k+) - rb_k.$$

Also, for $y \in (b - \delta, b) \subset (a, b)$, $g(y) \leq f(b-) - r(b - \delta)$ and, therefore,

$$G(b) \leq f(b-) - rb.$$

Thus,

$$r(b_k - a_k) \leq f(b_k+) - f(a_k+) \quad \text{if } b_k \neq b$$
$$r(b_k - a_k) \leq f(b-) - f(a_k+) \quad \text{if } b_k = b$$

and, hence,

$$\sum_k (b_k - a_k) \leq \frac{1}{r}(f(b-) - f(a+)).$$

This shows that $A \sim S$ is contained in the union of a countable number of open intervals the sum of whose lengths is arbitrarily small; that is, $A \sim S$ and, hence, A is of measure zero. ▮

3 · 5 Lemma. *If f is a bounded nondecreasing function on (a, b), then $D^+f(x) \leq D_-f(x)$ for a.a. $x \in (a, b)$.*

Proof. Since many details of this proof are similar to those in the proof of the previous lemma, they will be stated rather concisely. Let

$$A = \{x \in (a, b) : D_-f(x) < D^+f(x)\},$$

let $\langle s, r \rangle$ denote a pair of positive rational numbers with $s < r$, and let

$$A_{sr} = \{x \in (a, b) : D_-f(x) < s < r < D^+f(x)\}.$$

It is clear that if $x \in A$, then $x \in A_{sr}$ for some $\langle s, r \rangle$ and, hence, $A \subset \bigcup_{\langle s, r \rangle} A_{sr}$. Since this is a countable union, if A_{sr} is of measure zero for all $\langle s, r \rangle$, then A is of measure zero.

Now take a fixed pair $\langle s, r \rangle$ and let

$$A_s = \{x \in (a, b) : D_-f(x) < s\} \quad \text{and} \quad A_r = \{x \in (a, b) : D^+f(x) > r\}.$$

Then $A_{sr} = A_s \cap A_r$. If $D_- f(x) < s$, then, for some $y \in (a, x)$,

$$f(y) - sy > f(x) - sx.$$

Letting $g_s(x) = f(x) - sx$, we then have

$$A_s \subset B_s = \{x \in (a, b) : \text{for some } y \in (a, x), g_s(x) < g_s(y)\}.$$

Also, if $G_s = \overline{\mathrm{L}} g_s$ and S is the set of measure zero where f is not continuous, then

$$(A_s \sim S) \subset E_s = \{x \in (a, b) : \text{for some } y \in (a, x), G_s(x) < g_s(y)\}.$$

By Proposition 2 · 9, E_s is the union of a countable disjoint collection of open intervals (a_k, b_k) and

$$g_s(x) \leq G_s(a_k) \text{ for all } x \in (a_k, b_k) .$$

From this we can infer that, for all k,

$$f(b_k-) - sb_k = g_s(b_k-) \leq G_s(a_k) \leq f(a_k+) - sa_k$$

and, therefore,

$$f(b_k-) - f(a_k+) \leq s(b_k - a_k).$$

Let $g_r(x) = f(x) - rx$ and $G_r = \overline{\mathrm{L}} g_r$. Then

$$A_{sr} \sim S \subset \bigcup_k (a_k, b_k) \cap A_r \subset \bigcup_k E_r^k$$

where $E_r^k = \{x \in (a_k, b_k)$: for some $y \in (x, b_k)$, $G_r(x) < g_r(y)\}$. By Proposition 2 · 8, for each k, E_r^k is the union of a countable disjoint collection of open intervals (a_j^k, b_j^k) and

$$g_r(x) \leq G_r(b_j^k) \quad \text{for all } x \in (a_j^k, b_j^k).$$

Then

$$\begin{aligned} r(b_j^k - a_j^k) &\leq f(b_j^k+) - f(a_j^k+) \quad \text{if } b_j^k \neq b_k \\ r(b_j^k - a_j^k) &\leq f(b_k-) - f(a_j^k+) \quad \text{if } b_j^k = b_k \end{aligned}$$

and, therefore,

$$\sum_j (b_j^k - a_j^k) \leq \frac{1}{r}(f(b_k-) - f(a_k+)) \leq \frac{s}{r}(b_k - a_k).$$

Summing over k, we obtain

$$\sum_{k, j} (b_j^k - a_j^k) \leq \frac{s}{r}(b - a).$$

Now let the countable set $\{(a_j^k, b_j^k)\}_{k, j}$ be denoted by $\{I_i^1\}$ and let $(a, b) = I$. Then we have shown that

$$A_{sr} \sim S \subset \bigcup_i I_i^1 \quad \text{and} \quad \sum_i \mu(I_i^1) \leq \frac{s}{r}\mu(I).$$

If for each i we apply to $(A_{sr} \sim S) \cap I_i^1$ the process in the above two paragraphs, we will obtain a countable disjoint collection $\{J_{ij}^1\}$ of open intervals such that

$$(A_{sr} \sim S) \cap I_i^1 \subset \bigcup_j J_{ij}^1 \quad \text{and} \quad \sum_j \mu(J_{ij}^1) \leq \frac{s}{r} \mu(I_i^1).$$

Then

$$A_{sr} \sim S \subset \bigcup_{i,j} J_{ij}^1 \quad \text{and} \quad \sum_{i,j} \mu(J_{ij}^1) \leq \frac{s}{r} \sum_i \mu(I_i^1) \leq \left(\frac{s}{r}\right)^2 \mu(I).$$

Letting the countable set $\{J_{ij}^1\}$ be denoted by $\{I_i^2\}$, we can rewrite this as:

$$A_{sr} \sim S \subset \bigcup_i I_i^2 \quad \text{and} \quad \sum_i \mu(I_i^2) \leq \left(\frac{s}{r}\right)^2 \mu(I).$$

Continuing this process we will have, for any positive integer n,

$$A_{sr} \sim S \subset \bigcup_i I_i^n \quad \text{and} \quad \sum_i \mu(I_i^n) \leq \left(\frac{s}{r}\right)^n \mu(I).$$

Since $\lim (s/r)^n = 0$, this shows that $A_{sr} \sim S$ and, hence, A_{sr} is of measure zero. ▮

3 · 6 Lemma. *If f is a bounded nondecreasing function on (a, b), then $D^- f(x) \leq D_+ f(x)$ for a.a. $x \in (a, b)$.*

PROOF. If $g(x) = -f(-x)$ for $x \in (-b, -a)$, then g is nondecreasing on $(-b, -a)$ and, hence, by Lemma 3 · 5

$$D^+ g(-x) \leq D_- g(-x) \quad \text{for a.a.} -x \in (-b, -a).$$

Observing that

$$D^+ g(-x) = \overline{\lim_{h \to 0^+}} \frac{g(-x+h) - g(-x)}{h} = \overline{\lim_{h \to 0^+}} \frac{-f(x-h) + f(x)}{h}$$

$$= \overline{\lim_{h \to 0^+}} \frac{f(x-h) - f(x)}{-h} = D^- f(x)$$

and

$$D_- g(-x) = \underline{\lim_{h \to 0^-}} \frac{g(-x+h) - g(-x)}{h} = \underline{\lim_{h \to 0^-}} \frac{f(x-h) - f(x)}{-h} = D_+ f(x),$$

we see that

$$D^- f(x) \leq D_+ f(x) \quad \text{for a.a.} x \in (a, b). ▮$$

PROOF OF THEOREM 3 · 2. Using 3 · 3 and Lemmas 3 · 4, 3 · 5, and 3 · 6, we have for a.a. $x \in (a, b)$

$$0 \leq D^+ f(x) \leq D_- f(x) \leq D^- f(x) \leq D_+ f(x) \leq D^+ f(x) < \infty.$$

That is, for a.a.$x \in (a, b)$, the derivatives of f at x are all finite and equal and, hence, the derivative of f exists at x. ▌

It is clear that in Theorem 3 · 1 the restriction that the interval I be open can be dropped. Also, Theorem 3 · 1 still holds if "nondecreasing" is replaced by "nonincreasing." For if f is a nonincreasing function on I, then $-f$ is a nondecreasing function on I. A function which is either a nondecreasing function or a nonincreasing function is called a **monotone** function. Thus, we have:

3 · 7 Theorem. *A monotone function on an interval I is differentiable almost everywhere on I.*

Without the condition of monotonicity it is no longer true that a function defined on an interval I is differentiable almost everywhere on I. For a long time it was thought that continuity should be a sufficient condition for differentiability almost everywhere. However, Weierstrass found an example, which was first published by one of his pupils P. du Bois-Reymond in 1875 [11], of a function which was everywhere continuous but nowhere differentiable. Since then many more examples have been found. The following elementary example is due to the Dutch mathematician B. L. van der Waerden (1903–) [41]. It is based on the obvious fact that a sequence of integers can converge only if all terms are equal from some point on.

Let $\{x\}$ denote the distance from x to the nearest integer and define

$$f(x) = \sum_{n=0}^{\infty} \frac{\{10^n x\}}{10^n}.$$

Since

$$0 \leq \frac{\{10^n x\}}{10^n} \leq \frac{1}{2} \cdot \frac{1}{10^n}$$

and $\frac{1}{2} \sum 10^{-n}$ converges, the series defining f converges uniformly on $\mathbf{R}$. Moreover, each term in the series is continuous on $\mathbf{R}$. Therefore, f is continuous on $\mathbf{R}$.

Since f has period 1, $f(x + 1) = f(x)$ for all $x \in \mathbf{R}$, it is sufficient to consider $0 \leq x < 1$. Let the decimal expansion of x be

$$x = 0.a_1 a_2 \cdots a_n \cdots$$

where we agree not to use a representation ending in an infinite sequence of nines. Then

$$\{10^n x\} = 0.a_{n+1} a_{n+2} \cdots \quad \text{if } 0.a_{n+1} a_{n+2} \cdots \leq \tfrac{1}{2}$$

while

$$\{10^n x\} = 1 - 0.a_{n+1} a_{n+2} \cdots \quad \text{if } 0.a_{n+1} a_{n+2} \cdots > \tfrac{1}{2}.$$

Let $h_m = -10^{-m}$ if a_m equals 4 or 9 and $h_m = 10^{-m}$ otherwise. Then $x + h_m = 0.a_1a_2 \cdots a'_m \cdots$ where $a'_m = a_m - 1$ if $a_m = 4$ or 9 and $a'_m = a_m + 1$ if $a_m \neq 4$ or 9; i.e., $a'_m - a_m = 10^m h_m$. The expression $\{10^n(x + h_m)\} - \{10^n x\}$ is zero for $n \geq m$ while for $n < m$ we have $(a'_m - a_m)10^{n-m} = 10^n h_m$ if $a_{n+1} \leq 4$ and $-(a'_m - a_m)10^{n-m} = -10^n h_m$ if $a_{n+1} \geq 5$. Thus

$$\frac{f(x+h_m) - f(x)}{h_m} = \sum_{n=0}^{\infty} \frac{\{10^n(x+h_m)\} - \{10^n x\}}{10^n h_m}$$

$$= \sum_{n=0}^{m-1} \frac{\pm 10^n h_m}{10^n h_m} = \sum_{n=0}^{m-1} \pm 1.$$

For m even this sum is an even integer and for m odd it is an odd integer. Hence,

$$\lim_{m \to \infty} \frac{f(x+h_m) - f(x)}{h_m}$$

does not exist and f is not differentiable at x.

PROBLEMS

1. If f is a nondecreasing function on the open interval I, show that

a. for any $c \in I$, $f(c-)$ and $f(c+)$ exist.

b. for $c < d$ in I

$$f(c-) \leq f(c) \leq f(c+) \leq f(d-) \leq f(d) \leq f(d+).$$

c. the set of discontinuities of f is countable.

2. If f is a function from an open interval I of **R** into **R** such that $f(c-)$ and $f(c+)$ exist for all $c \in I$, show that f has a countable number of discontinuities.

Suggestion. Take $[a, b] \subset I$. Let $g(x) = \max\{|f(x) - f(x-)|, |f(x) - f(x+)|\}$ and let $A_n = \{x \in [a, b] : g(x) \geq 1/n\}$. Suppose A_n is infinite and let (x_k) be a sequence of distinct points in A_n which converges to a point $x \in [a, b]$. Take $\delta > 0$ such that $|f(y) - f(z)| < 1/(2n)$ whenever $y, z \in (x - \delta, x)$ or $y, z \in (x, x + \delta)$. Then, obtain a contradiction by considering a point $x_k \in A_n \cap ((x - \delta, x) \cup (x, x + \delta))$.

4 · FUNCTIONS OF BOUNDED VARIATION

If f_1 and f_2 are nondecreasing functions on a closed interval $[a, b]$, then it is clear that $f_1 + f_2$ is also a nondecreasing function but $f_1 - f_2$ need not be. The function $f_1 - f_2$ will be in the wider class of functions of bounded variation which we now define

Let f be a real-valued function defined on $[a, b]$, let $P = \{(x_{i-1}, x_i) : i = 1, \cdots, k\}$ be a partition of $[a, b]$, and let $\mathscr{P}$ be the set of all partitions of $[a, b]$. Then, the **(total) variation** $V_a^b(f)$ of f on $[a, b]$ is defined to be:

$$V_a^b(f) = \sup\left\{\sum_{i=1}^{k} |f(x_i) - f(x_{i-1})| : P \in \mathscr{P}\right\}.$$

If $V_a^b(f)$ is finite, then f is said to be of **bounded variation** on $[a, b]$. The set of all functions of bounded variation of $[a, b]$ is denoted by $BV[a, b]$.

It is easy to see that if a function is nondecreasing or nonincreasing on $[a, b]$, then it is of bounded variation on $[a, b]$. For example, if f is nondecreasing on $[a, b]$, then for any partition $P \in \mathscr{P}$ we have

$$\sum_{i=1}^{k} |f(x_i) - f(x_{i-1})| = \sum_{i=1}^{k} (f(x_i) - f(x_{i-1})) = f(b) - f(a)$$

and, hence, $V_a^b(f) = f(b) - f(a)$.

4 · 1 Proposition. *If f_1 and f_2 are nondecreasing functions on $[a, b]$, then $f_1 - f_2 \in BV[a, b]$.*

PROOF. If $f = f_1 - f_2$ and $P \in \mathscr{P}$, then

$$\sum_{i=1}^{k} |f(x_i) - f(x_{i-1})| \le f_1(b) - f_1(a) + f_2(b) - f_2(a)$$

and, hence, $V_a^b(f) \le f_1(b) - f_1(a) + f_2(b) - f_2(a)$. ∎

It is easy to show that $BV[a, b]$ is a linear space over $\mathbf{R}$ (Problem 1) and Proposition 4 · 1 also follows from this since nondecreasing functions belong to $BV[a, b]$. If $P = \{(x_{i-1}, x_i) : i = 1, \cdots, k\}$ is a partition of $[a, b]$, it is often convenient to use the notation:

$$\sum_P |\Delta f| = \sum_{i=1}^{k} |f(x_i) - f(x_{i-1})|.$$

4 · 2 Proposition. *If $f \in BV[a, b]$ and $c \in (a, b)$, then $f \in BV[a, c]$ and $f \in BV[c, b]$. Also,*

$$V_a^b(f) = V_a^c(f) + V_c^b(f).$$

PROOF. If P_1 and P_2 are partitions of $[a, c]$ and $[c, b]$, respectively, then $P = P_1 \cup P_2$ is a partition of $[a, b]$ and

$$\sum_{P_1} |\Delta f| + \sum_{P_2} |\Delta f| = \sum_P |\Delta f| \le V_a^b(f).$$

Therefore,

$$V_a^c(f) + V_c^b(f) \le V_a^b(f).$$

If $P = \{(x_{i-1}, x_i) : i = 1, \cdots, k\}$ is a partition of $[a, b]$, suppose $c \in [x_{j-1}, x_j]$. Then $P_1 = \{(x_{i-1}, x_i) : i = 1, \cdots, j-1\} \cup \{(x_{j-1}, c)\}$ and $P_2 = \{(c, x_j)\} \cup \{(x_{i-1}, x_i) : i = j+1, \cdots, k\}$ are partitions of $[a, c]$ and $[c, b]$, respectively. Also,

$$\begin{aligned}\sum_P |\Delta f| &= \sum_{i=1}^{j-1} |f(x_i) - f(x_{i-1})| + |f(x_j) - f(x_{j-1})| \\ &\quad + \sum_{i=j+1}^{k} |f(x_i) - f(x_{i-1})| \\ &\le \sum_{i=1}^{j-1} |f(x_i) - f(x_{i-1})| + |f(c) - f(x_{j-1})| \\ &\quad + |f(x_j) - f(c)| + \sum_{i=j+1}^{k} |f(x_i) - f(x_{i-1})| \\ &= \sum_{P_1} |\Delta f| + \sum_{P_2} |\Delta f| \\ &\le V_a^c(f) + V_c^b(f).\end{aligned}$$

Therefore, $V_a^b(f) \le V_a^c(f) + V_c^b(f)$ and, hence, $V_a^b(f) = V_a^c(f) + V_c^b(f)$. ∎

From this result we see that if $f \in BV[a, b]$, then we can define the function v_f on $[a, b]$ as follows:

$$v_f(a) = 0 \quad \text{and} \quad v_f(x) = V_a^x(f) \quad \text{for } x \in (a, b].$$

Also, if $x < y$ in $[a, b]$, then

$$v_f(y) = v_f(x) + V_x^y(f). \tag{4 · 3}$$

4 · 4 Theorem. *If $f \in BV[a, b]$, then f is the difference of two nondecreasing functions on $[a, b]$.*

PROOF. On $[a, b]$

$$f = v_f - (v_f - f).$$

If $x < y$ in $[a, b]$, then 4 · 3 implies

$$v_f(y) - v_f(x) = V_x^y(f) \ge 0$$

and, hence, v_f is nondecreasing on $[a, b]$. Also,

$$v_f(y) - v_f(x) = V_x^y(f) \ge |f(y) - f(x)| \ge f(y) - f(x);$$

that is,

$$v_f(y) - f(y) \ge v_f(x) - f(x).$$

Therefore, $v_f - f$ is also nondecreasing on $[a, b]$. ∎

If $f \in BV[a, b]$, we have shown that v_f is a nondecreasing function on $[a, b]$. Also, in the proof of Theorem 4 · 4 we obtained an inequality that shows that f is continuous at $x \in [a, b]$ if v_f is continuous at x. We now show that conversely v_f is continuous at x if f is continuous at x.

4 · 5 Proposition. *If $f \in BV[a, b]$ and f is continuous at $x \in [a, b]$, then v_f is continuous at x.*

PROOF. Take $x \in [a, b)$ and $\varepsilon > 0$. Let $P = \{(x_{k-1}, x_k) : k = 1, \cdots, n\}$ be a partition of $[a, b]$ such that

$$V_a^b(f) - \varepsilon < \sum_P |\Delta f| \leq V_a^b(f).$$

For some $k = 1, \cdots, n$, $x_{k-1} \leq x < x_k$. Take y such that $x < y < x_k$ and let P' be the refinement of P obtained by adding x and y as endpoints of subintervals. Then

$$V_a^b(f) - \varepsilon < \sum_{P'} |\Delta f| \leq V_a^b(f).$$

Since v_f is nondecreasing $V_a^b(f) = \sum_{P'} \Delta v_f$ and, hence,

$$0 \leq \sum_{P'} (\Delta v_f - |\Delta f|) < \varepsilon.$$

Since each term in this sum is nonnegative, we have

$$0 \leq v_f(y) - v_f(x) - |f(y) - f(x)| < \varepsilon.$$

Hence, if $f(x+) = f(x)$, we see that $v_f(x+) = v_f(x)$. Similarly, we can show that if $x \in (a, b]$ and $f(x-) = f(x)$, then $v_f(x-) = v_f(x)$. ∎

As a consequence of Theorem 4 · 4, some of the results obtained for nondecreasing functions can be extended to functions of bounded variation.

4 · 6 Theorem. *If $f \in BV[a, b]$, then f has only a countable number of discontinuities on $[a, b]$ and f is differentiable at almost all points of $[a, b]$.*

Thus, if $f \in BV[a, b]$, then f is bounded and continuous almost everywhere on $[a, b]$. Theorem 5 · 2 (p. 119) implies that f is Riemann and, hence, Lebesgue integrable on $[a, b]$. It is also true that if $f \in BV[a, b]$, then f' is integrable on $[a, b]$. First, we prove this for nondecreasing functions.

4 · 7 Theorem. *If f is a nondecreasing function on $[a, b]$, then $f' \in \mathscr{L}[a, b]$ and*

$$\int_a^b f' \leq f(b) - f(a).$$

PROOF. First extend f to $[a, b + 1]$ by defining $f(x) = f(b)$ if $x \in [b, b + 1]$. Then f is nondecreasing on $[a, b + 1]$ and, hence, is Riemann integrable on $[a, b + 1]$. For each positive integer n, let

$$g_n(x) = n\left(f\left(x + \frac{1}{n}\right) - f(x)\right) \quad \text{for } x \in [a, b].$$

Then $g_n(x) \geq 0$ for all $x \in [a, b]$ and $\lim g_n(x) = f'(x)$ for a.a.$x \in [a, b]$. Also, for each n, g_n is Riemann integrable on $[a, b]$ and

$$\int_a^b g_n = n\left(\int_a^b f\left(x + \frac{1}{n}\right) dx - \int_a^b f(x)\, dx\right) = n\left(\int_{a+1/n}^{b+1/n} f(x)\, dx - \int_a^b f(x)\, dx\right)$$

$$= n\left(\int_b^{b+1/n} f(x)\, dx - \int_a^{a+1/n} f(x)\, dx\right) = f(b) - n\int_a^{a+1/n} f(x)\, dx$$

$$\leq f(b) - f(a).$$

Then, by Fatou's Lemma (p. 129), $f' \in \mathscr{L}[a, b]$ and

$$\int_a^b f' \leq \underline{\lim} \int_a^b g_n \leq f(b) - f(a). \blacksquare$$

4 · 8 Corollary. *If $f \in BV[a, b]$, then $f' \in \mathscr{L}[a, b]$.*

PROOF. If $f \in BV[a, b]$, then there exist nondecreasing functions f_1 and f_2 such that $f = f_1 - f_2$ on $[a, b]$. By Theorem 4 · 7, $f_1', f_2' \in \mathscr{L}[a, b]$ and, hence, $f' = f_1' - f_2' \in \mathscr{L}[a, b]$. ▮

Note that in Theorem 4 · 7 and Corollary 4 · 8, the function f' may be undefined on a set of measure zero in $[a, b]$. However, since the value of a function on a set of measure zero does not affect the integral, we may consider that f' is extended in any way to all of $[a, b]$ and $\int_a^b f'$ will have a unique value. Thus, $\int_a^b f'$ is meaningful.

PROBLEMS

1. Show that $BV[a, b]$ is a linear space over $\mathbf{R}$. Also, if $f, g \in BV[a, b]$, then

$$V_a^b(f + g) \leq V_a^b(f) + V_a^b(g) \quad \text{and} \quad V_a^b(\alpha f) = |\alpha| V_a^b(f).$$

2. If f is continuous on $[a, b]$ and there exists an M such that $|f'(x)| \leq M$ for all $x \in (a, b)$, show that $f \in BV[a, b]$.

3. If $f \in BV[a, b]$, show that f is bounded on $[a, b]$.

4. If $f, g \in BV[a, b]$, show that $fg \in BV[a, b]$.

5. If $f \in BV[a, b]$ and there exists an m such that $0 < m \leq |f(x)|$ for all $x \in [a, b]$, show that $1/f \in BV[a, b]$ and $V_a^b(1/f) \leq (1/m^2) V_a^b(f)$.

6. If f is a step function on $[a, b]$, show that $f \in BV[a, b]$.

7. Let f be defined on $[0, 1]$ as follows: $f(0) = 0$ and $f(x) = x \sin(1/x)$ for $x \in (0, 1]$. Is f continuous on $[0, 1]$? Is f of bounded variation on $[0, 1]$?

8. Show that $BV[a, b]$ is a normed linear space if we define

$$\|f\| = |f(a)| + V_a^b(f).$$

9. If $f \in BV[a, b]$, define the positive variation $P_a^b(f)$ and negative variation $N_a^b(f)$ as follows:

$$P_a^b(f) = \sup\left\{\sum_P \max\{\Delta f, 0\} : P \in \mathscr{P}\right\}$$

$$N_a^b(f) = \sup\left\{-\sum_P \min\{\Delta f, 0\} : P \in \mathscr{P}\right\}$$

where $\mathscr{P}$ is the set of all partitions of $[a, b]$.

a. Show that

$$P_a^b(f) - N_a^b(f) = f(b) - f(a)$$

$$P_a^b(f) + N_a^b(f) = V_a^b(f).$$

b. Let $p_f(a) = 0$ and $p_f(x) = P_a^x(f)$ for $x \in (a, b]$ and let $n_f(a) = 0$ and $n_f(x) = N_a^x(f)$ for $x \in (a, b]$. Show that p_f and n_f are non-decreasing and

$$f(x) = f(a) + p_f(x) - n_f(x) \quad \text{for } x \in [a, b].$$

c. If f is continuous at $x \in [a, b]$, show that p_f and n_f are continuous at x.

5 · DIFFERENTIATION OF AN INDEFINITE INTEGRAL

Let f be a Lebesgue integrable function on the interval $[a, b]$. Then f is integrable on any interval $[a, x] \subset [a, b]$. We define the **indefinite integral** F of f as follows:

$$F(x) = \int_a^x f, \qquad x \in [a, b].$$

Our objective is to show that $F'(x) = f(x)$ for a.a. $x \in [a, b]$. We first show that F is of bounded variation and, hence, has a derivative almost everywhere on $[a, b]$.

5 · 1 Proposition. *If $f \in \mathscr{L}[a, b]$, then $F \in BV[a, b]$. In fact,*

$$V_a^b(F) = \int_a^b |f|.$$

PROOF. Recall that if $f \in \mathscr{L}[a, b]$, then $|f| \in \mathscr{L}[a, b]$. Let $P = \{(x_{i-1}, x_i) : i = 1, \cdots, k\}$ be a partition of $[a, b]$. Then

$$\sum_{i=1}^k |F(x_i) - F(x_{i-1})| = \sum_{i=1}^k \left|\int_a^{x_i} f - \int_a^{x_{i-1}} f\right| = \sum_{i=1}^k \left|\int_{x_{i-1}}^{x_i} f\right|$$

$$\leq \sum_{i=1}^k \int_{x_{i-1}}^{x_i} |f| = \int_a^b |f|.$$

Therefore, $V_a^b(F) \leq \int_a^b |f|$ and $F \in BV[a, b]$.

Since $f \in \mathscr{L}[a, b]$ there is a sequence (g_n) of step functions which are zero outside $[a, b]$ such that

$$\lim g_n(x) = f(x) \quad \text{for a.a.} x \in [a, b].$$

Let $h_n = (ng_n \wedge 1) \vee (-1)$. Then (h_n) is a sequence of step functions such that $|h_n| \leq 1$ and

$$\lim h_n(x) = 1 \quad \text{for a.a.} x \quad \text{where } f(x) > 0$$

$$\lim h_n(x) = -1 \quad \text{for a.a.} x \quad \text{where } f(x) < 0.$$

Then, $|h_n f| \leq |f|$ and

$$\lim h_n(x) f(x) = |f(x)| \quad \text{for a.a.} x \in [a, b].$$

Therefore, the Lebesgue Dominated Convergence Theorem implies that

$$\lim \int_a^b h_n f = \int_a^b |f|.$$

If the step function h_n has the representation $h_n(x) = c_i^n$ for $x \in (x_{i-1}, x_i)$ where $\{(x_{i-1}, x_i) : i = 1, \cdots, k\}$ is a partition of $[a, b]$, then

$$\int_a^b h_n f = \sum_{i=1}^k \int_{x_{i-1}}^{x_i} h_n f = \sum_{i=1}^k c_i^n(F(x_i) - F(x_{i-1}))$$

$$\leq \sum_{i=1}^k |F(x_i) - F(x_{i-1})| \leq V_a^b(F).$$

Therefore, $\int_a^b |f| \leq V_a^b(F)$ and, hence, $\int_a^b |f| = V_a^b(F)$. ▌

Thus, F is differentiable almost everywhere on $[a, b]$. We will show that $F'(x) = f(x)$ for a.a.$x \in [a, b]$ in three steps: for step functions, for functions in $\mathscr{I}$ and for integrable functions. In the second step we will need the following result.

5 · 2 Theorem. (Fubini Differentiation Theorem). *If (f_n) is a sequence of nondecreasing functions on $[a, b]$ and*

$$\sum f_n(x) = f(x) \quad \textit{for all } x \in [a, b],$$

then

$$\sum f_n'(x) = f'(x) \quad \textit{for a.a.} x \in [a, b].$$

PROOF. In the proof it is convenient to deal with nonnegative functions. If $g_n(x) = f_n(x) - f_n(a)$ and $g(x) = f(x) - f(a)$, then (g_n) is a sequence of nondecreasing nonnegative functions such that

$$\sum g_n(x) = g(x) \quad \text{for all } x \in [a, b].$$

And, if $\sum g_n'(x) = g'(x)$ for a.a.$x \in [a, b]$, then $\sum f_n'(x) = f'(x)$ for a.a.$x \in [a, b]$.

Let $s_k = \sum_{n=1}^{k} g_n$. Then, for each x and $x + h$ in $[a, b]$,

$$\frac{s_k(x+h) - s_k(x)}{h} \le \frac{s_{k+1}(x+h) - s_{k+1}(x)}{h} \le \frac{g(x+h) - g(x)}{h}.$$

If E is the set of measure zero where the nondecreasing functions s_k $(k = 1, 2, \cdots)$ and g fail to have derivatives, then

$$s_k'(x) \le s_{k+1}'(x) \le g'(x) \quad \text{for each } x \in [a, b] \sim E.$$

Thus, $(s_k'(x))$ is a bounded nondecreasing sequence and, hence, converges for each $x \in [a, b] \sim E$.

To show that (s_k') converges to g' on $[a, b] \sim E$ it is sufficient to prove that a subsequence (s_{k_j}') converges to g'. Take a subsequence (s_{k_j}) of (s_k) such that

$$g(b) - s_{k_j}(b) < 2^{-j}.$$

Since, for each j, $g - s_{k_j}$ is a nondecreasing function,

$$g(x) - s_{k_j}(x) < 2^{-j} \quad \text{for all } x \in [a, b]$$

and, hence, $\sum_{j=1}^{\infty}(g(x) - s_{k_j}(x))$ converges for all $x \in [a, b]$. Thus, $\sum_{j=1}^{\infty}(g(x) - s_{k_j}(x))$ has the same properties as $\sum g_n(x)$ and, therefore, by the first part of the proof $\sum_{j=1}^{\infty} (g'(x) - s_{k_j}'(x))$ converges for all $x \in [a, b] \sim E$. This implies that

$$\lim_{j \to \infty} (g'(x) - s_{k_j}'(x)) = 0 \quad \text{for all } x \in [a, b] \sim E$$

and, therefore, $\sum g_n'(x) = g'(x)$ for a.a.$x \in [a, b]$. ▮

5 · 3 Lemma. *If g is a step function on $[a, b]$ and $G(x) = \int_a^x g$, then*

$$G'(x) = g(x) \quad \textit{for a.a.} x \in [a, b].$$

Proof. Suppose g has the representation $g(x) = c_i$ for $x \in (x_{i-1}, x_i)$ where $\{(x_{i-1}, x_i) : i = 1, \cdots, k\}$ is a partition of $[a, b]$. If $x \in (x_{j-1}, x_j)$, then

$$G(x) = \int_a^x g = \sum_{i=1}^{j-1} c_i(x_i - x_{i-1}) + c_j(x - x_{j-1})$$

and, therefore,

$$G'(x) = c_j = g(x). ▮$$

5 · 4 Lemma. *If $f \in \mathscr{P}$ on $[a, b]$ and $F(x) = \int_a^x f$, then*

$$F'(x) = f(x) \quad \textit{for a.a.} x \in [a, b].$$

Proof. Let (g_n) be a nondecreasing sequence of step functions such that $\lim g_n(x) = f(x)$ for a.a.$x \in [a, b]$ and $(\int_a^b g_n)$ is bounded. Then for all $x \in [a, b]$

$$F(x) = \int_a^x f = \lim \int_a^x g_n = \lim G_n(x)$$

where $G_n(x) = \int_a^x g_n$. Lemma 5 · 3 implies that $G_n'(x) = g_n(x)$ for a.a.$x \in [a, b]$.

For $n \geq 2$, $(G_n - G_{n-1})$ is a sequence of nondecreasing functions on $[a, b]$ and

$$\sum_{n=2}^{\infty} (G_n(x) - G_{n-1}(x)) = F(x) - G_1(x) \quad \text{for all } x \in [a, b].$$

Then, by the Fubini Differentiation Theorem

$$\sum_{n=2}^{\infty} (G_n'(x) - G_{n-1}'(x)) = F'(x) - G_1'(x) \quad \text{for} \quad \text{a.a.} x \in [a, b];$$

that is,

$$\sum_{n=2}^{\infty} (g_n(x) - g_{n-1}(x)) = F'(x) - g_1(x) \quad \text{for a.a.} x \in [a, b].$$

Therefore,

$$f(x) = \lim g_n(x) = F'(x) \quad \text{for a.a.} x \in [a, b]. \blacksquare$$

5 · 5 Theorem. *If f is integrable on $[a, b]$ and $F(x) = \int_a^x f$, then*

$$F'(x) = f(x) \quad \textit{for a.a.} x \in [a, b].$$

PROOF. If $f \in \mathscr{L}[a, b]$, then there are functions f_1 and f_2 in $\mathscr{G}$ such that $f = f_1 - f_2$ almost everywhere on $[a, b]$. Let $F_i(x) = \int_a^x f_i$, $i = 1, 2$. Then

$$F(x) = \int_a^x f = \int_a^x f_1 - \int_a^x f_2 = F_1(x) - F_2(x)$$

and

$$F'(x) = F_1'(x) - F_2'(x) = f_1(x) - f_2(x) = f(x) \quad \text{for a.a.} x \in [a, b]. \blacksquare$$

PROBLEMS

1. If f is the greatest integer function and $F(x) = \int_0^x f$, determine F on $[0, 5]$ and verify that $F'(x) = f(x)$ for a.a.$x \in [0, 5]$.

2. Let

$$f(x) = \begin{cases} 1 & \text{for } x \text{ rational} \\ 0 & \text{for } x \text{ irrational} \end{cases}$$

and $F(x) = \int_0^x f$. Determine F on $[0, 1]$ and verify that $F'(x) = f(x)$ for a.a.$x \in [0, 1]$.

3. If $f \in \mathscr{L}[a, b]$ and $F(x) = \int_a^x f = 0$ for all $x \in [a, b]$, show that $f(x) = 0$ for a.a.$x \in [a, b]$.

4. If $f \in \mathscr{L}[a, b]$ and $\alpha \in \mathbf{R}$, show that for a.a.$x \in [a, b]$

$$\lim_{h \to 0^+} \frac{1}{h} \int_0^h |f(x + t) - \alpha| \, dt = |f(x) - \alpha| = \lim_{h \to 0^+} \frac{1}{h} \int_0^h |f(x - t) - \alpha| \, dt.$$

5. If $f \in \mathscr{L}[a, b]$, show that for a.a.$x \in [a, b]$

$$(5 \cdot 6) \qquad \lim_{h \to 0^+} \frac{1}{h} \int_0^h |f(x+t) + f(x-t) - 2f(x)|\, dt = 0.$$

The set of all $x \in (a, b)$ for which $5 \cdot 6$ holds is called the **Lebesgue set** of f.

Suggestion. For each rational number r let E_r be the set of all $x \in [a, b]$ such that

$$\lim_{h \to 0^+} \frac{1}{h} \int_0^h |f(x+t) - r|\, dt \neq |f(x) - r|$$

or

$$\lim_{h \to 0^+} \frac{1}{h} \int_0^h |f(x-t) - r|\, dt \neq |f(x) - r|.$$

Let $E = \bigcup E_r$. If $x \notin E$, then for each $\varepsilon > 0$ there exists a rational number r such that $|f(x) - r| < \varepsilon/4$. Conclude that $5 \cdot 6$ holds if $x \notin E$.

6 · INTEGRATION OF A DERIVATIVE

We have shown that if f is of bounded variation on $[a, b]$, then f' is integrable on $[a, b]$. Also if $g(x) = \int_a^x f'$ then $g'(x) = f'(x)$ for a.a.$x \in [a, b]$. However, this tells us little about the relationship of g to f. For example, any two step functions on $[a, b]$ will have equal derivatives almost everywhere on $[a, b]$. Even requiring that the functions be continuous is of no help since there is a nonconstant function which is nondecreasing and continuous on $[a, b]$ with derivative zero almost everywhere (Problem 6). Of course, if we require that f' be continuous on $[a, b]$ then the Second Fundamental Theorem of Calculus states that $\int_a^x f' = f(x) - f(a)$; i.e., $g(x) = f(x) - f(a)$.

We would like to obtain this result under less stringent conditions. It turns out that the concept of absolute continuity is what we want.

6 · 1 Definition. *A real-valued function f defined on an interval I is **absolutely continuous** if for each $\varepsilon > 0$ there exists a $\delta > 0$ such that*

$$\sum_{i=1}^{n} |f(b_i) - f(a_i)| < \varepsilon$$

for every finite pairwise disjoint collection $\{(a_i, b_i) : i = 1, \cdots, n\}$ of open intervals in I with

$$\sum_{i=1}^{n} (b_i - a_i) < \delta.$$

It is clear that an absolutely continuous function on I is uniformly continuous on I. We will see shortly that the converse is not true. However, if a function f satisfies the Lipschitz condition

$$|f(x) - f(y)| \leq c\,|x - y| \quad \text{for all } x, y \in I,$$

then f is absolutely continuous on I. Also, it is easy to show that the set of absolutely continuous functions on an interval I is a linear space (Problem 2).

6 · 2 Proposition. *If f is absolutely continuous on $[a, b]$, then $f \in BV[a, b]$.*

PROOF. Since f is absolutely continuous on $[a, b]$, there exists a $\delta > 0$ such that

$$\sum_{i=1}^{n} |f(b_i) - f(a_i)| < 1 \tag{6 · 3}$$

for any finite pairwise disjoint collection $\{(a_i, b_i) : i = 1, \cdots, n\}$ of open intervals with $\sum_{i=1}^{n} (b_i - a_i) < \delta$. Let M be a positive integer such that $M\delta > b - a$.

For any partition P of $[a, b]$ let P' be the refinement of P formed by adjoining to P the points $a + i(b - a)/M$ $(i = 0, \cdots, M)$. Then $P' = \bigcup_{k=1}^{M} P_k$, where P_k is a partition of $[a + (k - 1)(b - a)/M, a + k(b - a)/M]$. Using 6 · 3 we have $\sum_{P_k} |\Delta f| < 1$ and, therefore,

$$\sum_{P} |\Delta f| \le \sum_{P'} |\Delta f| = \sum_{k=1}^{M} \sum_{P_k} |\Delta f| < M. \blacksquare$$

Since the function f defined by $f(0) = 0$ and $f(x) = x \sin(1/x)$ for $x \in (0, 1]$ is continuous on $[0, 1]$ but not of bounded variation on $[0, 1]$, this is an example of a continuous function which is not absolutely continuous (Problem 7, p. 233).

Proposition 6 · 2 implies that if f is absolutely continuous on $[a, b]$, then f is differentiable at almost all points of $[a, b]$ and f' is integrable on $[a, b]$. Also, if $g(x) = \int_a^x f'$, then $g'(x) = f'(x)$ for a.a. $x \in [a, b]$. In the case when f is absolutely continuous we will be able to prove that $g'(x) = f'(x)$ for a.a. $x \in [a, b]$ implies that, for some constant c, $g(x) = f(x) + c$ for all $x \in [a, b]$ and, hence, $\int_a^x f' = f(x) - f(a)$. For this proof we will need a number of preliminary results.

Since an absolutely continuous function f on $[a, b]$ is of bounded variation on $[a, b]$ it can be written as the difference of two nondecreasing functions:

$$f = v_f - (v_f - f)$$

where v_f is the variation function defined by $v_f(x) = V_a^x(f)$. We now show that v_f is absolutely continuous and, hence, f is the difference of two nondecreasing absolutely continuous functions.

6 · 4 Proposition. *If f is absolutely continuous on $[a, b]$, then v_f is absolutely continuous on $[a, b]$.*

PROOF. Take $\varepsilon > 0$. There exists a $\delta > 0$ such that

$$\sum_{i=1}^{n} |f(b_i) - f(a_i)| < \varepsilon$$

for every finite pairwise disjoint collection $\{(a_i, b_i) : i = 1, \cdots, n\}$ with $\sum_{i=1}^{n} (b_i - a_i) < \delta$. For any such collection let $P_i = \{(x_{k-1}^i, x_k^i) : k = 1, \cdots, r_i\}$ be a partition of $[a_i, b_i]$. Then $\{(x_{k-1}^i, x_k^i) : i = 1, \cdots, n; k = 1, \cdots, r_i\}$ is a finite disjoint collection of open intervals with

$$\sum_{i=1}^{n} \sum_{k=1}^{r_i} (x_k^i - x_{k-1}^i) = \sum_{i=1}^{n} (b_i - a_i) < \delta$$

and, therefore,

$$\sum_{i=1}^{n} \sum_{k=1}^{r_i} |f(x_k^i) - f(x_{k-1}^i)| < \varepsilon.$$

Taking the least upper bound over all partitions P_i of $[a_i, b_i]$ for $i = 1, \cdots, n$ we obtain

$$\sum_{i=1}^{n} V_{a_i}^{b_i}(f) = \sum_{i=1}^{n} |v_f(b_i) - v_f(a_i)| \leq \varepsilon.$$

This shows that v_f is absolutely continuous on $[a, b]$. ∎

In Proposition 5 · 1 we showed that if $f \in \mathscr{L}[a, b]$ and $F(x) = \int_a^x f$, then $F \in BV[a, b]$. We now show that F is absolutely continuous on $[a, b]$.

6 · 5 Proposition. *If $f \in \mathscr{L}[a, b]$ and $F(x) = \int_a^x f$, for $x \in [a, b]$, then F is absolutely continuous on $[a, b]$.*

PROOF. For each positive integer k let

$$g_k = |f| \wedge k.$$

Since $f \in \mathscr{L}[a, b]$, $|f|$ and g_k are integrable on $[a, b]$. Also, $|g_k| \leq |f|$ and $\lim g_k(x) = |f(x)|$ for all $x \in [a, b]$. Therefore, by the Lebesgue Dominated Convergence Theorem

$$\lim \int_a^b g_k = \int_a^b |f|.$$

Take $\varepsilon > 0$. Then there is a positive integer m such that

$$\int_a^b (|f| - g_m) < \frac{\varepsilon}{2}.$$

If we let $\delta = \varepsilon/(2m)$ and take any finite pairwise disjoint collection $\{(a_i, b_i) : i = 1, \cdots, n\}$ with $\sum_{i=1}^{n} (b_i - a_i) < \delta$, then we have

$$\sum_{i=1}^{n} |F(b_i) - F(a_i)| \leq \sum_{i=1}^{n} \int_{a_i}^{b_i} |f| = \sum_{i=1}^{n} \int_{a_i}^{b_i} (|f| - g_m) + \sum_{i=1}^{n} \int_{a_i}^{b_i} g_m$$

$$< \frac{\varepsilon}{2} + \sum_{i=1}^{n} \int_{a_i}^{b_i} m = \frac{\varepsilon}{2} + m \sum_{i=1}^{n} (b_i - a_i) < \varepsilon. \blacksquare$$

6 · 6 Proposition. *If f is a nondecreasing absolutely continuous function on $[a, b]$ and $f'(x) = 0$ for a.a. $x \in [a, b]$, then f is a constant on $[a, b]$.*

PROOF. Take $\varepsilon > 0$. Suppose that $f'(x) = 0$ for all $x \in [a, b] \sim E$ where E is of measure zero. It is convenient to extend the definition of f to the interval $[a-1, b+1]$ by letting $f(x) = f(a)$ for $x \in [a-1, a]$ and $f(x) = f(b)$ for $x \in [b, b+1]$. Then f is a nondecreasing absolutely continuous function on $[a-1, b+1]$. For the given $\varepsilon > 0$ there exists a $\delta > 0$ such that

$$\sum_{i=1}^{n} (f(b_i) - f(a_i)) < \varepsilon \tag{6 · 7}$$

for every finite pairwise disjoint collection $\{(a_i, b_i) : i = 1, \cdots, n\}$ of subintervals of $[a-1, b+1]$ with $\sum_{i=1}^{n} (b_i - a_i) < \delta$.

Since E is of measure zero there is a countable collection $\{I_j\}$ of open intervals such that $E \subset \bigcup_{j=1}^{\infty} I_j$ and $\sum_{j=1}^{\infty} \mu(I_j) < \delta$. We may assume that, for all j, $I_j \subset [a - 1, b + 1]$.

For each point $x \in [a, b] \sim E$ there exists an $\eta > 0$ such that

$$\frac{f(y) - f(x)}{y - x} < \varepsilon \quad \text{for } 0 < |y - x| < \eta.$$

Then, for $0 < x - c < \eta, f(x) - f(c) < \varepsilon(x - c)$ and for $0 < d - x < \eta$, $f(d) - f(x) < \varepsilon(d - x)$. Thus, there exists an open interval $J_x = (c, d)$ such that $x \in J_x \subset [a-1, b+1]$ and

$$f(d) - f(c) < \varepsilon(d - c). \tag{6 · 8}$$

The collection $\{I_j : j = 1, 2, \cdots\} \cup \{J_x : x \in [a, b] \sim E\}$ is an open covering of $[a, b]$ and, hence, by the Heine-Borel Theorem has a finite subcovering $\{I_j : j = 1, \cdots, m\} \cup \{J_k : k = 1, \cdots, n\}$. We may assume that no interval in this finite collection is contained in the union of other intervals in the collection since it would then be superfluous. Let $I_j = (a_j, b_j)$ and $J_k = (c_k, d_k)$ and assume that these intervals are labeled in such a way that

$$a_1 < a_2 < \cdots < a_m \quad \text{and} \quad c_1 < c_2 < \cdots < c_n.$$

Then, we also have

$$b_1 < b_2 < \cdots < b_m \quad \text{and} \quad d_1 < d_2 < \cdots < d_n.$$

For, if $b_{j+1} \leq b_j$, then $(a_{j+1}, b_{j+1}) \subset (a_j, b_j)$.

Observe that the collections $\{I_j : 1 \leq j \leq m, j \text{ odd}\}$, $\{I_j : 1 \leq j \leq m, j \text{ even}\}$, $\{J_k : 1 \leq k \leq n, k \text{ odd}\}$, and $\{J_k : 1 \leq k \leq n, k \text{ even}\}$ are each pairwise disjoint. For example, if $I_j \cap I_{j+2} \neq \varnothing$, then $a_{j+2} < b_j$ and $(a_{j+1}, b_{j+1}) \subset (a_j, b_j) \cup (a_{j+2}, b_{j+2})$ but we have assumed that no interval of the finite collection is contained in the union of other intervals in the collection.

Using 6 · 7 and 6 · 8 we have for $1 \le j \le m$ and $1 \le k \le n$

$$f(b) - f(a) \le \sum_{j \text{ odd}} (f(b_j) - f(a_j)) + \sum_{j \text{ even}} (f(b_j) - f(a_j))$$
$$+ \sum_{k \text{ odd}} (f(d_k) - f(c_k)) + \sum_{k \text{ even}} (f(d_k) - f(c_k))$$
$$< \varepsilon + \varepsilon + \varepsilon(b - a + 2) + \varepsilon(b - a + 2) = 2\varepsilon(3 + b - a).$$

Thus, $f(b) = f(a)$ and f is constant on $[a, b]$. ▮

We now prove the Second Fundamental Theorem of Calculus for the Lebesgue integral.

6 · 9 Theorem. *If f is absolutely continuous on $[a, b]$, then*

$$\int_a^x f' = f(x) - f(a) \quad \textit{for all } x \in [a, b].$$

PROOF. From Proposition 6 · 4 and the discussion preceding it we know that f can be written in the form $f = f_1 - f_2$ where f_1 and f_2 are nondecreasing absolutely continuous functions on $[a, b]$. For $i = 1, 2$, let $g_i(x) = \int_a^x f_i'$ for $x \in [a, b]$. Then, g_i is absolutely continous on $[a, b]$ by Proposition 6 · 5. Also, Theorem 5 · 5 implies that $g_i'(x) = f_i'(x)$ for a.a.$x \in [a, b]$. Let $h_i = f_i - g_i$. Then h_i is absolutely continuous on $[a, b]$ and $h_i'(x) = 0$ for a.a.$x \in [a, b]$. Also, for any x and y in $[a, b]$ with $x < y$, Theorem 4 · 7 implies that

$$h_i(y) - h_i(x) = f_i(y) - \int_a^y f_i' - f_i(x) + \int_a^x f_i'$$
$$= f_i(y) - f_i(x) - \int_x^y f_i' \ge 0$$

and, therefore, h_i is nondecreasing. Then, h_i is a constant by Proposition 6 · 6; that is,

$$\int_a^x f' = \int_a^x f_1' - \int_a^x f_2' = g_1(x) - g_2(x) = f_1(x) + c_1 - f_2(x) - c_2$$
$$= f(x) + c.$$

Letting $x = a$ we see that $c = -f(a)$ and, hence,

$$\int_a^x f' = f(x) - f(a). \ ▮$$

Observe that Proposition 6 · 5 and Theorem 6 · 9 imply that a function is absolutely continuous on $[a, b]$ if and only if it is the sum of a constant and the indefinite integral of a function in $\mathscr{L}[a, b]$.

PROBLEMS ·

1. If $f(x) = x^2 \sin(1/x)$ for $x \in (0, 1]$ and $f(0) = 0$, show that f is absolutely continuous on $[0, 1]$.

2. Show that the set of absolutely continuous functions on an interval I is a linear space.

3. If f and g are bounded and absolutely continuous on I, show that fg is absolutely continuous on I.

4. If g is a nondecreasing absolutely continuous function on $[a, b]$ and f is absolutely continuous on $[g(a), g(b)]$, show that $f \circ g$ is absolutely continuous on $[a, b]$.

5. If f is absolutely continuous on $[a, b]$ and $f'(x) \geq 0$ for a.a. $x \in [a, b]$, show that f is nondecreasing on $[a, b]$.

6. The Cantor function f (p. 20) was defined from the Cantor set C onto $[0, 1]$ by the rule

$$f\left(\sum_{k=1}^{\infty} \frac{x_k}{3^k}\right) = \sum_{k=1}^{\infty} \frac{r_k}{2^k} \quad \text{where } r_k = \tfrac{1}{2}x_k.$$

If (a, b) is one of the countable number of disjoint open intervals whose union is $[0, 1] \sim C$, then, in ternary notation,

$$a = 0.a_1a_2 \cdots a_{n-1}0222 \cdots = 0.a_1a_2 \cdots a_{n-1}1000 \cdots$$

$$b = 0.a_1a_2 \cdots a_{n-1}2000 \cdots = 0.a_1a_2 \cdots a_{n-1}1222 \cdots$$

where $a_k = 0$ or 2 for $k \leq n - 1$ and, hence, in binary notation,

$$f(a) = f(b) = 0.c_1c_2 \cdots c_{n-1}1000 \cdots \text{ where } c_k = \tfrac{1}{2}a_k.$$

Thus, we can extend f to all of $[0, 1]$ by defining f to have the value $f(a)$ on (a, b); that is, if $x \in [0, 1] \sim C$ and

$$x = 0.x_1x_2 \cdots x_{n-1}1x_{n+1} \cdots \text{ (ternary)} \quad \text{where } x_k = 0 \text{ or } 2 \text{ for } k \leq n - 1$$

then we define

$$f(x) = 0.r_1r_2 \cdots r_{n-1}1000 \cdots \text{ (binary) where } r_k = \tfrac{1}{2}x_k \text{ for } k \leq n - 1.$$

Show that f is a nonconstant function which is nondecreasing and continuous on $[0, 1]$ with derivative zero almost everywhere in $[0, 1]$.

Suggestion. To prove continuity use Problem 1, p. 229, and the fact that f maps $[0, 1]$ onto $[0, 1]$.

7. Show that the Cantor function (Problem 6) is of bounded variation on $[0, 1]$ but is not absolutely continuous on $[0, 1]$.

7 · INTEGRATION BY PARTS AND BY SUBSTITUTION

For the Lebesgue integral we can prove a result on integration by parts which is similar to that for the Riemann integral.

7 · 1 Theorem. *If f and g are in* $\mathscr{L}[a, b]$ *and* $F(x) = \int_a^x f + c_1$ *and* $G(x) = \int_a^x g + c_2$ *for* $x \in [a, b]$, *then*

$$\int_a^b Fg + \int_a^b fG = F(b)G(b) - F(a)G(a).$$

PROOF. By Proposition 6 · 5, F and G are absolutely continuous on $[a, b]$ and, hence, FG is absolutely continuous on $[a, b]$ (Problem 3, §6). Then, Theorem 6 · 9 implies that

$$\int_a^b (FG)' = F(b)G(b) - F(a)G(a).$$

Also, by Theorem 5 · 5, $F'(x) = f(x)$ and $G'(x) = g(x)$ for a.a. $x \in [a, b]$ and, hence,

$$(FG)' = FG' + F'G = Fg + fG \quad \text{a.e. on } [a, b].$$

Thus,

$$\int_a^b Fg + \int_a^b fG = \int_a^b (FG)' = F(b)G(b) - F(a)G(a). \blacksquare$$

7 · 2 Corollary. *If f and g are absolutely continuous on* $[a, b]$, *then*

$$\int_a^b fg' + \int_a^b f'g = f(b)g(b) - f(a)g(a).$$

PROOF. Since f and g are absolutely continuous on $[a, b]$, Theorem 6 · 9 implies that f' and g' are in $\mathscr{L}[a, b]$. Also, $f(x) = \int_a^x f' + f(a)$ and $g(x) = \int_a^x g' + g(a)$. Then, Theorem 7 · 1 implies

$$\int_a^b fg' + \int_a^b f'g = f(b)g(b) - f(a)g(a). \blacksquare$$

We now prove a theorem on substitution or change of variable for the Lebesgue integral. This result could be improved considerably but the result obtained is sufficient for our purposes. First we prove some preliminary results.

7 · 3 Lemma. *Let g be a nondecreasing absolutely continuous function on* $[a, b]$ *and let h be a step function on* $[g(a), g(b)]$. *Then,* $(h \circ g)g' \in \mathscr{L}[a, b]$ *and*

$$\int_a^b (h \circ g)g' = \int_{g(a)}^{g(b)} h.$$

PROOF. Although g' may fail to be defined on a set S of measure zero in $[a, b]$, we may assume that g' is defined arbitrarily on S since this will not affect integration.

First assume that h is the characteristic function of a closed interval $[c, d] \subset [g(a), g(b)]$. Since g is continuous on $[a, b]$, it takes on the values c and d. Let

$$a_1 = \min\{t \in [a, b] : g(t) = c\} \quad \text{and} \quad b_1 = \max\{t \in [a, b] : g(t) = d\}.$$

Then

$$\int_a^b (h \circ g)g' = \int_{a_1}^{b_1} g' = g(b_1) - g(a_1) = d - c = \int_{g(a)}^{g(b)} h.$$

Since a step function on $[g(a), g(b)]$ is a linear combination of characteristic functions of closed subintervals of $[g(a), g(b)]$ except at a finite number of points, the linearity of the integral implies the stated result. ∎

7·4 Lemma. *If g is a nondecreasing absolutely continuous function on $[a, b]$ and E is a set of measure zero in $[g(a), g(b)]$, then $T = \{t \in [a, b] : g(t) \in E \text{ and } g'(t) > 0\}$ is a set of measure zero.*

PROOF. Since E is of measure zero there exists a nondecreasing sequence (h_n) of nonnegative step functions such that $(h_n(x))$ diverges to ∞ for each $x \in E$ but $\left(\int_{g(a)}^{g(b)} h_n\right)$ converges. Then, $((h_n \circ g)g')$ is an a.e. nondecreasing sequence of integrable functions which diverges to ∞ for each $t \in T$. Since

$$\int_a^b (h_n \circ g)g' = \int_{g(a)}^{g(b)} h_n,$$

$\left(\int_a^b (h_n \circ g)g'\right)$ converges. The Monotone Convergence Theorem implies that $(h_n(g(t))g'(t))$ converges for a.a. $t \in [a, b]$ and, hence, T is of measure zero. ∎

7·5 Theorem. *Let g be a nondecreasing absolutely continuous function on $[a, b]$ and let f be integrable on $[g(a), g(b)]$. Then, $(f \circ g)g' \in \mathscr{L}[a, b]$ and*

$$\int_a^b (f \circ g)g' = \int_{g(a)}^{g(b)} f.$$

PROOF. First take $f \in \mathscr{I}$. Then there is a nondecreasing sequence (h_n) of step functions on $[g(a), g(b)]$ such that

$$\lim h_n(x) = f(x) \quad \text{for all } x \in [g(a), g(b)] \sim E,$$

where E is of measure zero, and

$$\lim \int_{g(a)}^{g(b)} h_n = \int_{g(a)}^{g(b)} f.$$

Therefore, $((h_n \circ g)g')$ is an a.e. nondecreasing sequence of integrable functions such that

$$\lim h_n(g(t))g'(t) = f(g(t))g'(t) \quad \text{for all } t \in [a, b] \sim T$$

where $T = \{t \in [a, b] : g(t) \in E \text{ and } g'(t) > 0\}$ and

$$\int_a^b (h_n \circ g)g' = \int_{g(a)}^{g(b)} h_n \le \int_{g(a)}^{g(b)} f.$$

Since T is of measure zero, the Monotone Convergence Theorem implies that

$$\int_a^b (f \circ g)g' = \lim \int_a^b (h_n \circ g)g' = \lim \int_{g(a)}^{g(b)} h_n = \int_{g(a)}^{g(b)} f.$$

If f is integrable on $[g(a), g(b)]$, then $f = f_1 - f_2$ a.e. on $[g(a), g(b)]$, where $f_1, f_2 \in \mathscr{P}$. Then $(f \circ g)g' = (f_1 \circ g)g' - (f_2 \circ g)g'$ a.e. on $[a, b]$ and

$$\int_a^b (f \circ g)g' = \int_a^b (f_1 \circ g)g' - \int_a^b (f_2 \circ g)g'$$

$$= \int_{g(a)}^{g(b)} f_1 - \int_{g(a)}^{g(b)} f_2 = \int_{g(a)}^{g(b)} f. \blacksquare$$

PROBLEMS

1. If $f \in \mathscr{L}$ and $F(x) = \int_{-\infty}^{x} f$, show that F is absolutely continuous on **R**.

2. Let $f \in \mathscr{L}$ and let t be any number. Show that

$$\int_{-\infty}^{\infty} f(x)\, dx = \int_{-\infty}^{\infty} f(x - t)\, dx.$$

3. Let f and g be in $\mathscr{L}$. Then the **convolution** of f and g, denoted by $f * g$, is defined by

$$(f * g)(x) = \int_{-\infty}^{\infty} f(t)g(x - t)\, dt.$$

Show that:

a. $(f * g)(x)$ is defined for a.a.x and $f * g \in \mathscr{L}$.

Suggestion. Use Problem 2 to show that

$$\int_{-\infty}^{\infty} \int_{-\infty}^{\infty} |f(t)g(x - t)|\, dx\, dt = \int_{-\infty}^{\infty} |g(x)|\, dx \int_{-\infty}^{\infty} |f(t)|\, dt$$

and then use the Tonelli and Fubini Theorems.

b. $\|f * g\|_1 \le \|f\|_1 \|g\|_1$

c. $f * g = g * f$

d. $(f * g) * h = f * (g * h)$.

8 · THE RIESZ REPRESENTATION THEOREM

Take a finite interval $[a, b] \subset \mathbf{R}$. For $1 < p < \infty$ if $f \in \mathscr{L}_p[a, b]$ and $g \in \mathscr{L}_q[a, b]$, where $1/p + 1/q = 1$, then Hölder's Inequality implies that $fg \in \mathscr{L}_1[a, b]$ and

$$\int_a^b |fg| \le \left[\int_a^b |f|^p\right]^{1/p} \left[\int_a^b |g|^q\right]^{1/q} = \|f\|\, \|g\|.$$

For $g \in \mathscr{L}_q[a, b]$, let

$$F_g(f) = \int_a^b fg, \qquad f \in \mathscr{L}_p[a, b].$$

Then, F_g is a linear functional on $\mathscr{L}_p[a, b]$. In fact, since

$$|F_g(f)| \leq \int_a^b |fg| \leq \|f\| \, \|g\|,$$

F_g is a bounded linear functional on $\mathscr{L}_p[a, b]$ and

$$(8 \cdot 1) \qquad \|F_g\| = \sup_{\|f\|=1} |F_g(f)| \leq \|g\|.$$

If we define $T(g) = F_g$ for $g \in \mathscr{L}_q[a, b]$, then T is a linear transformation from $\mathscr{L}_q[a, b]$ into $\mathscr{L}_p^*$ $[a, b]$, the dual of $\mathscr{L}_p[a, b]$. We now show that T is an isometry. Let

$$f = (\operatorname{sgn} g)\, |g|^{q-1}$$

where $\operatorname{sgn} g(x) = -1$, 0, or 1 according as $g(x) < 0$, $g(x) = 0$, or $g(x) > 0$. Since $|f(x)|^p = |g(x)|^{pq-p} = |g(x)|^q$, then $f \in \mathscr{L}_p$ $[a, b]$ and

$$\|f\| = \left[\int_a^b |g|^q\right]^{1/p}.$$

Also,

$$|F_g(f)| = \left|\int_a^b fg\right| = \int_a^b |g|^q = \left[\int_a^b |g|^q\right]^{1/q} \left[\int_a^b |g|^q\right]^{1/p}$$
$$= \|g\| \, \|f\|.$$

Therefore, $\|F_g\| \geq \|g\|$ and this combined with $8 \cdot 1$ gives

$$\|F_g\| = \|g\|.$$

Thus, T is an isometry of $\mathscr{L}_q[a, b]$ into $\mathscr{L}_p^*$ $[a, b]$.

The following theorem due to the Hungarian mathematician F. Riesz shows that T is *onto* and, hence, $\mathscr{L}_q[a, b]$ and $\mathscr{L}_p^*[a, b]$ are isometrically isomorphic normed linear spaces.

8 · 2 Theorem (Riesz). *For each $F \in \mathscr{L}_p^*$ $[a, b]$, $1 < p < \infty$, there is a function $g \in \mathscr{L}_q[a, b]$ such that*

$$F(f) = \int_a^b fg, \qquad f \in \mathscr{L}_p[a, b]$$

and $\|F\| = \|g\|$.

We break the proof of this theorem up into a sequence of lemmas.

8 · 3 Lemma. *If $F \in \mathscr{L}_p^*[a, b]$ and if*

$$G(a) = 0 \quad \textit{and} \quad G(x) = F(\chi_{[a, x)}) \quad \textit{for } x \in (a, b],$$

then G is absolutely continuous on $[a, b]$.

PROOF. Let $\{(a_i, b_i) : i = 1, \cdots, n\}$ be a pairwise disjoint collection of open intervals in $[a, b]$. Let

$$\sigma_i = \operatorname{sgn}[G(b_i) - G(a_i)] = -1, 0, \text{ or } 1$$

according as $G(b_i) - G(a_i)$ is negative, zero, or positive. Then

$$|G(b_i) - G(a_i)| = F(\sigma_i \chi_{[a_i, b_i)})$$

and, hence,

$$\begin{aligned}
\sum_{i=1}^{n} |G(b_i) - G(a_i)| = F\left(\sum_{i=1}^{n} \sigma_i \chi_{[a_i, b_i)}\right) &\leq \|F\| \left\| \sum_{i=1}^{n} \sigma_i \chi_{[a_i, b_i)} \right\|_p \\
&= \|F\| \left[\int_a^b \left| \sum_{i=1}^{n} \sigma_i \chi_{[a_i, b_i)} \right|^p \right]^{1/p} \\
&\leq \|F\| \left[\int_a^b \sum_{i=1}^{n} \chi_{[a_i, b_i)} \right]^{1/p} \\
&= \|F\| \left[\sum_{i=1}^{n} (b_i - a_i)\right]^{1/p}
\end{aligned}$$

Thus, G is absolutely continuous on $[a, b]$. ▮

Since G is absolutely continuous, G' exists a.e. on $[a, b]$, and if $g = G'$ a.e. on $[a, b]$ then $g \in \mathscr{L}_1[a, b]$.

8 · 4 Lemma. *If $g = G'$ a.e. on $[a, b]$, then for each step function h and each bounded measurable function f on $[a, b]$*

$$F(h) = \int_a^b hg \quad \textit{and} \quad F(f) = \int_a^b fg.$$

PROOF. For each interval $[c, d) \subset [a, b]$,

$$F(\chi_{[c, d)}) = G(d) - G(c) = \int_c^d g = \int_a^b \chi_{[c, d)}\, g.$$

Thus, for each step function h on $[a, b]$

$$F(h) = \int_a^b hg.$$

If f is measurable and $|f(x)| \leq M$ for all $x \in [a, b]$, then $f \in \mathscr{L}_p[a, b]$. Let (h_n) be a sequence of step functions such that $\lim h_n(x) = f(x)$ for a.a.$x \in [a, b]$ (Problem 10, p. 140). We may assume that $|h_n| \leq M$ for all n; otherwise take $(-M) \vee (h_n \wedge M)$. Since $|f - h_n|^p \leq (2M)^p$ for all n and

$$\lim |f(x) - h_n(x)|^p = 0 \quad \text{for a.a.} x \in [a, b],$$

the Lebesgue Dominated Convergence Theorem implies that

$$\lim \int_a^b |f - h_n|^p = 0$$

and, hence, $\lim \|f - h_n\|_p = 0$. Since $|F(f) - F(h_n)| \le \|F\| \, \|f - h_n\|_p$,

$$\lim F(h_n) = F(f).$$

Also, $|h_n g| \le M|g|$ and

$$\lim h_n(x)g(x) = f(x)g(x) \quad \text{for a.a.} x \in [a, b].$$

Again applying the Lebesgue Dominated Convergence Theorem we have

$$\lim \int_a^b h_n g = \int_a^b fg.$$

Thus,

$$F(f) = \lim F(h_n) = \lim \int_a^b h_n g = \int_a^b fg. \quad \blacksquare$$

8 · 5 Lemma. *If $g = G'$ a.e. on $[a, b]$, then $g \in \mathscr{L}_q[a, b]$.*

PROOF. For each positive integer n let

$$g_n = (\operatorname{sgn} g)[|g|^{q-1} \wedge n].$$

Then, g_n is a bounded measurable function and, hence, $g_n \in \mathscr{L}_p[a, b]$. Also $(|g_n|^p)$ is a nondecreasing sequence of integrable functions such that

$$\lim |g_n(x)|^p = |g(x)|^{pq-p} = |g(x)|^q \quad \text{for all } x \in [a, b].$$

Since

$$|g_n|^p = |g_n| \, |g_n|^{p-1} \le |g_n| \, |g| = g_n g,$$

we have

$$\int_a^b |g_n|^p \le \int_a^b g_n g = F(g_n) \le \|F\| \, \|g_n\|_p = \|F\| \left[\int_a^b |g_n|^p\right]^{1/p}$$

and, hence,

$$\left[\int_a^b |g_n|^p\right]^{1/q} \le \|F\|.$$

The Monotone Convergence Theorem implies that $|g|^q$ is integrable on $[a, b]$ and

$$\|g\| = \left[\int_a^b |g|^q\right]^{1/q} = \lim \left[\int_a^b |g_n|^p\right]^{1/q} \le \|F\|.$$

Thus, $g \in \mathscr{L}_q[a, b]$. $\blacksquare$

8 · 6 Lemma. *If* $f \in \mathscr{L}_p[a, b]$, *then* $F(f) = \int_a^b fg$.

PROOF. By Theorem 3 · 6, p. 197, there is a sequence (h_n) of step functions such that

$$\lim \|f - h_n\|_p = 0.$$

Since $|F(f) - F(h_n)| \le \|F\| \, \|f - h_n\|$,

$$\lim F(h_n) = F(f).$$

Also,

$$\left| \int_a^b fg - \int_a^b h_n g \right| = \left| \int_a^b (f - h_n)g \right| \le \|f - h_n\| \, \|g\|$$

and, hence,

$$\lim \int_a^b h_n g = \int_a^b fg.$$

Therefore,

$$F(f) = \lim F(h_n) = \lim \int_a^b h_n g = \int_a^b fg. \quad ∎$$

This completes the proof of the Riesz Representation Theorem. We have shown that, for $1 < p < \infty$, $\mathscr{L}_p^*[a, b]$ is isometrically isomorphic to $\mathscr{L}_q[a, b]$ which we may indicate by writing $\mathscr{L}_p^*[a, b] = \mathscr{L}_q[a, b]$. We leave it to the reader (Problem 1) to prove that $\mathscr{L}_1^*[a, b] = \mathscr{L}_\infty[a, b]$. This proof is similar to what has been given but the proof that $g \in \mathscr{L}_\infty[a, b]$ is much simpler and consideration of bounded measurable functions is not necessary.

Also, from the results obtained it is not difficult to show that $\mathscr{L}_p^* = \mathscr{L}_q$ and $\mathscr{L}_1^* = \mathscr{L}_\infty$ (Problems 2 and 3). However, it is not true that $\mathscr{L}_\infty^*$ is $\mathscr{L}_1$ (Problem 4).

PROBLEMS

1. Show that $\mathscr{L}_1^*[a, b] = \mathscr{L}_\infty[a, b]$.

2. Show that $\mathscr{L}_p^* = \mathscr{L}_q$, $1 < p < \infty$.

3. Show that $\mathscr{L}_1^* = \mathscr{L}_\infty$.

4. Show that $\mathscr{L}_\infty^*[0, 1] \neq \mathscr{L}_1[0, 1]$.

Suggestion. For $f \in C[0, 1]$, let $F(f) = f(1)$. By the Hahn-Banach Theorem extend F to $\mathscr{L}_\infty[0, 1]$. Let

$$f_n(x) = \begin{cases} 0, & 0 \le x < 1 - 1/n \\ nx - n + 1, & 1 - 1/n \le x \le 1. \end{cases}$$

Then $F(f_n) - 1$. But, for all $g \in \mathscr{L}_1[0, 1]$, $\lim \int_0^1 f_n g = 0$.

10 | *The Stieltjes Integrals*

1 · INTRODUCTION

In 1894 the Dutch astronomer and mathematician Thomas Joannes Stieltjes (1856-1894) introduced an integral, now known as the Riemann-Stieltjes integral, in conjunction with his work on continued fractions and what is called the problem of moments. The concept of the Stieltjes integral received little attention until 1909 when the Hungarian mathematician Frigyes Riesz showed that every bounded linear functional on the space of continuous functions on $[a, b]$ can be represented in terms of a Riemann-Stieltjes integral. The Stieltjes integral is particularly useful in probability theory where it makes possible the simultaneous treatment of continuous and discrete random variables. It is also of great importance in many concepts and problems of analysis, mechanics, and theoretical physics, for example, in curvilinear and surface integrals. In this chapter, we will consider some integrals of the Stieltjes type and study relations among them.

2 · THE RIEMANN-STIELTJES INTEGRAL

Stieltjes' generalization of the Riemann integral involves two functions. Let f and g be two real-valued functions defined on the interval $[a, b] \subset \mathbf{R}$. For any partition $P = \{I_1, \cdots, I_n\}$ of $[a, b]$, where $I_k = (x_{k-1}, x_k)$, choose any point $z_k \in [x_{k-1}, x_k]$ and form the sum

$$S(f, g, P) = \sum_{k=1}^{n} f(z_k)(g(x_k) - g(x_{k-1})).$$

2 · 1 Definition. *If* $\lim_{|P|\to 0} S(f, g, P)$ *exists, then this limit is called the* ***Riemann-Stieltjes integral*** *of f with respect to g and is denoted by* $\int_a^b f\,dg$.

The function f is called the **integrand** and the function g is called the **integrator**.

The statement that $\lim_{|P|\to 0} S(f, g, P) = \int_a^b f\,dg$ means that for each $\varepsilon > 0$ there exists a $\delta > 0$ such that for all partitions P with $|P| < \delta$ and for all choices of $z_k \in [x_{k-1}, x_k]$

$$\left| \int_a^b f\,dg - \sum_{k=1}^{n} f(z_k)(g(x_k) - g(x_{k-1})) \right| < \varepsilon.$$

The following two propositions are direct consequences of the definition of the integral.

2·2 Proposition. *If* $\int_a^b f_1\, dg$ *and* $\int_a^b f_2\, dg$ *exist and* α *and* β *are any real numbers, then*

$$\int_a^b (\alpha f_1 + \beta f_2)\, dg = \alpha \int_a^b f_1\, dg + \beta \int_a^b f_2\, dg.$$

2·3 Proposition. *If* $\int_a^b f\, dg_1$ *and* $\int_a^b f\, dg_2$ *exist and* α *and* β *are any real numbers, then*

$$\int_a^b f\, d(\alpha g_1 + \beta g_2) = \alpha \int_a^b f\, dg_1 + \beta \int_a^b f\, dg_2.$$

The following Cauchy type convergence criterion is useful in many instances.

2·4 Proposition. *The integral* $\int_a^b f\, dg$ *exists if and only if for each* $\varepsilon > 0$ *there exists a* $\delta > 0$ *such that for any partitions* P *and* P' *with mesh less than* δ *and any sums* $S(f, g, P)$ *and* $S(f, g, P')$

$$|S(f, g, P) - S(f, g, P')| < \varepsilon.$$

PROOF. If $\int_a^b f\, dg = \lim_{|P| \to 0} S(f, g, P)$ exists, then clearly the condition holds.

Assume now that the condition holds. Take $\varepsilon > 0$ and let $\delta > 0$ correspond to $\varepsilon/2$. Let (P_n) be a sequence of partitions such that $\lim |P_n| = 0$ and for each P_n take a specific sum $S_n(f, g, P_n)$. Then, there exists a positive integer n_0 such that $|P_n| < \delta$ whenever $n \geq n_0$ and, hence,

$$|S_m(f, g, P_m) - S_n(f, g, P_n)| < \varepsilon/2 \quad \text{whenever } m, n \geq n_0.$$

Therefore, $(S_n(f, g, P_n))$ is a Cauchy sequence in $\mathbf{R}$ and, hence, converges. Let

$$\lim S_n(f, g, P_n) = r.$$

We now show that $\lim_{|P| \to 0} S(f, g, P) = r$. Take $n \geq n_0$ such that

$$|S_n(f, g, P_n) - r| < \varepsilon/2.$$

Then, for all partitions P with $|P| < \delta$ and all sums $S(f, g, P)$,

$$|S(f, g, P) - r| \leq |S(f, g, P) - S_n(f, g, P_n)| + |S_n(f, g, P_n) - r| < \varepsilon. \blacksquare$$

We can use this Cauchy criterion to show that if $\int_a^b f\, dg$ exists and $[c, d] \subset [a, b]$, then $\int_c^d f\, dg$ exists (Problem 2). Then, it follows easily that if $\int_a^b f\, dg$ exists and $c \in (a, b)$ then $\int_a^c f\, dg$ and $\int_c^b f\, dg$ exist and

$$\int_a^b f\, dg = \int_a^c f\, dg + \int_c^b f\, dg \qquad \text{(Problem 3).}$$

However, it is not always true that if $\int_a^c f\, dg$ and $\int_c^b f\, dg$ exist, then $\int_a^b f\, dg$ exists (Problem 4).

2 · 5 Example. Let f be a function from $[a, b]$ into **R** which is continuous at $c \in [a, b]$ and let χ_c be the function from **R** into **R** such that

$$\chi_c(x) = \begin{cases} 1 & \text{if } x = c \\ 0 & \text{if } x \neq c. \end{cases}$$

Show that

$$\int_a^b f\, d\chi_c = \begin{cases} 0 & \text{if } c \in (a, b) \\ -f(a) & \text{if } c = a \\ f(b) & \text{if } c = b. \end{cases}$$

SOLUTION. Let $P = \{(x_{k-1}, x_k) : k = 1, \cdots, n\}$ be a partition of $[a, b]$. Suppose $c \in (a, b)$. If $x_k \neq c$ for all $k = 0, \cdots, n$, then $S(f, \chi_c, P) = 0$. If $x_i = c$ for some $i = 1, \cdots, n-1$ then

$$\begin{aligned} S(f, \chi_c, P) &= f(z_i)(\chi_c(x_i) - \chi_c(x_{i-1})) + f(z_{i+1})(\chi_c(x_{i+1}) - \chi_c(x_i)) \\ &= f(z_i) - f(z_{i+1}). \end{aligned}$$

Since f is continuous at c, for each $\varepsilon > 0$ there exists a $\delta > 0$ such that $|f(c) - f(x)| < \varepsilon/2$ whenever $|c - x| < \delta$. Then, if $|P| < \delta$,

$$|S(f, \chi_c, P)| \le |f(z_i) - f(c)| + |f(z_{i+1}) - f(c)| < \varepsilon.$$

Therefore,

$$\int_a^b f\, d\chi_c = \lim_{|P| \to 0} S(f, \chi_c, P) = 0.$$

The cases $c = a$ and $c = b$ are similar and are left to the reader. ▮

2 · 6 Example. Suppose $(c, d) \subset [a, b]$, f is a function from $[a, b]$ into **R** which is continuous at c and d and $\chi_{(c,d)}$ is the characteristic function of (c, d). Show that

$$\int_a^b f\, d\chi_{(c,d)} = f(c) - f(d).$$

SOLUTION. Let $P = \{(x_{k-1}, x_k) : k = 1, \cdots, n\}$ be a partition of $[a, b]$ with $|P| < d - c$. If $c \in [x_{i-1}, x_i)$ and $d \in (x_{j-1}, x_j]$, then

$$\begin{aligned} S(f, \chi_{(c,d)}, P) &= \sum_{k=1}^{n} f(z_k)(\chi_{(c,d)}(x_k) - \chi_{(c,d)}(x_{k-1})) \\ &= f(z_i) - f(z_j). \end{aligned}$$

Therefore,

$$\int_a^b f\, d\chi_{(c,d)} = \lim_{|P| \to 0} S(f, \chi_{(c,d)}, P) = f(c) - f(d). \ ▮$$

2 · 7 Example. Let g be a step function with representation

$$g(x) = c_i \quad \text{for } x \in I_i, \quad i = 1, \cdots, r$$

where $I_i = (a_{i-1}, a_i)$ and $\{I_1, \cdots, I_r\}$ is a partition of $[a, b]$. If f is a function from $[a, b]$ into $\mathbf{R}$ which is continuous at a_i $(i = 0, \cdots, r)$, show that

$$\int_a^b f\,dg = f(a)(c_1 - g(a)) + \sum_{i=1}^{r-1} f(a_i)(c_{i+1} - c_i) + f(b)(g(b) - c_r).$$

SOLUTION. Since

$$g = \sum_{i=1}^{r} c_i \chi_{I_i} + \sum_{i=0}^{r} g(a_i)\chi_{a_i},$$

using 2 · 3, 2 · 5, and 2 · 6 we obtain

$$\begin{aligned}\int_a^b f\,dg &= \sum_{i=1}^{r} c_i \int_a^b f\,d\chi_{I_i} + \sum_{i=0}^{r} g(a_i) \int_a^b f\,d\chi_{a_i}\\ &= \sum_{i=1}^{r} c_i(f(a_{i-1}) - f(a_i)) - g(a)f(a) + g(b)f(b)\\ &= f(a)(c_1 - g(a)) + \sum_{i=1}^{r-1} f(a_i)(c_{i+1} - c_i) + f(b)(g(b) - c_r). \blacksquare\end{aligned}$$

Remark. If the step function g of Example 2 · 7 is the greatest integer function, then

$$\int_0^n f\,dg = \sum_{i=1}^{n} f(i)$$

since $c_i = i - 1$, $g(0) = 0$, and $g(n) = n$. Thus, the Riemann-Stieltjes integral includes sums as a special case. If we define the improper Riemann-Stieltjes integral

$$\int_0^\infty f\,dg = \lim_{b\to\infty} \int_0^b f\,dg,$$

then for g equal to the greatest integer function we obtain an infinite series:

$$\int_0^\infty f\,dg = \sum_{i=1}^{\infty} f(i).$$

For the Riemann-Stieltjes integral the following integration by parts result holds.

2 · 8 Proposition. *If f is integrable with respect to g, then g is integrable with respect to f and*

$$\int_a^b g\,df = \Big[fg\Big]_a^b - \int_a^b f\,dg.$$

PROOF. Let $P = \{(x_{k-1}, x_k) : k = 1, \cdots, n\}$ be a partition of $[a, b]$. Then,

$$S(g, f, P) = \sum_{k=1}^{n} g(z_k)(f(x_k) - f(x_{k-1}))$$

$$= g(z_n)f(b) - g(z_1)f(a) + \sum_{k=1}^{n-1} f(x_k)(g(z_k) - g(z_{k+1}))$$

$$= f(b)g(b) - f(a)g(a) - \sum_{k=0}^{n} f(x_k)(g(z_{k+1}) - g(z_k))$$

where $z_0 = a$ and $z_{n+1} = b$. Observe that $P' = \{(z_{k-1}, z_k) : k = 1, \cdots, n + 1\}$ is a partition of $[a, b]$ such that $|P'| \to 0$ as $|P| \to 0$. Thus,

$$\lim_{|P| \to 0} S(g, f, P) = f(b)g(b) - f(a)g(a) - \lim_{|P'| \to 0} S(f, g\ P')$$

$$= \Big[fg \Big]_a^b - \int_a^b f\ dg. \blacksquare$$

Using Proposition 2 · 8 and Example 2 · 7 we obtain:

2 · 9 Proposition. *Let f be a step function with representation*

$$f(x) = c_i \quad \textit{for} \quad x \in I_i, \quad i = 1, \cdots, r$$

where $I_i = (a_{i-1}, a_i)$ and $\{I_1, \cdots, I_r\}$ is a partition of $[a, b]$. If g is a function from $[a, b]$ into **R** *which is continuous at a_i $(i = 0, \cdots, r)$, then*

$$\int_a^b f\ dg = \sum_{i=1}^{r} c_i(g(a_i) - g(a_{i-1})).$$

Now consider the case where f is Riemann integrable and the integrator g is absolutely continuous. In this case, the Stieltjes integral of f with respect to g exists and is equal to the Lebesgue integral of fg'.

2 · 10 Proposition. *If f is Riemann integrable on $[a, b]$ and g is absolutely continuous on $[a, b]$, then f is Stieltjes integrable with respect to g on $[a, b]$ and*

$$\int_a^b f\ dg = \int_a^b fg'.$$

PROOF. Since f is Riemann integrable on $[a, b]$, it is bounded on $[a, b]$, say $|f(x)| \le M$ for all $x \in [a, b]$, and it is continuous a.e. on $[a, b]$. And, since g is absolutely continuous on $[a, b]$, by the Second Fundamental Theorem of Calculus

$$g(x) = g(a) + \int_a^x g' \quad \text{for all } x \in [a, b].$$

Let $P = \{(x_{k-1}, x_k) : k = 1, \cdots, n\}$ be a partition of $[a, b]$. Then,

$$S(f, g, P) = \sum_{k=1}^{n} f(z_k)(g(x_k) - g(x_{k-1})) = \sum_{k=1}^{n} f(z_k) \int_{x_{k-1}}^{x_k} g'$$
$$= \int_a^b hg'$$

where $h(x) = f(z_k)$ for all $x \in (x_{k-1}, x_k)$.

If we now take a sequence (P_i) of partitions of $[a, b]$ such that $\lim |P_i| = 0$, then we obtain a sequence (h_i) of step functions such that $\lim h_i(x) = f(x)$ for a.a.$x \in [a, b]$. Since $|h_i g'| \leq M|g'|$, the Lebesgue Dominated Convergence Theorem (p. 130) implies that $fg' \in \mathscr{L}[a, b]$ and

$$\lim \int_a^b h_i g' = \int_a^b fg'.$$

Thus, for any sequence (P_i) such that $\lim |P_i| = 0$,

$$\lim S(f, g, P_i) = \int_a^b fg'$$

and, hence,

$$\lim_{|P| \to 0} S(f, g, P) = \int_a^b fg'. \quad \blacksquare$$

From Proposition 2 · 10 we infer that if f and g' are Riemann integrable on $[a, b]$, then,

$$\int_a^b f \, dg = \int_a^b fg'$$

where the integral on the right is a Riemann integral. For, if g' is Riemann integrable on $[a, b]$, it is defined and bounded on $[a, b]$ and, hence, g is absolutely continuous on $[a, b]$. Also, fg' is Riemann integrable on $[a, b]$ and its Riemann integral is the same as its Lebesgue integral.

PROBLEMS

1. Evaluate the Riemann-Stieltjes integral $\int_0^2 f \, dg$ if

a. $f(x) = x^2$
$g(x) = [x]$, the greatest integer function

b. $f(x) = [x]$
$g(x) = x^2$

c. $f(x) = x^2$

$$g(x) = \begin{cases} 3x, & x \in [0, 1) \\ 3x + 1, & x \in [1, 2) \\ 8, & x = 2 \end{cases}$$

2. If $\int_a^b f \, dg$ exists and $[c, d] \subset [a, b]$, show that $\int_c^d f \, dg$ exists.

3. If $\int_a^b f\,dg$ exists and $c \in (a, b)$, show that $\int_a^c f\,dg$ and $\int_c^b f\,dg$ exist and

$$\int_a^b f\,dg = \int_a^c f\,dg + \int_c^b f\,dg.$$

4. Let $c \in (a, b)$ and

$$f(x) = \begin{cases} 0, & x \in [a, c] \\ 1, & x \in (c, b] \end{cases}, \qquad g(x) = \begin{cases} 0, & x \in [a, c) \\ 1, & x \in [c, b] \end{cases}.$$

Show that $\int_a^c f\,dg$ and $\int_c^b f\,dg$ exist but $\int_a^b f\,dg$ does not exist.

5. If $g \in BV[a, b]$, $|f(x)| \leq M$ for all $x \in [a, b]$, and $\int_a^b f\,dg$ exists, show that

$$\left|\int_a^b f\,dg\right| \leq M V_a^b(g).$$

3 · THE DARBOUX–STIELTJES INTEGRAL

Suppose f is bounded on $[a, b]$ and g is nondecreasing on $[a, b]$. In this case, using the method employed in Chapter 6 to define the Riemann integral we can define an integral called the Darboux-Stieltjes integral and denoted by (DS) $\int_a^b f\,dg$. We will show that, unless f and g have a common point of discontinuity, this integral has the same value as the Riemann-Stieltjes integral $\int_a^b f\,dg$ which we denote by (RS) $\int_a^b f\,dg$ whenever we wish to distinguish it from the Darboux-Stieltjes integral.

For any partition $P = \{(x_{k-1}, x_k) : k = 1, \cdots, n\}$ of $[a, b]$, let

$$m_k(f) = \inf\{f(x) : x \in [x_{k-1}, x_k]\},$$

$$M_k(f) = \sup\{f(x) : x \in [x_{k-1}, x_k]\},$$

$$L(f, g, P) = \sum_{k=1}^{n} m_k(f)(g(x_k) - g(x_{k-1})),$$

$$U(f, g, P) = \sum_{k=1}^{n} M_k(f)(g(x_k) - g(x_{k-1})).$$

Since $g(x_k) - g(x_{k-1})$ is nonnegative, this development requires only minor modifications in that given for the Riemann integral in §2 of Chapter 6.

If P' is a refinement of P, then

$$L(f, g, P) \leq L(f, g, P') \quad \text{and} \quad U(f, g, P') \leq U(f, g, P).$$

Then, for any two partitions P_1 and P_2 of $[a, b]$,

$$m(g(b) - g(a)) \leq L(f, g, P_1) \leq U(f, g, P_2) \leq M(g(b) - g(a))$$

where $m \leq f(x) \leq M$ for all $x \in [a, b]$. If $\mathscr{P}$ denotes the set of all partitions of $[a, b]$, let the lower and upper integrals be defined by

$$\underline{\int_a^b} f\,dg = \sup\{L(f, g, P) : P \in \mathscr{P}\} \quad \text{and} \quad \overline{\int_a^b} f\,dg = \inf\{U(f, g, P) : P \in \mathscr{P}\}.$$

Then, for each $P \in \mathscr{P}$

$$L(f, g, P) \leq \underline{\int_a^b} f\, dg \leq \overline{\int_a^b} f\, dg \leq U(f, g, P).$$

If the lower and upper integrals are equal, then we say that the **Darboux-Stieltjes integral** of f with respect to g exists and

$$(\mathrm{DS}) \int_a^b f\, dg = \underline{\int_a^b} f\, dg = \overline{\int_a^b} f\, dg.$$

Note that if $g(x) = x$, then this integral is the Riemann integral $\int_a^b f$. It is clear that (DS) $\int_a^b f\, dg$ exists if and only if for each $\varepsilon > 0$ there is a partition P of $[a, b]$ such that

$$U(f, g, P) - L(f, g, P) < \varepsilon.$$

We now compare the Riemann-Stieltjes and Darboux-Stieltjes integrals.

3 · 1 Proposition. *Let f be bounded on $[a, b]$ and g be nondecreasing on $[a, b]$. If* (RS) $\int_a^b f\, dg$ *exists, then* (DS) $\int_a^b f\, dg$ *exists and*

$$(\mathrm{DS}) \int_a^b f\, dg = (\mathrm{RS}) \int_a^b f\, dg.$$

PROOF. Assume $g(a) < g(b)$; otherwise the result is trivial. For each $\varepsilon > 0$ there exists a partition $P = \{(x_{k-1}, x_k) : k = 1, \cdots, n\}$ such that for any $z_k \in [x_{k-1}, x_k]$

$$(\mathrm{RS}) \int_a^b f\, dg - \frac{\varepsilon}{2} < \sum_{k=1}^{n} f(z_k)(g(x_k) - g(x_{k-1})) < (\mathrm{RS}) \int_a^b f\, dg + \frac{\varepsilon}{2}.$$

For each $k = 1, \cdots, n$ we can choose $z_k \in [x_{k-1}, x_k]$ such that

$$f(z_k) < m_k(f) + \frac{\varepsilon}{2(g(b) - g(a))}.$$

Then,

$$\sum_{k=1}^{n} f(z_k)(g(x_k) - g(x_{k-1})) < L(f, g, P) + \frac{\varepsilon}{2}$$

and, hence,

$$(\mathrm{RS}) \int_a^b f\, dg - \varepsilon < L(f, g, P) \leq \underline{\int_a^b} f\, dg.$$

Similarly,

$$\overline{\int_a^b} f\, dg \leq U(f, g, P) < (\mathrm{RS}) \int_a^b f\, dg + \varepsilon.$$

Since for each $\varepsilon > 0$

$$(\mathrm{RS}) \int_a^b f\, dg - \varepsilon < \underline{\int_a^b} f\, dg \le \overline{\int_a^b} f\, dg \not< (\mathrm{RS}) \int_a^b f\, dg + \varepsilon,$$

$$(\mathrm{RS}) \int_a^b f\, dg = \underline{\int_a^b} f\, dg = \overline{\int_a^b} f\, dg = (\mathrm{DS}) \int_a^b f\, dg. \blacksquare$$

3 · 2 Proposition. *Let f be bounded on $[a, b]$ and g be nondecreasing on $[a, b]$. If f and g have no common points of discontinuity and* (DS) $\int_a^b f\,dg$ *exists, then* (RS) $\int_a^b f\,dg$ *exists and*

$$(\mathrm{RS}) \int_a^b f\, dg = (\mathrm{DS}) \int_a^b f\, dg.$$

PROOF. Take $\varepsilon > 0$. Let $P_1 = \{I_j : j = 1, \cdots, r\}$, where $I_j = (x_{j-1}, x_j)$, be a partition of $[a, b]$ such that

$$U(f, g, P_1) - L(f, g, P_1) < \varepsilon/2.$$

If P' is any refinement of P_1, then,

$$\left| S(f, g, P') - (\mathrm{DS}) \int_a^b f\, dg \right| < \frac{\varepsilon}{2}.$$

Since f and g are both bounded, we can choose a number M such that $|f(x)| \le M$ and $|g(x)| \le M$ for all $x \in [a, b]$. At the points x_j $(j = 0, \cdots, r)$ either f or g is continuous. Hence, there exists a $\delta > 0$ such that, for all $j = 0, \cdots, r$, if $|x - x_j| < \delta$, then,

$$|f(x) - f(x_j)| < \frac{\varepsilon}{8rM} \quad \text{or} \quad |g(x) - g(x_j)| < \frac{\varepsilon}{8rM}.$$

For any partition $P = \{J_k : k = 1, \cdots, s\}$, where $J_k = (y_{k-1}, y_k)$, such that $|P| < \delta$, take a sum

$$S(f, g, P) = \sum_{k=1}^{s} f(z_k)(g(y_k) - g(y_{k-1})).$$

The partition $P' = \{I_j \cap J_k : j = 1, \cdots, r;\ k = 1, \cdots, s\}$ is a common refinement of P_1 and P. If $I_j \cap J_k \ne \varnothing$, let $I_j \cap J_k = (a_{jk}, b_{jk})$ and form the sum

$$S(f, g, P') = \sum_{j,k} f(z_{jk})(g(b_{jk}) - g(a_{jk})),$$

where

$$z_{jk} = z_k \quad \text{if} \quad I_j \cap J_k = J_k$$

$$z_{jk} = x_{j-1} \text{ or } x_j \quad \text{if} \quad I_j \cap J_k \ne J_k.$$

Then,

$$S(f, g, P') - S(f, g, P) = \sum_{j,k} (f(z_{jk}) - f(z_k))(g(b_{jk}) - g(a_{jk})).$$

This sum has less than $2r$ nonzero terms and for each nonzero term we have $z_{jk} = x_{j-1}$ or x_j and $a_{jk} = x_{j-1}$ or $b_{jk} = x_j$. Thus,

$$|S(f, g, P') - S(f, g, P)| < (2r)\left(\frac{\varepsilon}{8rM}\right)(2M) = \frac{\varepsilon}{2}$$

and, hence, for any partition P with $|P| < \delta$,

$$\left|S(f, g, P) - (\text{DS}) \int_a^b f\, dg\right| \leq |S(f, g, P) - S(f, g, P')|$$
$$+ \left|S(f, g, P') - (\text{DS}) \int_a^b f\, dg\right| < \varepsilon.$$

That is, (RS) $\int_a^b f\, dg =$ (DS) $\int_a^b f\, dg$. ▮

Propositions 3 · 1 and 3 · 2 show that if f is bounded and g is nondecreasing, then the Riemann-Stieltjes and Darboux-Stieltjes integrals coincide except possibly when f and g have a common point of discontinuity. In the following example f and g have a common point of discontinuity and (DS) $\int_a^b f\, dg$ exists but (RS) $\int_a^b f\, dg$ does not exist. In fact, it is true in general that (RS) $\int_a^b f\, dg$ does not exist whenever f and g have a common discontinuity [15].

Let

$$f(x) = \begin{cases} 0, & x \in [0, 1] \\ 1, & x \in (1, 2] \end{cases} \quad \text{and} \quad g(x) = \begin{cases} 0, & x \in [0, 1) \\ 1, & x \in [1, 2]. \end{cases}$$

Then, for $P = \{(0, 1), (1, 2)\}$,

$$L(f, g, P) = 0 = U(f, g, P)$$

and, hence,

$$(\text{DS}) \int_0^2 f\, dg = 0.$$

To show that (RS) $\int_0^2 f\, dg$ does not exist, for each $\delta > 0$, take a partition $P = \{(x_{k-1}, x_k) : k = 1, \cdots, n\}$ of $[0, 2]$ such that $|P| < \delta$ and $1 \in (x_{j-1}, x_j)$ for some $j = 1, \cdots, n$. Let

$$S(f, g, P) = \sum_{k=1}^{n} f(z_k)(g(x_k) - g(x_{k-1})) \quad \text{where} \quad z_j > 1$$

and

$$S'(f, g, P) = \sum_{k=1}^{n} f(z_k')(g(x_k) - g(x_{k-1}))$$

where $z_k' = z_k$ for $k \neq j$ and $z_j' \leq 1$. Then,

$$S(f, g, P) - S'(f, g, P) = (f(z_j) - f(z_j'))(g(x_j) - g(x_{j-1})) = 1.$$

Therefore, (RS) $\int_0^2 f\, dg$ does not exist.

PROBLEMS

1. If g is nondecreasing, $c \in (a, b)$, and (DS) $\int_a^c f\,dg$ and (DS) $\int_c^b f\,dg$ exist, show that (DS) $\int_a^b f\,dg$ exists and

$$\text{(DS)} \int_a^b f\ dg = \text{(DS)} \int_a^c f\ dg + \text{(DS)} \int_c^b f\ dg.$$

2. If g is nondecreasing, $f_1 \leq f_2$, and (RS) $\int_a^b f_1\ dg$ and (RS) $\int_a^b f_2\ dg$ exist, show that

$$\text{(RS)} \int_a^b f_1\ dg \leq \text{(RS)} \int_a^b f_2\ dg.$$

3. If f is continuous on $[a, b]$ and g is nondecreasing on $[a, b]$, prove that $\int_a^b f\,dg$ exists.

4. If f is continuous on $[a, b]$ and g is of bounded variation on $[a, b]$, prove that $\int_a^b f\,dg$ exists.

5. If $f \in BV[a, b]$ and $g \in C[a, b]$, show that $\int_a^b f\,dg$ exists.

4 · STEP FUNCTIONS

Let g be a fixed nondecreasing function on $\mathbf{R}$. An open interval (a, b) will be called **admissible** (with respect to g) if g is continuous at a and b. We define a measure μ_g on admissible open intervals by the following rule:

$$\mu_g((a, b)) = g(b) - g(a).$$

Let f be a step function with representation $f(x) = c_i$ for $x \in I_i$ and $f(x) = 0$ for $x \notin [a, b]$ where $\{I_1, \cdots, I_r\}$ is a partition of $[a, b]$ into admissible open intervals $I_i = (a_{i-1}, a_i)$. Define an integral for such step functions, called **admissible step functions** as follows:

$$\int f\ dg = \sum_{i=1}^{r} c_i\,\mu_g(I_i). \tag{4 · 1}$$

Since f and g have no common discontinuity, Proposition 2 · 9 implies that

$$\int_a^b f\ dg = \sum_{i=1}^{r} c_i(g(a_i) - g(a_{i-1})) = \sum_{i=1}^{r} c_i\,\mu_g(I_i).$$

Thus, the integral $\int f\,dg$ defined for admissible step functions coincides with the Riemann-Stieltjes integral $\int_a^b f\,dg$.

The following result is an immediate consequence of the definition 4 · 1.

4 · 2 Proposition. *If f is an admissible step function such that $f(x) \geq 0$ for all x, then $\int f\,dg \geq 0$.*

Also, minor modifications in the proofs of Propositions 3 · 4 and 3 · 5 of Chapter 6 yield:

4 · 3 Proposition. *The set $\mathscr{S}^g$ of admissible step functions is a vector lattice and the integral is a linear functional on $\mathscr{S}^g$.*

A set E in $\mathbf{R}$ is said to be of **μ_g-measure zero** if for each $\varepsilon > 0$ there is a countable collection $\{I_k\}$ of admissible open intervals such that

$$E \subset \bigcup_{k=1}^{\infty} I_k \quad \text{and} \quad \sum_{k=1}^{\infty} \mu_g(I_k) \leq \varepsilon.$$

In what follows "almost everywhere" and "almost all" will refer to μ_g-measure zero. It is easy to see that a countable union of sets of μ_g-measure zero is a set of μ_g-measure zero. However, a set consisting of a single point need not be of μ_g-measure zero. If g is discontinuous at c, then $g(c-) < g(c+)$ and, for any admissible open interval (a, b) containing c, we have

$$\mu_g((a, b)) = g(b) - g(a) \geq g(c+) - g(c-).$$

Hence, if c is a point of discontinuity of g, then $\{c\}$ is not of μ_g-measure zero. Clearly, if g is continuous at c, then $\{c\}$ is of μ_g-measure zero.

A set E in $\mathbf{R}$ is said to be a **μ_g-null set** if there is a nondecreasing sequence (f_n) of nonnegative admissible step functions such that $(f_n(x))$ diverges to ∞ for each $x \in E$ and $(\int f_n \, dg)$ converges.

4 · 4 Proposition. *A set E is a μ_g-null set if and only if it is of μ_g-measure zero.*

The proof of this proposition is essentially the same as that of Proposition 4 · 3 of Chapter 6. Also, minor modifications in the proofs of Propositions 4 · 4 and 4 · 5 of Chapter 6 yield:

4 · 5 Proposition. *If (f_n) is a nonincreasing sequence of nonnegative admissible step functions and* $\lim \int f_n \, dg = 0$, *then*

$$\lim f_n(x) = 0 \quad \textit{for a.a.x.}$$

4 · 6 Proposition. *If (f_n) is a nonincreasing sequence of nonnegative admissible step functions and* $\lim f_n(x) = 0$ *for a.a.x, then*

$$\lim \int f_n \, dg = 0.$$

5 · EXISTENCE OF THE RIEMANN–STIELTJES INTEGRAL

If f is a bounded function on $[a, b]$ and g is nondecreasing on $[a, b]$, then we can obtain a criterion for the existence of the Riemann-Stieltjes integral $\int_a^b f \, dg$ which is similar to that for the Riemann integral. First, we will take care of a few details.

If $\int_a^b f\,dg$ exists, then f and g cannot both be discontinuous (from the right) at a or (from the left) at b. For example, if f and g are both discontinuous at a, then $g(a) < g(a+)$. Also, there exists an $\varepsilon > 0$ such that for each partition $P = \{(x_{k-1}, x_k) : k = 1, \cdots, n\}$ of $[a, b]$ there exists a point $y \in (a, x_1)$ such that

$$|f(a) - f(y)| \geq \varepsilon.$$

For a partition P take

$$S(f, g, P) = \sum_{k=1}^{n} f(z_k)(g(x_k) - g(x_{k-1})) \quad \text{where } z_1 = a$$

and

$$S'(f, g, P) = \sum_{k=1}^{n} f(z_k')(g(x_k) - g(x_{k-1}))$$

where $z_1' = y$ and $z_k' = z_k$ for $k = 2, \cdots, n$. Then

$$\begin{aligned} |S(f, g, P) - S'(f, g, P)| &= |f(a) - f(y)|\,(g(x_1) - g(a)) \\ &\geq \varepsilon\,(g(a+) - g(a)). \end{aligned}$$

Thus, $\int_a^b f\,dg$ does not exist.

If $\int_a^b f\,dg$ exists, then we may assume that g is continuous at a and b. For, if g is discontinuous at a, then f is continuous at a and $\int_a^b f\,d\chi_a$ exists. If we let

$$g_1(a) = g(a+) \quad \text{and} \quad g_1(x) = g(x) \quad \text{for} \quad x \in (a, b],$$

then g_1 is continuous at a and

$$g_1 = g + (g(a+) - g(a))\chi_a.$$

Since both $\int_a^b f\,dg$ and $\int_a^b f\,d\chi_a$ exist, $\int_a^b f\,dg_1$ exists.

A partition $P = \{(x_{j-1}, x_j) : j = 1, \cdots, s\}$ is said to be **admissible** with respect to g if g is continuous at each of the points x_j. Now assume that f and g have no common points of discontinuity and g is continuous at a and b. For each partition $P^1 = \{(x_{i-1}^1, x_i^1) : i = 1, \cdots, r\}$ of $[a, b]$ such that $|P^1| < \delta$ and for each $\varepsilon > 0$ we will show that there exists an admissible partition P such that $|P| < \delta$ and

$$\text{(5 · 1)} \qquad \begin{aligned} L(f, g, P^1) &\leq L(f, g, P) + \varepsilon(g(b) - g(a)) \\ U(f, g, P^1) &\geq U(f, g, P) - \varepsilon(g(b) - g(a)). \end{aligned}$$

In constructing an admissible partition P from a given partition P^1 we will replace each partition point of P^1 at which g is discontinuous by another point at which g is continuous. This can change the corresponding lower sum and the change can even be negative. However, what we show is that if we make the choice of the new partition points correctly the lower sum will not be reduced very much if at all.

If g is discontinuous at x_i^1, then f is continuous at x_i^1 and, hence, there exists a $\delta_i > 0$ such that

$$|f(x) - f(x_i^1)| < \varepsilon/3 \quad \text{whenever} \quad |x - x_i^1| < \delta_i.$$

Since the set of discontinuities of g is countable, by adding points at which g is continuous we can obtain a refinement $P^2 = \{(x_{j-1}^2, x_j^2) : j = 1, \cdots, s\}$ of P^1 with mesh less than each δ_i. We now modify the partition P^2 by replacing any point x_j^2 at which g is discontinuous by a point x_j at which g is continuous and which is slightly to the right of x_j^2. That is, if g is discontinuous at x_j^2, we can choose a point $x_j \in (x_{j-1}^2, x_{j-1}^2 + \delta) \cap (x_j^2, x_{j+1}^2)$ at which g is continuous. If g is continuous at x_j^2, choose $x_j = x_j^2$. Then, the partition $P = \{(x_{j-1}, x_j) : j = 1, \cdots, s\}$ is admissible and $|P| < \delta$. Let

$$m_j^2 = \inf \{f(x) : x \in [x_{j-1}^2, x_j^2]\}$$

and

$$m_j = \inf \{f(x) : x \in [x_{j-1}, x_j]\}$$

and for brevity let $g_j = g(x_j)$, $g_j^2 = g(x_j^2)$, and $\Delta g_j = g_j - g_j^2$. Since $x_j \geq x_j^2$ and g is nondecreasing $\Delta g_j \geq 0$. If $x_j > x_j^2$ then f is continuous at x_j^2 and

$$f(x_j^2) - \varepsilon/3 < f(x) < f(x_j^2) + \varepsilon/3 \quad \text{for all } x \in [x_{j-1}^2, x_{j+1}^2].$$

This implies that $|m_j^2 - m_{j+1}^2| \leq 2\varepsilon/3$. Also since $[x_{j-1}, x_j] \subset [x_{j-1}^2, x_{j+1}^2]$ we have

$$m_j^2 \leq f(x_j^2) < f(x) + \varepsilon/3 \quad \text{for all } x \in [x_{j-1}, x_j]$$

which implies that

$$m_j^2 \leq m_j + \varepsilon/3.$$

If $x_j = x_j^2$, then $\Delta g_j = 0$ and, moreover, since $[x_{j-1}, x_j] \subset [x_{j-1}^2, x_{j+1}^2]$

$$m_j^2 \leq m_j < m_j + \varepsilon/3.$$

Thus, for all $j = 1, \cdots, s$ we have $m_j^2 \leq m_j + \varepsilon/3$. Now

$$\begin{aligned} g_j^2 - g_{j-1}^2 &= g_j - g_{j-1} + (g_{j-1} - g_{j-1}^2) - (g_j - g_j^2) \\ &= g_j - g_{j-1} - \Delta g_j + \Delta g_{j-1}. \end{aligned}$$

Hence,

$$\begin{aligned} L(f, g, P^1) \leq L(f, g, P^2) &= \sum_{j=1}^{s} m_j^2 (g_j^2 - g_{j-1}^2) \\ &\leq \sum_{j=1}^{s} \left(m_j + \frac{\varepsilon}{3}\right)(g_j - g_{j-1}) - \sum_{j=1}^{s} m_j^2 (\Delta g_j - \Delta g_{j-1}) \\ &\leq L(f, g, P) + \frac{\varepsilon}{3}(g(b) - g(a)) + \sum_{j=1}^{s-1} |m_j^2 - m_{j+1}^2| \, \Delta g_j \\ &\leq L(f, g, P) + \varepsilon(g(b) - g(a)) \end{aligned}$$

since for those terms in the last sum in which $\Delta g_j \neq 0$ we have $|m_j^2 - m_{j+1}^2| \leq 2\varepsilon/3$ while $\sum \Delta g_j \leq g(b) - g(a)$.

The proof of the second of Inequalities 5 · 1 is similar.

If $P = \{I_1, \cdots, I_r\}$ is an admissible partition of $[a, b]$, let

$$(5 \cdot 2)\quad h(x) = \begin{cases} m_i(f) & \text{for } x \in I_i \\ f(x) & \text{for } x \in \bar{I}_i \sim I_i \\ 0 & \text{for } x \notin [a, b] \end{cases} \quad \text{and} \quad H(x) = \begin{cases} M_i(f) & \text{for } x \in I_i \\ f(x) & \text{for } x \in \bar{I}_i \sim I_i \\ 0 & \text{for } x \notin [a, b], \end{cases}$$

where $m_i(f) = \inf\{f(x) : x \in \bar{I}_i\}$ and $M_i(f) = \sup\{f(x) : x \in \bar{I}_i\}$. Then h and H are admissible step functions and

$$\int h\, dg = L(f, g, P) \quad \text{and} \quad \int H\, dg = U(f, g, P).$$

5 · 3 Theorem. *Let f be bounded and g be nondecreasing on $[a, b]$. If $\int_a^b f\, dg$ exists, then the set of discontinuities of f on $[a, b]$ is of μ_g-measure zero.*

PROOF. Assume that g is continuous at a and b. Take a sequence (P_n) of admissible partitions of $[a, b]$ such that $P_n \subset P_{n+1}$ and $\lim |P_n| = 0$. Using 5 · 2 we obtain sequences (h_n) and (H_n) of admissible step functions which are, respectively, nondecreasing and nonincreasing. Since $\lim S(f, g, P_n) = \int_a^b f\, dg$, it follows that

$$\lim \int h_n\, dg = \int_a^b f\, dg = \lim \int H_n\, dg.$$

Let $k_n = H_n - h_n$. Then, (k_n) is a nonincreasing sequence of nonnegative admissible step functions with

$$\lim \int k_n\, dg = 0.$$

Hence, by Proposition 4 · 5

$$\lim k_n(x) = 0 \quad \text{for a.a.} x.$$

Since $h_n(x) \leq f(x) \leq H_n(x)$ for all $x \in [a, b]$, this implies that

$$\lim h_n(x) = f(x) = \lim H_n(x) \quad \text{for a.a.} x \in [a, b].$$

The rest of the proof is the same as that of Theorem 5 · 1, p. 118. ▌

5 · 4 Theorem. *Let f be bounded and g be nondecreasing on $[a, b]$. If the set of discontinuities of f is of μ_g-measure zero, then $\int_a^b f\, dg$ exists.*

PROOF. Since the set of discontinuities of f is of μ_g-measure zero, f and g have no common discontinuities. Thus, we may assume as before that g is continuous at a and b. Also, if (DS) $\int_a^b f\, dg$ exists, then $\int_a^b f\, dg =$ (RS) $\int_a^b f\, dg$ exists (Proposition 3 · 2).

From the proof of Proposition $2 \cdot 1$ of Chapter 6 and the discussion preceding Proposition $5 \cdot 3$ we see that there exists a sequence (P_n) of admissible partitions of $[a, b]$ such that $P_n \subset P_{n+1}$, $\lim |P_n| = 0$, and

$$\lim L(f, g, P_n) = \underline{\int_a^b} f\, dg \quad \text{and} \quad \lim U(f, g, P_n) = \overline{\int_a^b} f\, dg.$$

Let (h_n) and (H_n) be the corresponding sequences of admissible step functions and $k_n = H_n - h_n$. If E_1 is the set of μ_g-measure zero where f is not continuous and E_2 is the set of μ_g-measure zero comprised of the endpoints of intervals in the partitions P_n, let $E = E_1 \cup E_2$. Then, for each $x \in [a, b] \sim E$ and each $\varepsilon > 0$ there exists a $\delta > 0$ such that

$$|f(x) - f(y)| < \varepsilon/2 \quad \text{whenever } y \in (x - \delta, x + \delta) \cap [a, b].$$

Also, there exists a positive integer n_0 such that $|P_n| < \delta$ whenever $n \geq n_0$. Thus, for each $n \geq n_0$, if x is in the interval I_i^n in P_n then

$$k_n(x) = H_n(x) - h_n(x) = M_i(f) - m_i(f) \leq \varepsilon.$$

This shows that $\lim k_n(x) = 0$ for a.a.x and, therefore, by Proposition $4 \cdot 6$

$$\lim \int k_n\, dg = 0;$$

that is,

$$\underline{\int_a^b} f\, dg = \lim \int h_n\, dg = \lim \int H_n\, dg = \overline{\int_a^b} f\, dg. \quad \blacksquare$$

The above criterion for the existence of the Riemann-Stieltjes integral $\int_a^b f\, dg$ can be extended to the case where g is a function of bounded variation on $[a, b]$. If $g \in BV[a, b]$, then $g = g_1 - g_2$ where g_1 and g_2 are nonincreasing functions on $[a, b]$ and g_1 and g_2 are continuous where g is continuous. Also, if $\int_a^b f\, dg_1$ and $\int_a^b f\, dg_2$ exist, then $\int_a^b f\, dg$ exists and

$$\int_a^b f\, dg = \int_a^b f\, dg_1 - \int_a^b f\, dg_2.$$

5 · 5 Corollary. *If $f \in C[a, b]$ and $g \in BV[a, b]$, then $\int_a^b f\, dg$ exists.*

5 · 6 Corollary. *If $f \in BV[a, b]$ and $g \in C[a, b]$, then $\int_a^b f\, dg$ exists.*

PROOF. By Corollary $5 \cdot 5$, g is integrable with respect to f. Then, by Proposition $2 \cdot 8$, f is integrable with respect to g and

$$\int_a^b f\, dg = \Big[fg\Big]_a^b - \int_a^b g\, df. \quad \blacksquare$$

PROBLEMS

1. If f_1 and f_2 are bounded and Riemann-Stieltjes integrable with respect to the nondecreasing function g on $[a, b]$, show that $f_1 f_2$ is integrable with respect to g on $[a, b]$.

2. If f is bounded and Riemann-Stieltjes integrable with respect to the nondecreasing function g on $[a, b]$, show that $|f|$ is integrable with respect to g on $[a, b]$ and

$$\left|\int_a^b f\,dg\right| \leq \int_a^b |f|\,dg.$$

3. If f is continuous and g is nondecreasing on $[a, b]$, show that there exists a point $c \in [a, b]$ such that

$$\int_a^b f\,dg = f(c)(g(b) - g(a)).$$

4. If $f \in C[a, b]$, $g \in BV[a, b]$, and $F(x) = \int_a^x f\,dg$ for $x \in [a, b]$, show that $F \in BV[a, b]$. Also, if g is continuous at x then F is continuous at x.

6 · THE RIESZ REPRESENTATION THEOREM

Corollary 5 · 5 shows that if $g \in BV[a, b]$, then we can define a linear functional F_g on $C[a, b]$ as follows:

$$F_g(f) = \int_a^b f\,dg.$$

In fact, F_g is a bounded linear functional, since (Problem 5, p. 257)

$$|F_g(f)| = \left|\int_a^b f\,dg\right| \leq V_a^b(g)\,\|f\|.$$

Thus, $\|F_g\| \leq V_a^b(g)$.

We now show that every bounded linear functional on $C[a, b]$ can be represented as a Riemann-Stieltjes integral with respect to some function $g \in BV[a, b]$. This result, known as the *Riesz Representation Theorem*, is due to F. Riesz who published it in 1909 [31].

6 · 1 Theorem. (F. Riesz). *If F is a bounded linear functional on $C[a, b]$, then there exists a function $g \in BV[a, b]$ such that*

$$F(f) = \int_a^b f\,dg \quad \textit{for all } f \in C[a, b]$$

and $\|F\| = V_a^b(g)$.

PROOF. Since $C[a, b]$ is a subspace of the normed linear space $B[a, b]$ of all bounded functions on $[a, b]$ with the sup-norm, $\|f\| = \sup\{|f(x)| : x \in [a, b]\}$, the Hahn-Banach Theorem implies that F can be extended to $B[a, b]$. The extended linear functional, which we still denote by F, will have the same norm as the original F.

Define a function g on $[a, b]$ by the rule:

$$g(a) = 0 \quad \text{and} \quad g(x) = F(\chi_{[a, x)}) \quad \text{for } x \in (a, b].$$

If $P = \{(x_{i-1}, x_i) : i = 1, \cdots, k\}$ is a partition of $[a, b]$, then, letting $\sigma_i = \operatorname{sgn}(g(x_i) - g(x_{i-1}))$, we have

$$\sum_{i=1}^{k} |g(x_i) - g(x_{i-1})| = F\left[\sum_{i=1}^{k} \sigma_i \chi_{[x_{i-1}, x_i)}\right]$$
$$\leq \|F\| \left\|\sum_{i=1}^{k} \sigma_i \chi_{[x_{i-1}, x_i)}\right\| = \|F\|.$$

This shows that $g \in BV[a, b]$ and $V_a^b(g) \leqq \|F\|$.

For any $f \in C[a, b]$ and any partition P of $[a, b]$ take $z_i \in [x_{i-1}, x_i]$ and let

$$s = \sum_{i=1}^{k} f(z_i)\chi_{[x_{i-1}, x_i)}.$$

Then

$$F(s) = \sum_{i=1}^{k} f(z_i)(g(x_i) - g(x_{i-1}))$$

and

$$\lim_{|P| \to 0} F(s) = \int_a^b f \, dg.$$

Since $f \in C[a, b]$, f is uniformly continuous on $[a, b]$; that is, for each $\varepsilon > 0$ there exists a $\delta > 0$ such that

$$|f(x) - f(y)| < \varepsilon \quad \text{whenever } x, y \in [a, b] \text{ and } |x - y| < \delta.$$

Thus, if $|P| < \delta$, then

$$|F(f) - F(s)| \leq \|F\| \, \|f - s\| < \varepsilon \, \|F\|.$$

Therefore,

$$F(f) = \lim_{|P| \to 0} F(s) = \int_a^b f \, dg.$$

Thus, as noted at the beginning of this section, $\|F\| \leq V_a^b(g)$ and, hence, $\|F\| = V_a^b(g)$. ∎

We now show that $C^*[a, b]$, the dual of $C[a, b]$, is isometrically isomorphic to a subspace of the normed linear space $BV[a, b]$, where $\|g\| = |g(a)| + V_a^b(g)$ (Problem 8, p. 233). Let T be the transformation from $BV[a, b]$ into $C^*[a, b]$ defined as follows: for $g \in BV[a, b]$,

$$(6 \cdot 2) \qquad T(g) = F_g \quad \text{where } F_g(f) = \int_a^b f \, dg, \quad f \in C[a, b].$$

Clearly, T is a linear transformation but it is not one-to-one. For example, if $c \in (a, b)$ and $r \in \mathbf{R}$, then $T(g) = T(g + r\chi_c)$. The following result describes the relationship between g_1 and g_2 when $T(g_1) = T(g_2)$; note that if $g = g_1 - g_2$, then $T(g_1) = T(g_2)$ if and only if $T(g) = 0$.

6 · 3 Proposition. *If $g \in BV[a, b]$, then $T(g) = 0$ if and only if $g(a) = g(a+) = g(x-) = g(x+) = g(b-) = g(b)$ for all $x \in (a, b)$.*

Proof. Suppose $T(g) = 0$; that is, $\int_a^b f\,dg = 0$ for all $f \in C[a, b]$. For $f = 1$ we obtain

$$0 = \int_a^b dg = g(b) - g(a)$$

and, hence, $g(b) = g(a)$.

Now take $y \in (a, b)$ such that g is continuous at y. For h sufficiently small define

$$f(x) = \begin{cases} 1, & x \in [a, y) \\ 1 - \dfrac{1}{h}(x - y), & x \in [y, y + h) \\ 0, & x \in [y + h, b]. \end{cases}$$

Then, $f \in C[a, b]$ and

$$0 = g(y) - g(a) + \int_y^{y+h} \left[1 - \frac{1}{h}(x - y)\right] dg(x).$$

Thus,

$$|g(y) - g(a)| \leq V_y^{y+h}(g) = v(y + h) - v(y)$$

and, since the variation function v is continuous at y, if we let $h \to 0$, we obtain

$$g(y) = g(a).$$

Since g is continuous a.e. on $[a, b]$, for any $x \in [a, b)$, we can take a sequence (y_n) of points converging to x from the right such that g is continuous at each y_n. Then

$$g(x+) = \lim g(y_n) = g(a).$$

Similarly,

$$g(x-) = g(a) \quad \text{for } x \in (a, b].$$

This completes the "only if" part of the proof.

Now suppose that $g \in BV[a, b]$ and

$$g(a) = g(a+) = g(x-) = g(x+) = g(b-) = g(b) \quad \text{for all } x \in (a, b).$$

Since g is continuous a.e. on $[a, b]$, $g(x) = g(a)$ for a.a.$x \in [a, b]$. If $P = \{(x_{i-1}, x_i) : i = 1, \cdots, k\}$ is a partition of $[a, b]$ such that g is continuous at x_i, $i = 0, \cdots, k$, then for each $f \in C[a, b]$

$$S(f, g, P) = \sum_{i=1}^{k} f(z_i)(g(x_i) - g(x_{i-1})) = 0.$$

If we take a sequence (P_n) of such partitions with $|P_n| \to 0$, we see that

$$\int_a^b f\,dg = 0. \blacksquare$$

Let $BVN[a, b]$ denote the set of functions $g \in BV[a, b]$ such that $g(a) = 0$ and g is continuous from the right in (a, b); that is, $g(x) = g(x+)$ for all $x \in (a, b)$. Clearly, $BVN[a, b]$ is a subspace of $BV[a, b]$ and, for $g \in BVN[a, b]$, $\|g\| = V_a^b(g)$. This space $BVN[a, b]$ is called the space of normalized functions of bounded variation on $[a, b]$.

Now consider T, defined by 6 · 2, to be a linear transformation from $BVN[a, b]$ into $C^*[a, b]$. Proposition 6 · 3 implies that T is one-to-one. We can use the Riesz Representation Theorem to show that T is *onto*. The Riesz Theorem implies that, for $F \in C^*[a, b]$, there exists a $g \in BV[a, b]$ such that $T(g) = F$. Also, $g(a) = 0$ and $V_a^b(g) \le \|F\|$. But g is not necessarily in $BVN[a, b]$ since it may fail to be continuous from the right at some points. If we define $\bar{g}$ by

$$\bar{g}(a) = 0, \quad \bar{g}(b) = g(b), \quad \text{and} \quad \bar{g}(x) = g(x+) \quad \text{for } x \in (a, b),$$

then $\bar{g} \in BVN[a, b]$, $T(\bar{g}) = T(g) = F$, and $V_a^b(\bar{g}) \le V_a^b(g) \le \|F\|$. Therefore, $\|\bar{g}\| = V_a^b(\bar{g}) = \|F\|$.

Thus, we have shown that T is a linear isometry of $BVN[a, b]$ onto $C^*[a, b]$ and, hence, $C^*[a, b]$ and $BVN[a, b]$ are isometrically isomorphic.

7 · THE SECOND MEAN VALUE THEOREM

Later we will have use for the Second Mean Value Theorem for Lebesgue integrals. We first prove a preliminary result.

7 · 1 Proposition. *If F is continuous and g is nondecreasing on $[a, b]$, then there exists a $c \in [a, b]$ such that*

$$\int_a^b F\,dg = F(c)(g(b) - g(a)).$$

PROOF. If $m = \min\{F(x) : x \in [a, b]\}$ and $M = \max\{F(x) : x \in [a, b]\}$, then

$$m(g(b) - g(a)) \le \int_a^b F\,dg \le M(g(b) - g(a)).$$

Thus, for some $N \in [m, M]$,

$$\int_a^b F\,dg = N(g(b) - g(a)).$$

Since F is continuous on $[a, b]$ there exists a $c \in [a, b]$ such that $F(c) = N$ and, hence,

$$\int_a^b F\,dg = F(c)(g(b) - g(a)). \blacksquare$$

7 · 2 Theorem (Second Mean Value Theorem). *If f is Lebesgue integrable on $[a, b]$ and g is nondecreasing on $[a, b]$, then there exists a $c \in [a, b]$ such that*

$$\int_a^b gf = g(a) \int_a^c f + g(b) \int_c^b f.$$

PROOF. Let $F(x) = \int_a^x f$ for $x \in [a, b]$. Then F is absolutely continuous on $[a, b]$. Since g is nondecreasing on $[a, b]$, g is Riemann integrable on $[a, b]$ and Proposition 2 · 10 implies that

$$\int_a^b g\, dF = \int_a^b gF' = \int_a^b gf.$$

Using integration by parts and Proposition 7 · 1, we obtain

$$\begin{aligned} \int_a^b gf = \int_a^b g\, dF &= g(b)F(b) - g(a)F(a) - \int_a^b F\, dg \\ &= g(b) \int_a^b f - F(c)(g(b) - g(a)) \quad \text{for some } c \in [a, b] \\ &= g(a) \int_a^c f + g(b) \int_c^b f. \ \blacksquare \end{aligned}$$

It is clear that the Second Mean Value Theorem holds if g is nonincreasing; consider $-g$. Also, the following result is easily obtained.

7 · 3 Corollary. *If f is Lebesgue integrable on $[a, b]$ and g is nondecreasing on $[a, b]$, then there exists a $c \in [a, b]$ such that*

$$\int_a^b gf = g(a+) \int_a^c f + g(b-) \int_c^b f.$$

PROOF. Let $g_1(a) = g(a+)$, $g_1(b) = g(b-)$, and $g_1(x) = g(x)$ for $x \in (a, b)$. Then g_1 is nondecreasing on $[a, b]$ and for some $c \in [a, b]$

$$\begin{aligned} \int_a^b gf = \int_a^b g_1 f &= g_1(a) \int_a^c f + g_1(b) \int_c^b f \\ &= g(a+) \int_a^c f + g(b-) \int_c^b f. \ \blacksquare \end{aligned}$$

PROBLEM

Show that

a. if $1 \le a < b$, then $\left| \int_a^b \frac{\sin x}{x}\, dx \right| \le 3$

b. $\left| \int_0^1 \frac{\sin x}{x}\, dx \right| \le 1$

c. if $0 \le a < b$, then $\left| \int_a^b \frac{\sin x}{x}\, dx \right| \le 4.$

8 · THE LEBESGUE–STIELTJES INTEGRAL

Let g be a fixed nondecreasing function on $\mathbf{R}$. In §4 we pointed out that the set $\mathscr{S}^g$ of admissible step functions is a vector lattice and the integral defined on $\mathscr{S}^g$ is a linear functional with the properties:

(1) if $f \in \mathscr{S}^g$ and $f \geq 0$, then $\int f\,dg \geq 0$

(2) if (f_n) is a nonincreasing sequence of nonnegative functions in $\mathscr{S}^g$ and $\lim f_n(x) = 0$ for all $x \in \mathbf{R}$, then

$$\lim \int f_n\,dg = 0.$$

Thus, we have the prerequisites for the Daniell method of developing an integral, in this instance called the **Lebesgue-Stieltjes integral.** Therefore, we have here a special case of the general development given in §15 of Chapter 6 and we need only restate those results for this special case.

Let $\bar{\mathscr{S}}^g$ be the set of functions f such that there is a nondecreasing sequence (f_n) of functions in $\mathscr{S}^g$ such that $\lim f_n(x) = f(x)$ for a.a.x and $(\int f_n\,dg)$ is bounded. In this case "a.a.x" refers to the μ_g-measure discussed in §4. For $f \in \bar{\mathscr{S}}^g$ we define

$$\int f\,dg = \lim \int f_n\,dg.$$

Let $\mathscr{L}^g$ be the set of functions f such that $f = f_1 - f_2$ a.e., where f_1 and f_2 are in $\bar{\mathscr{S}}^g$, and define

$$\int f\,dg = \int f_1\,dg - \int f_2\,dg.$$

Then $\mathscr{L}^g$, the set of functions which are Lebesgue-Stieltjes integrable with respect to g, is a vector lattice and the integral is a linear functional on $\mathscr{L}^g$; that is,

$$\int (\alpha_1 f_1 + \alpha_2 f_2)\,dg = \alpha_1 \int f_1\,dg + \alpha_2 \int f_2\,dg.$$

Also, Levi's Theorem, the Monotone Convergence Theorem, Fatou's Lemma, and the Lebesgue Dominated Convergence Theorem hold for the Lebesgue-Stieltjes integral. In addition we have:

$$\text{if } f \in \mathscr{L}^g \text{ and } f \geq 0 \text{ a.e., then } \int f\,dg \geq 0$$

and

$$\text{if } f \in \mathscr{L}^g \text{ and } \int |f|\,dg = 0, \text{ then } f = 0 \text{ a.e.}$$

A function f is **μ_g-measurable** if, for each $h \in \mathscr{L}^g$ such that $h \geq 0$, $(-h) \vee (f \wedge h) \in \mathscr{L}^g$. The set of μ_g- measurable functions is denoted by $\mathscr{M}^g$. Clearly, Stone's Axiom holds in this case; if $f \in \mathscr{S}^g$, then $1 \wedge f \in \mathscr{S}^g$. Thus, $\mathscr{M}^g$ is an algebra and a lattice with the property: if (f_n) is a sequence of functions in $\mathscr{M}^g$ and $f(x) = \lim f_n(x)$ for a.a.x, then $f \in \mathscr{M}^g$. Also, $\mathscr{L}^g \subset \mathscr{M}^g$.

A subset E of $\mathbf{R}$ is called μ_g-measurable if χ_E is a μ_g-measurable function. If $\mathscr{M}^g$ denotes the set of μ_g-measurable subsets of $\mathbf{R}$, then $\mathscr{M}^g$ is a σ-algebra. The measure μ_g is a nonnegative extended real-valued function defined on $\mathscr{M}^g$ by the rule:

$$\mu_g(E) = \begin{cases} \int \chi_E \, dg & \text{if } \chi_E \text{ is integrable with respect to } g \\ \infty & \text{otherwise.} \end{cases}$$

Then, μ_g is a countably additive function on $\mathscr{M}^g$.

If (a, b) is an admissible open interval, then by $4 \cdot 1$

$$\int \chi_{(a,b)} \, dg = g(b) - g(a) = \mu_g(a, b).$$

Thus, the above definition for μ_g agrees with that given in §4 for admissible open intervals. If (a, b) is not admissible, then there is a sequence $((a_n, b_n))$ of admissible open intervals such that (a_n) is nonincreasing and converges to a and (b_n) is nondecreasing and converges to b. Then, letting $\chi_n = \chi_{(a_n, b_n)}$ we see that (χ_n) is a nondecreasing sequence of functions in $\mathscr{L}^g$ such that $(\int \chi_n \, dg)$ is bounded and $\lim \chi_n(x) = \chi_{(a,b)}(x)$ for all x. By the Monotone Convergence Theorem, $\chi_{(a,b)} \in \mathscr{L}^g$ and

$$\int \chi_{(a,b)} \, dg = \lim \int \chi_n \, dg = \lim(g(b_n) - g(a_n)).$$

Thus,

$$\mu_g(a, b) = g(b-) - g(a+).$$

Since $(a, b] = (a, c) \sim (b, c)$ for $c > b$, then $(a, b] \in \mathscr{M}^g$. Also, for any $c > b$,

$$\begin{aligned} \mu_g(a, b] &= \mu_g(a, c) - \mu_g(b, c) = g(c-) - g(a+) - g(c-) + g(b+) \\ &= g(b+) - g(a+). \end{aligned}$$

Similarly, we can show that any interval is μ_g-measurable (Problem 2).

If E is a set in $\mathbf{R}$ such that $f\chi_E \in \mathscr{L}^g$, then the Lebesgue-Stieltjes integral of f over E is defined to be:

$$\int_E f \, dg = \int f \chi_E \, dg.$$

In case f is bounded on $[a, b]$ we can relate the Lebesgue-Stieltjes integral, (LS) $\int_a^b f \, dg$, to the Riemann-Stieltjes integral, (RS) $\int_a^b f \, dg$, where g is nondecreasing.

8 · 1 Theorem. *Let f be bounded and g be nondecreasing on $[a, b]$. If* (RS) $\int_a^b f \, dg$ *exists, then* (LS) $\int_a^b f \, dg$ *exists and*

$$\text{(LS)} \int_a^b f \, dg = \text{(RS)} \int_a^b f \, dg.$$

PROOF. In the proof of Theorem 5 · 3 we showed that if (RS) $\int_a^b f\,dg$ exists, then there is a nondecreasing sequence (h_n) of functions in $\mathscr{S}^g$ such that

$$\lim h_n(x) = f(x)\chi_{[a,\,b]}(x) \quad \text{for a.a.}x.$$

and

$$\lim \int h_n\,dg = (\text{RS}) \int_a^b f\,dg.$$

Thus, $f \in \mathscr{P}^g$ and

$$(\text{LS}) \int_a^b f\,dg = \lim \int h_n\,dg = (\text{RS}) \int_a^b f\,dg. \;\blacksquare$$

Under certain conditions we can also relate a Lebesgue-Stieltjes integral to a Lebesgue integral.

8 · 2 Theorem. *If g is nondecreasing and absolutely continuous on* $\mathbf{R}$ *and* $f \in \mathscr{L}^g$, *then* $fg' \in \mathscr{L}$ *and*

$$\int f\,dg = \int fg'.$$

PROOF. Suppose first that f is an admissible step function with representation $f(x) = 0$ for $x \notin [a, b]$ and

$$f(x) = c_i \quad \text{for} \quad x \in (a_{i-1}, a_i), \quad i = 1, \cdots, r.$$

Then, $fg' \in \mathscr{L}[a, b]$ and

$$\int fg' = \int_a^b fg' = \sum_{i=1}^r \int_{a_{i-1}}^{a_i} fg' = \sum_{i=1}^r c_i \int_{a_{i-1}}^{a_i} g'.$$

Also, the absolute continuity of g implies by the Second Fundamental Theorem of Calculus (p. 242) that

$$\int f\,dg = \sum_{i=1}^r c_i(g(a_i) - g(a_{i-1})) = \sum_{i=1}^r c_i \int_{a_{i-1}}^{a_i} g' = \int fg'.$$

Next, suppose that $f \in \mathscr{P}^g$; that is, there is a nondecreasing sequence (f_n) of admissible step functions such that $\lim f_n(x) = f(x)$ for all $x \in \mathbf{R} \sim E$ where E is of μ_g-measure zero and $(\int f_n\,dg)$ is bounded. Then,

$$\lim \int f_n\,dg = \int f\,dg.$$

Also, $\int f_n\,dg = \int f_n g'$. Let $E' = \{x \in E\colon g'(x) \neq 0\}$ and let E'' be the set of Lebesgue-measure zero where g' does not exist. Then, $(f_n g')$ is a nondecreasing sequence of functions in $\mathscr{L}$ such that

$$\lim f_n(x)g'(x) = f(x)g'(x) \quad \text{for all } x \in \mathbf{R} \sim (E' \cup E'')$$

and $(\int f_n g')$ is bounded. If E' is of Lebesgue-measure zero, then the Monotone Convergence Theorem implies that

$$\int fg' = \lim \int f_n g' = \lim \int f_n \, dg = \int f \, dg.$$

To complete this case we now show that E' is of Lebesgue-measure zero. Since E is of μ_g-measure zero there is a nondecreasing sequence of nonnegative admissible step functions such that $(h_n(x))$ diverges to ∞ for $x \in E$ and $(\int h_n \, dg)$ converges. Then, $(h_n g')$ is a nondecreasing sequence of nonnegative functions in $\mathscr{L}$ such that $(h_n(x)g'(x))$ diverges to ∞ for $x \in E'$ and $(\int h_n g')$ converges. By the Monotone Convergence Theorem, $(h_n(x)g'(x))$ converges except for x in a set of Lebesgue-measure zero. Thus, E' is of Lebesgue-measure zero.

Finally, if $f \in \mathscr{L}^g$, then μ_g-a.e. we have $f = f_1 - f_2$ where f_1 and f_2 are in $\mathscr{P}^g$. Using the same argument as above we see that $fg' = (f_1 - f_2)g'$ Lebesgue-a.e. Thus,

$$\int f \, dg = \int f_1 \, dg - \int f_2 \, dg = \int f_1 g' - \int f_2 g' = \int (f_1 - f_2)g' = \int fg'. \blacksquare$$

If E is a μ_g-measurable set in $\mathbf{R}$, let $\mathscr{L}^g(E)$ denote the set of functions f which are Lebesgue-Stieltjes integrable over E with respect to g; i.e., $\int_E f \, dg = \int f \chi_E \, dg$ exists. Then with obvious modifications the Lebesgue-Stieltjes theory holds for integration over E.

Also, we can extend this theory to complex-valued functions. The function g is still taken to be real-valued and nondecreasing. If $f = f_1 + if_2$ is a complex-valued function defined on $\mathbf{R}$ and if $\int f_1 \, dg$ and $\int f_2 \, dg$ exist, then we say that f is Lebesgue-Stieltjes integrable with respect to g and

$$\int f \, dg = \int f_1 \, dg + i \int f_2 \, dg.$$

PROBLEMS

1. If g is a nondecreasing function on $\mathbf{R}$, show that $\int dg$ exists if and only if g is bounded. If g is bounded and we let $g(\infty) = \lim_{b \to \infty} g(b)$ and $g(-\infty) = \lim_{a \to -\infty} g(a)$, show that

$$\int dg = \int_{-\infty}^{\infty} dg = g(\infty) - g(-\infty).$$

2. Show that $[a, b]$ and $[a, b)$ are μ_g-measurable and

$$\mu_g[a, b] = g(b+) - g(a-)$$
$$\mu_g[a, b) = g(b-) - g(a-).$$

3. If $f \in \tilde{\mathscr{M}}^g$ and $|f| \leqq h$ for some $h \in \mathscr{L}^g$, show that $f \in \mathscr{L}^g$.

4. If f is continuous, show that $f \in \tilde{\mathscr{M}}^g$.

5. If $f \in \tilde{\mathscr{M}}^g$ show that $|f| \in \tilde{\mathscr{M}}^g$.

9 · THE SPACE $\mathscr{L}_2^g(E)$

In Chapter 7 we considered some important normed linear spaces associated with Lebesgue integration. Similarly there are normed linear spaces associated with Lebesgue-Stieltjes integration and we now consider one such space.

For a nondecreasing function g on **R** let $\mathscr{L}_2^g(E)$ be the set of μ_g-measurable functions f on the μ_g-measurable set E such that f^2 is Lebesgue-Stieltjes integrable over E with respect to g. Then, $\mathscr{L}_2^g(E)$ is a linear space over **R**. The following analogues of Hölder's and Minkowski's Inequalities (for the case $p = 1/2 = q$) hold:

(9 · 1) If $f_1, f_2 \in \mathscr{L}_2^g(E)$, then $f_1 f_2 \in \mathscr{L}^g(E)$ and

$$\int_E |f_1 f_2| \, dg \leqq \left[\int_E f_1^2 \, dg\right]^{1/2} \left[\int_E f_2^2 \, dg\right]^{1/2}.$$

(9 · 2) If $f_1, f_2 \in \mathscr{L}_2^g(E)$, then $f_1 + f_2 \in \mathscr{L}_2^g(E)$ and

$$\left[\int_E (f_1 + f_2)^2 \, dg\right]^{1/2} \leqq \left[\int_E f_1^2 \, dg\right]^{1/2} + \left[\int_E f_2^2 \, dg\right]^{1/2}$$

On $\mathscr{L}_2^g(E)$ let

$$s(f) = \left[\int_E f^2 \, dg\right]^{1/2};$$

then s is a semi-norm on $\mathscr{L}_2^g(E)$. Identifying functions in $\mathscr{L}_2^g(E)$ which differ on a subset of E of μ_g-measure zero, we consider $\mathscr{L}_2^g(E)$ to be a normed linear space with

$$\|f\| = \left[\int_E f^2 \, dg\right]^{1/2}.$$

By a proof that is the same as that for $\mathscr{L}_p$ (Theorem 8 · 6, Chapter 7) we can show that $\mathscr{L}_2^g(E)$ is a Banach space.

PROBLEM

If $f \in \mathscr{L}_2^g[a, b]$, show that for each $\varepsilon > 0$ there exists a continuous function h such that

$$\|f - h\| < \varepsilon.$$

11 | *Inner Product Spaces*

1 · INTRODUCTION

If $x = (\alpha^1, \cdots, \alpha^n)$ and $y = (\beta^1, \cdots, \beta^n)$ are points in $\mathbf{R}^n$, then the scalar, dot, or inner product of x and y is

$$x \cdot y = \sum_{k=1}^{n} \alpha^k \beta^k.$$

In this chapter, this product, generally called the inner product, will be generalized so as to apply to an abstract linear space X. An inner product will be a scalar-valued function defined on $X \times X$ which has certain properties. A linear space with a given inner product defined on it is called an inner product space.

In an inner product space a norm can be defined in terms of the given inner product. Thus, an inner product space is a normed linear space, but it is more. For example, in an inner product space the notion of orthogonality of vectors can be defined and this has important consequences.

2 · INNER PRODUCT SPACES

Let X be a linear space over the field $\mathbf{F}$ where $\mathbf{F}$ is either $\mathbf{R}$ or $\mathbf{C}$.

2 · 1 Definition. *An* ***inner product*** *on X is a scalar-valued function on $X \times X$, whose values are denoted by $\langle x, y\rangle$, which has the following properties:*

(1) $\langle x, x\rangle \geq 0$; $\langle x, x\rangle = 0$ *if and only if* $x = 0$;
(2) $\langle x, y\rangle = \overline{\langle y, x\rangle}$;
(3) $\langle x + y, z\rangle = \langle x, z\rangle + \langle y, z\rangle$;
(4) $\langle \alpha x, y\rangle = \alpha\langle x, y\rangle$;

where the bar in property 2 denotes the complex conjugate. A linear space with inner product defined on it is called an ***inner product space.***

If X is a real linear space, property 2 reduces to $\langle x, y\rangle = \langle y, x\rangle$.

The following properties are immediate consequences of the definition of an inner product:

$$\langle x, y + z\rangle = \langle x, y\rangle + \langle x, z\rangle \tag{2 · 2}$$

$$\langle x, \alpha y\rangle = \bar{\alpha}\langle x, y\rangle \tag{2 · 3}$$

$$\langle x, 0\rangle = 0 = \langle 0, y\rangle. \tag{2 · 4}$$

Another important property of the inner product is given in the inequality of the following proposition. This inequality was first introduced by the French mathematician Augustin-Louis Cauchy (1789–1857) for the sums of products of real or complex numbers and was generalized in 1859 by the Russian mathematician Victor Jacovlevich Bunyakovski (1804–1889) for integrals. Bunyakovski's contribution was overlooked in the west and was later independently rediscovered by the German mathematician Hermann Amandus Schwarz (1843–1921). In English language publications the inequality is variously known as the Schwarz, the Cauchy–Schwarz, or the Cauchy–Bunyakovski–Schwarz (CBS) Inequality.

2 · 5 Proposition (The Schwarz Inequality). *If X is an inner product space, then for all $x, y \in X$*

$$|\langle x, y\rangle| \leq \sqrt{\langle x, x\rangle} \cdot \sqrt{\langle y, y\rangle}$$

where equality holds if and only if x and y are linearly dependent.

PROOF. If x and y are linearly independent vectors in X then for each $\lambda \in \mathbf{F}$

$$0 < \langle x - \lambda y, x - \lambda y\rangle = \langle x, x\rangle - \lambda\langle y, x\rangle - \bar{\lambda}\langle x, y\rangle + |\lambda|^2\langle y, y\rangle.$$

If we let $\lambda = \langle x, y\rangle / \langle y, y\rangle$, we obtain:

$$0 < \langle x, x\rangle - \frac{|\langle x, y\rangle|^2}{\langle y, y\rangle} - \frac{|\langle x, y\rangle|^2}{\langle y, y\rangle} + \frac{|\langle x, y\rangle|^2}{\langle y, y\rangle}.$$

Therefore,

$$|\langle x, y\rangle|^2 < \langle x, x\rangle\langle y, y\rangle$$

from which the Schwarz Inequality follows.

Suppose now that x and y are linearly dependent. If $y = 0$, then it is clear that

$$|\langle x, y\rangle| = 0 = \sqrt{\langle x, x\rangle} \cdot \sqrt{\langle y, y\rangle}.$$

If $y \neq 0$, then $x = \lambda y$ and

$$\begin{aligned} |\langle x, y\rangle| = |\langle \lambda y, y\rangle| &= |\lambda|\,|\langle y, y\rangle| = |\lambda|\sqrt{\langle y, y\rangle} \cdot \sqrt{\langle y, y\rangle} \\ &= \sqrt{\langle x, x\rangle} \cdot \sqrt{\langle y, y\rangle}. \ \blacksquare \end{aligned}$$

2 · 6 Theorem. *If X is an inner product space, then $\|x\| = \sqrt{\langle x, x\rangle}$ defines a norm on X.*

PROOF. With the exception of the Triangle Inequality, the verification that this function has the basic properties of a norm is a direct consequence of the defining properties of the inner product. The Schwarz Inequality is used to verify the Triangle Inequality:

$$\begin{aligned}\|x + y\|^2 &= \langle x + y, x + y\rangle = \langle x, x\rangle + \langle x, y\rangle + \langle y, x\rangle + \langle y, y\rangle \\ &= \|x\|^2 + 2\text{Rl}(\langle x, y\rangle) + \|y\|^2 \\ &\le \|x\|^2 + 2|\langle x, y\rangle| + \|y\|^2 \\ &\le \|x\|^2 + 2\|x\|\,\|y\| + \|y\|^2 = (\|x\| + \|y\|)^2.\end{aligned}$$

Thus,

$$\|x + y\| \le \|x\| + \|y\|. \blacksquare$$

From Theorem 2 · 6 we see that an inner product space can be considered to be a normed linear space with norm defined by $\|x\| = \sqrt{\langle x, x\rangle}$. This norm is said to be the norm associated with the given inner product. If an inner product space is complete in the associated norm, then it is called a **Hilbert space** after the German mathematician David Hilbert (1862–1943) who introduced the space l_2 in 1906 in conjunction with his study of integral equations.

Another important property of the inner product in a space X is that it is a continuous function on the product space $X \times X$. Recall that if X is a normed linear space, then we can define a norm on $X \times X$ by

$$\|(x, y)\|_\infty = \max\{\|x\|, \|y\|\}$$

and the corresponding topology is the product topology on $X \times X$. We could also use the norms

$$\|(x, y)\|_p = [\|x\|^p + \|y\|^p]^{1/p}, \quad 1 \le p < \infty$$

on $X \times X$. Since

$$\|(x, y)\|_\infty \le \|(x, y)\|_p \le 2^{1/p}\|(x, y)\|_\infty,$$

these p-norms are all equivalent to the infinity-norm and, hence, they all yield the same topological space.

2 · 7 Proposition. *If X is an inner product space, then the inner product is a continuous function on $X \times X$.*

PROOF. Take $(x_0, y_0) \in X \times X$. Then, for $(x, y) \in X \times X$, we have

$$|\langle x, y\rangle - \langle x_0, y_0\rangle| \le |\langle x_0, y - y_0\rangle| + |\langle x - x_0, y_0\rangle| + |\langle x - x_0, y - y_0\rangle|$$

and by the Schwarz Inequality

$$|\langle x, y\rangle - \langle x_0, y_0\rangle| \le \|x_0\|\,\|y - y_0\| + \|x - x_0\|\,\|y_0\| + \|x - x_0\|\,\|y - y_0\|.$$

Thus,

$$\begin{aligned}|\langle x, y\rangle - \langle x_0, y_0\rangle| \le [\|x_0\| + \|y_0\|]\,\|(x, y) - (x_0, y_0)\|_\infty \\ + \|(x, y) - (x_0, y_0)\|_\infty^2.\end{aligned}$$

The continuity of the inner product at (x_0, y_0) follows from this inequality. $\blacksquare$

The Euclidean spaces $\mathbf{R}^n$, the unitary spaces $\mathbf{C}^n$ the sequence spaces l_2, and the function spaces $\mathscr{L}_2(E)$ and $\mathscr{L}_2{}^g(E)$ are all examples of inner product spaces. Since these spaces are complete, they are Hilbert spaces.

(2 · 8) As already noted, an inner product can be defined in $\mathbf{R}^n$ by

$$\langle x, y\rangle = \sum_{k=1}^{n} \alpha^k \beta^k \quad \text{where } x = (\alpha^1, \ldots, \alpha^n),\ y = (\beta^1, \ldots, \beta^n).$$

(2 · 9) In $\mathbf{C}^n$ we can define an inner product by

$$\langle x, y\rangle = \sum_{k=1}^{n} \alpha^k \overline{\beta^k} \quad \text{where } x = (\alpha^1, \ldots, \alpha^n),\ y = (\beta^1, \ldots, \beta^n).$$

The fact that this satisfies the properties of an inner product is easily verified. Also, it is clear that the associated norm is the norm of $\mathbf{C}^n$.

(2 · 10) We now consider the complex l_2 space in more detail. Let $x = (\alpha^k)$ and $y = (\beta^k)$ be points in l_2 and define

$$\langle x, y\rangle = \sum_{k=1}^{\infty} \alpha^k \overline{\beta^k}.$$

The convergence of this series follows from the Schwarz Inequality in $\mathbf{R}^n$: for each positive integer n,

$$\sum_{k=1}^{n} \left|\alpha^k \overline{\beta^k}\right| = \sum_{k=1}^{n} |\alpha^k|\, |\beta^k| \le \left[\sum_{k=1}^{n} |\alpha^k|^2\right]^{1/2} \left[\sum_{k=1}^{n} |\beta^k|^2\right]^{1/2}$$
$$\le \left[\sum_{k=1}^{\infty} |\alpha^k|^2\right]^{1/2} \left[\sum_{k=1}^{\infty} |\beta^k|^2\right]^{1/2}.$$

The fact that property 2 of the inner product is satisfied follows from elementary properties of the complex conjugate including continuity of the operation of taking the complex conjugate:

$$\langle x, y\rangle = \sum_{k=1}^{\infty} \alpha^k \overline{\beta^k} = \lim_{n\to\infty} \sum_{k=1}^{n} \alpha^k \overline{\beta^k}$$
$$= \lim_{n\to\infty} \overline{\left(\sum_{k=1}^{n} \overline{\alpha^k} \beta^k\right)} = \overline{\left(\sum_{k=1}^{\infty} \beta^k \overline{\alpha^k}\right)} = \overline{\langle y, x\rangle}.$$

The verification of the other basic properties of an inner product is routine and is left to the reader.

(2 · 11) In the complex space $\mathscr{L}_2(E)$, where E is a Lebesgue-measurable set in $\mathbf{R}^n$, let

$$\langle f, g\rangle = \int_E f\bar{g}.$$

Hölder's Inequality shows that if f and g are in $\mathscr{L}_2(E)$, then $f\bar{g}$ is integrable over E. If we take equality in $\mathscr{L}_2(E)$ to mean equality almost everywhere, then we can easily verify that this defines an inner product and that the associated norm is the norm of $\mathscr{L}_2(E)$.

(2 · 12) In the real space $\mathscr{L}_2{}^g(E)$ where g is a nondecreasing function on $\mathbf{R}$ and E is a Stieltjes-measurable set in $\mathbf{R}$, let

$$\langle f_1, f_2 \rangle = \int_E f_1 f_2 \, dg.$$

(2 · 13) The space $C_2[a, b]$ of continuous real-valued functions on $[a, b]$ with

$$\langle f, g \rangle = \int_a^b fg$$

is an example of an inner product space which is not a Hilbert space (Problem 3, p. 180).

Not all normed linear spaces are inner product spaces; that is, it is not always possible to define an inner product on a normed linear space in such a way that the norm associated with the inner product is the original norm of the space. The norm associated with an inner product satisfies the **parallelogram law:**

(2 · 14) $$\|x + y\|^2 + \|x - y\|^2 = 2(\|x\|^2 + \|y\|^2).$$

This is easily verified:

$$\begin{aligned}\|x + y\|^2 + \|x - y\|^2 &= \langle x + y, x + y \rangle + \langle x - y, x - y \rangle \\ &= \langle x, x \rangle + \langle x, y \rangle + \langle y, x \rangle + \langle y, y \rangle \\ &\quad + \langle x, x \rangle - \langle x, y \rangle - \langle y, x \rangle + \langle y, y \rangle \\ &= 2(\|x\|^2 + \|y\|^2).\end{aligned}$$

Geometrically the parallelogram law expresses the fact that in a parallelogram the sum of the squares of the lengths of the diagonals is equal to the sum of the squares of the lengths of the four sides.

If in the normed linear space $l_\infty{}^2$ we take $x = (1, 0)$ and $y = (0, 1)$, then

$$\|x + y\|^2 + \|x - y\|^2 = 1 + 1 = 2$$

while

$$2(\|x\|^2 + \|y\|^2) = 2(1 + 1) = 4.$$

Since the norm of $l_\infty{}^2$ does not satisfy the parallelogram law, $l_\infty{}^2$ is not an inner product space.

We have shown that if a normed linear space is an inner product space then its norm satisfies the parallelogram law. It is also true that if the parallelogram law holds in a normed linear space, then the space is an inner product space (Problem 7).

PROBLEMS

1. Prove 2 · 2, 2 · 3, and 2 · 4.

2. Let X be the space of continuous complex-valued functions on $[a, b]$ with $\mathbf{F} = \mathbf{C}$ and define

$$\langle x, y \rangle = \int_a^b x\bar{y}.$$

Verify that this is an inner product on X.

3. If $\sum_{k=1}^{\infty} x_k$ converges in the inner product space X, show that

$$\left\langle \sum_{k=1}^{\infty} x_k, y \right\rangle = \sum_{k=1}^{\infty} \langle x_k, y \rangle.$$

4. If X is a real inner product space and $\|x\| = \sqrt{\langle x, x \rangle}$, show that

$$\langle x, y \rangle = \tfrac{1}{4}[\|x + y\|^2 - \|x - y\|^2].$$

5. If X is a complex inner product space and $\|x\| = \sqrt{\langle x, x \rangle}$, show that

$$\langle x, y \rangle = \tfrac{1}{4}[\|x + y\|^2 - \|x - y\|^2 + i\,\|x + iy\|^2 - i\,\|x - iy\|^2].$$

6. Assuming that $\mathbf{C}^n$ is an inner product space use Problem 5 to obtain a formula for the inner product which has the norm of $\mathbf{C}^n$ as its associated norm.

7. Let X be a complex normed linear space whose norm satisfies the parallelogram law. If we define

$$\langle x, y \rangle = \tfrac{1}{4}[\|x + y\|^2 - \|x - y\|^2 + i\,\|x + iy\|^2 - i\,\|x - iy\|^2],$$

show that this is an inner product on X.

Suggestion. To establish property 3, let $\langle x + y, z \rangle - \langle x, z \rangle - \langle y, z \rangle = \tfrac{1}{4}\varphi(x, y, z) + (i/4)\varphi(x, y, iz)$ where

$$\varphi(x, y, z) = \|(x + z) + y\|^2 - \|(x - z) + y\|^2 - \|x + z\|^2 + \|x - z\|^2 - \|y + z\|^2 + \|y - z\|^2.$$

Apply the parallelogram law to each of the first two terms and obtain another expression for $\varphi(x, y, z)$. Average these two expressions and show that $\varphi(x, y, z) = 0$.

To establish property 4, let $\varphi(\alpha) = \langle \alpha x, y \rangle - \alpha\langle x, y \rangle$. Show that $\varphi(\alpha) = 0$ where α is successively a positive integer, an integer, a rational number, i, a complex number $r + is$ where r and s are rational, and finally a complex number.

8. If X is an inner product space, show that its completion $\tilde{X}$ can be made an inner product space.

9. A square matrix $H = [h_{ik}]$ with complex entries is said to be **Hermitian** if $h_{ik} = \bar{h}_{ki}$ $(i, k = 1, \cdots, n)$. (A real Hermitian matrix is said to be **symmetric**.)

An Hermitian (or real symmetric) $n \times n$ matrix is called **positive definite** if

$$\sum_{i=1}^{n} \sum_{k=1}^{n} h_{ik} \alpha^i \overline{\alpha^k} > 0$$

for all $(\alpha^1, \cdots, \alpha^n) \neq (0, \cdots, 0)$.

a. Let $H = [h_{ik}]$ be a positive definite Hermitian matrix. Show that

$$\langle x, y \rangle = \sum_{i=1}^{n} \sum_{k=1}^{n} h_{ik} \alpha^i \overline{\beta^k}$$

where $x = (\alpha^1, \cdots, \alpha^n)$ and $y = (\beta^1, \cdots, \beta^n)$ defines an inner product on $V_n(\mathbf{C})$, the space of n-tuples of complex numbers.

b. Let $(e_1, \cdots, e_n)$ be a basis for a complex (real) n-dimensional inner product space. If $x = \sum_{i=1}^{n} \alpha^i e_i$ and $y = \sum_{i=1}^{n} \beta^i e_i$, show that

$$\langle x, y \rangle = \sum_{i=1}^{n} \sum_{k=1}^{n} h_{ik} \alpha^i \overline{\beta^k}$$

where $H = [h_{ik}]$ is a positive definite Hermitian (symmetric) matrix.

3 · ORTHOGONALITY

The existence of an inner product in a vector space provides a means of introducing the notion of orthogonality in such spaces. The vectors x and y in an inner product space X are said to be **orthogonal** if $\langle x, y \rangle = 0$. Note that according to this definition the zero vector is orthogonal to every vector in X. If a vector x is orthogonal to every vector in a set E, then we say that x is orthogonal to E. The set of all vectors orthogonal to a set E is called the **orthogonal complement** of E, denoted by $E^\perp$.

3·1 Proposition. *If E is a set in an inner product space X, then $E^\perp$ is a closed linear subspace of X.*

PROOF. Take $x, y \in E^\perp$ and $\alpha, \beta \in \mathbf{F}$. Then, for all $z \in E$,

$$\langle \alpha x + \beta y, z \rangle = \alpha \langle x, z \rangle + \beta \langle y, z \rangle = 0.$$

This shows that $E^\perp$ is a linear subspace. Now take $x \in \overline{E^\perp}$ and let (x_n) be a sequence of points in $E^\perp$ such that $\lim x_n = x$. Then, for all $z \in E$,

$$\langle x, z \rangle = \lim \langle x_n, z \rangle = 0.$$

Thus, $x \in E^\perp$. ∎

We will generalize to inner product spaces the familiar result that a vector x in $\mathbf{R}^2$ has a unique representation as the sum of two vectors, one parallel to a given nonzero vector u and the other orthogonal to u. Let E be the subspace spanned by u. We call the vector parallel to u in this representation the **projection** of x on E and denote it by $P_E(x)$. The projection $P_E(x)$ can be characterized as the point in E closest to x. This aspect of $P_E(x)$ will be used as the starting point for our generalization.

Recall that a set E in the linear space X is said to be convex if, for each pair of points x and y in E, the line segment $[x, y]$ joining x and y lies in E; i.e., if $x, y \in E$ then $[x, y] = \{\tau x + (1 - \tau)y : \tau \in [0, 1]\} \subset E$. Clearly, any subspace of a linear space is convex.

3 · 2 Proposition. *If E is a convex set which is complete in an inner product space X and x is any point in X, then there is a unique point $P_E(x)$ in E such that*

$$\|x - P_E(x)\| = \inf\{\|x - y\| : y \in E\}.$$

PROOF. Let $\delta = \inf\{\|x - y\| : y \in E\}$ and let (y_n) be a sequence of points in E such that

$$\lim \|x - y_n\| = \delta.$$

If we show that (y_n) is a Cauchy sequence, then (y_n) converges to some point in E since E is complete. Note that if y_m and y_n are in E, then $\frac{1}{2}(y_m + y_n) \in E$ since E is convex.

Take $\varepsilon > 0$. There exists a positive integer n_0 such that

$$\|x - y_n\|^2 < \delta^2 + \tfrac{1}{4}\varepsilon^2 \quad \text{whenever } n \geq n_0.$$

Using the parallelogram law, we have for $n, m \geq n_0$

$$\begin{aligned}\|y_m - y_n\|^2 &= \|(y_m - x) + (x - y_n)\|^2 \\ &= 2\,\|y_m - x\|^2 + 2\,\|x - y_n\|^2 - \|y_m + y_n - 2x\|^2 \\ &< 2(\delta^2 + \tfrac{1}{4}\varepsilon^2) + 2(\delta^2 + \tfrac{1}{4}\varepsilon^2) - 4\,\|\tfrac{1}{2}(y_m + y_n) - x\|^2 \\ &\leq 4\delta^2 + \varepsilon^2 - 4\delta^2 = \varepsilon^2.\end{aligned}$$

Thus, (y_n) is a Cauchy sequence in E and, hence, converges to some point $y \in E$. Then $\|x - y\| = \delta$.

To show uniqueness, suppose $y, y' \in E$, $\|x - y\| = \delta$, and $\|x - y'\| = \delta$. Then,

$$\begin{aligned}\|y - y'\|^2 &= \|(y - x) + (x - y')\|^2 \\ &= 2\,\|y - x\|^2 + 2\,\|x - y'\|^2 - \|y + y' - 2x\|^2 \\ &= 4\delta^2 - 4\,\|\tfrac{1}{2}(y + y') - x\|^2 \leq 0.\end{aligned}$$

Therefore, $y = y'$. ∎

If Y is a complete subspace of X and x is any point in X, then Proposition 3 · 2 states that there is a unique point $P_Y(x)$ in Y which is closest to x. We now show that $x - P_Y(x)$ is orthogonal to Y.

3 · 3 Proposition. *If Y is a complete subspace of the inner product space X and x is any point in X, then $x - P_Y(x) \in Y^{\perp}$.*

PROOF. Take $y \in Y$. Then, for each real number α, $P_Y(x) + \alpha y \in Y$ and, hence,

$$\begin{aligned}\|x - P_Y(x)\|^2 &\leq \|x - P_Y(x) - \alpha y\|^2 = \langle x - P_Y(x) - \alpha y, x - P_Y(x) - \alpha y\rangle \\ &= \|x - P_Y(x)\|^2 - 2\alpha \mathrm{Rl}(\langle x - P_Y(x), y\rangle) + \alpha^2 \|y\|^2.\end{aligned}$$

Thus,

$$2\alpha \mathrm{Rl}(\langle x - P_Y(x), y\rangle) \leqq \alpha^2 \|y\|^2.$$

For each positive number α, we have

$$\mathrm{Rl}(\langle x - P_Y(x), y\rangle) \leq \tfrac{1}{2}\alpha \|y\|^2$$

and, therefore, $\mathrm{Rl}(\langle x - P_Y(x), y\rangle) \leq 0$. Similarly, for each negative number α we have

$$\mathrm{Rl}(\langle x - P_Y(x), y\rangle) \geq \tfrac{1}{2}\alpha \|y\|^2$$

and, therefore, $\mathrm{Rl}(\langle x - P_Y(x), y\rangle) \geq 0$. Thus, $\mathrm{Rl}(\langle x - P_Y(x), y\rangle) = 0$. Replacing α by $i\alpha$ in the above calculations, we obtain $\mathrm{Im}(\langle x - P_Y(x), y\rangle) = 0$ and, hence,

$$\langle x - P_Y(x), y\rangle = 0. \blacksquare$$

If Y and Z are subspaces of a linear space X with the property that every element x in X has a unique representation of the form

$$x = y + z, \quad \text{where } y \in Y \text{ and } z \in Z,$$

then we say that X is the **direct sum** of Y and Z and we write $X = Y \oplus Z$. We know that if Y is a complete subspace of an inner product space X, then any point x in X can be represented in the form

$$x = P_Y(x) + (x - P_Y(x)), \quad \text{where } P_Y(x) \in Y \text{ and } x - P_Y(x) \in Y^{\perp}.$$

To see that this representation is unique, suppose that

$$x = y + z, \quad \text{where } y \in Y \text{ and } z \in Y^{\perp}.$$

Then, $y - P_Y(x) = (x - P_Y(x)) - z$. Since $y - P_Y(x) \in Y$, $(x - P_Y(x)) - z \in Y^{\perp}$, and $Y \cap Y^{\perp} = \{0\}$, we have $y = P_Y(x)$ and $z = x - P_Y(x)$. Thus, $X = Y \oplus Y^{\perp}$ and we have proved the following:

3 · 4 Theorem. *If Y is a complete subspace of the inner product space X, then*

$$X = Y \oplus Y^{\perp}.$$

Now assume X is a Hilbert space and Y is a closed subspace of X. Then Y is complete and, hence, $X = Y \oplus Y^{\perp}$. Also, $Y^{\perp}$ is a complete subspace and $Y^{\perp\perp} = Y$ (Problem 5). Thus, if $x = y + z$ where $y \in Y$ and $z \in Y^{\perp}$, then $y = P_Y(x)$ and $z = P_{Y^{\perp}}(x)$.

3 · 5 Proposition. *If Y is a closed subspace of the Hilbert space X, then the projection P_Y of X onto Y is a bounded linear transformation from X onto Y with the properties:*

(1) $P_Y{}^2 = P_Y \circ P_Y = P_Y$
(2) $P_{Y^\perp} \circ P_Y = 0$
(3) $\langle P_Y(x), y\rangle = \langle x, P_Y(y)\rangle$ *for all* $x, y \in X$.

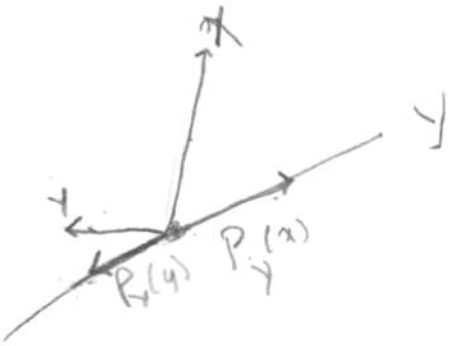

PROOF. For any $x, y \in X$ and $\alpha, \beta \in \mathbf{F}$,

$$\alpha x + \beta y = [\alpha P_Y(x) + \beta P_Y(y)] + [\alpha(x - P_Y(x)) + \beta(y - P_Y(y))]$$

where $\alpha P_Y(x) + \beta P_Y(y) \in Y$ and $\alpha(x - P_Y(x)) + \beta(y - P_Y(y)) \in Y^\perp$. Therefore,

$$P_Y(\alpha x + \beta y) = \alpha P_Y(x) + \beta P_Y(y).$$

Also, since $P_Y(x)$ and $x - P_Y(x)$ are orthogonal, by the Pythagorean Theorem (Problem 4)

$$\langle x, x\rangle = \|x\|^2 = \|P_Y(x) + (x - P_Y(x))\|^2 = \|P_Y(x)\|^2 + \|x - P_Y(x)\|^2$$
$$\geq \|P_Y(x)\|^2$$

and, therefore,

$$\|P_Y\| = \sup_{x \neq 0} \frac{\|P_Y(x)\|}{\|x\|} \leq 1.$$

Thus, $P_Y \in BL(X, Y)$. Since, for each $x \in X, P_Y(x) = P_Y(x) + 0$ where $P_Y(x) \in Y$ and $0 \in Y^\perp$, we have $P_Y{}^2(x) = P_Y(x)$ and $P_{Y^\perp}(P_Y(x)) = 0$. Also, for each $x, y \in X$,

$$\langle P_Y(x), y\rangle = (P_Y(x), P_Y(y) + (y - P_Y(y))\rangle = \langle P_Y(x), P_Y(y)\rangle$$
$$= \langle x - (x - P_Y(x)), P_Y(y)\rangle = \langle x, P_Y(y)\rangle. \blacksquare$$

In a Hilbert space a bounded linear functional can be represented in terms of the inner product. First, we note that, if for a fixed vector a in an inner product space X, we define

$$f_a(x) = \langle x, a\rangle, x \in X,$$

then f_a is a linear functional on X. Also, since

$$|f_a(x)| = |\langle x, a\rangle| \leq \|a\| \, \|x\|$$

and $|f_a(a)| = \|a\|^2$, f_a is a bounded linear functional with $\|f_a\| = \|a\|$.

Thus, if for all $a \in X$ we define

$$(3 \cdot 6) \qquad T(a) = f_a \quad \text{where } f_a(x) = \langle x, a\rangle,$$

then T is an isometry from X into its dual X^*. Since

$$T(\alpha a + \beta b) = \bar{\alpha} T(a) + \bar{\beta} T(b),$$

T is linear if X is a real space but it is not linear if X is complex. The following result shows that if X is a Hilbert space, then T is *onto.*

3 · 7 Theorem (Riesz Representation Theorem). *If f is a bounded linear functional on a Hilbert space X, then there exists a unique vector a in X such that $\|a\| = \|f\|$ and*

$$f(x) = \langle x, a\rangle, \text{ for all } x \in X.$$

PROOF. We prove only the existence of a vector a such that $f(x) = \langle x, a\rangle$; the rest follows easily.

If $f = 0$, take $a = 0$. Suppose now that $f \neq 0$ and let N be the null space of f: $N = \{x : f(x) = 0\}$. Then N is a closed subspace of X and hence is complete. Since $N^{\perp}$ is a subspace containing a nonzero vector (Theorem 3 · 4), we may choose $a_1 \in N^{\perp}$ such that $f(a_1) = 1$. Then, $f(x) = 0$ implies $\langle x, a_1\rangle = 0$. For each $x \in X$, $f(x - f(x)a_1) = f(x) - f(x)f(a_1) = 0$ and, therefore, $\langle x - f(x)a_1, a_1\rangle = 0$. That is,

$$\langle x, a_1\rangle - f(x)\langle a_1, a_1\rangle = 0$$

and, hence,

$$f(x) = \langle x, a_1/\|a_1\|^2\rangle.$$

Thus, if $a = \|a_1\|^{-2}a_1$, then $f(x) = \langle x, a\rangle$ for all $x \in X$. ▮

For a Hilbert space X we have shown that the transformation T defined by 3 · 6 is an isometry from X onto X^*. If, for $f, g \in X^*$, we let

$$\langle f, g\rangle = \overline{\langle T^{-1}(f), T^{-1}(g)\rangle},$$

it is easy to show that this is an inner product on X^* and the associated norm is the same as the norm of X^*. Hence, X^* is a Hilbert space.

If X is a real Hilbert space, then T is an isometric isomorphism of X onto X^* which preserves the inner product. Therefore, in this case X and X^* may be identified.

If X is a complex Hilbert space, then T is not linear. In this case we can identify X with its second dual $X^{**} = (X^*)^*$. Since X^* is a Hilbert space there is an isometry S from X^* onto X^{**} such that

$$S(\alpha f + \beta g) = \bar{\alpha}S(f) + \bar{\beta}S(g).$$

Thus, $S \circ T$ is an isometric isomorphism of X onto X^{**}. Also, if we define an inner product in X^{**} by

$$\langle F, G\rangle = \overline{\langle S^{-1}(F), S^{-1}(G)\rangle}$$

then $S \circ T$ preserves the inner product and, hence, X and X^{**} may be identified.

PROBLEMS

1. If E is a set in an inner product space X, show that $E \cap E^{\perp} = \{0\}$.

2. If the set E is dense in the inner product space X, show that $E^{\perp} = \{0\}$.

3. If Y and Z are subspaces of a linear space X, show that $X = Y \oplus Z$ if and only if every element x in X can be written as a sum $x = y + z$ where $y \in Y$ and $z \in Z$ and $Y \cap Z = \{0\}$.

4. In an inner product space X prove the Pythagorean Theorem: if x and y are orthogonal, then $\|x + y\|^2 = \|x\|^2 + \|y\|^2$.

5. If X is a Hilbert space and Y is a closed linear subspace of X, show that $Y^{\perp\perp} = Y$.

6. Let Y be the closure of the linear subspace spanned by a set E in a Hilbert space X. Show that $E^{\perp} = Y^{\perp}$.

7. If E is a set on a Hilbert space X, show that $E^{\perp\perp}$ is the closure of the linear subspace spanned by E.

8. If Y is a closed linear subspace of the Hilbert space X and P_Y is the projection of X onto Y, show that:
   a. $Y = \{x \in X : P_Y(x) = x\}$
   b. $Q = I - P_Y$ is the projection of X onto $Y^{\perp}$.

4 · ORTHOGONAL FAMILIES

A family $(x_s)_{s \in S}$ of points in X is said to be **orthogonal** if $x_s \neq 0$ for all $s \in S$ and $\langle x_s, x_t \rangle = 0$ for $s \neq t$. If, in addition, $\|x_s\| = 1$ for all $s \in S$, then (x_s) is called an **orthonormal family**. For example, in $\mathbf{R}^3$ the triple (i, j, k), where $i = (1, 0, 0)$, $j = (0, 1, 0)$, and $k = (0, 0, 1)$, is orthonormal. Clearly, an orthogonal family can be made orthonormal by dividing each term by its norm.

4 · 1 Proposition. *An orthogonal family* $(x_s)_{s \in S}$ *of points in* X *is linearly independent.*

PROOF. Let $(x_s)_{s \in S'}$ be a finite subfamily. Then $\sum_{s \in S'} \alpha^s x_s = 0$ implies that, for any $t \in S'$,

$$\alpha^t \langle x_t, x_t \rangle = \sum_{s \in S'} \alpha^s \langle x_s, x_t \rangle = \left\langle \sum_{s \in S'} \alpha^s x_s, x_t \right\rangle = 0$$

and, hence, $\alpha^t = 0$. ▮

From Proposition 4 · 1 we see that if $(x_1, \cdots, x_n)$ is an orthogonal n-tuple in an inner product space X of dimension n, then $(x_1, \cdots, x_n)$ is a Hamel basis and therefore, any element x in X can be expressed in the form:

$$x = \sum_{k=1}^{n} \alpha^k x_k .$$

Then, for any $j = 1, \cdots, n$,

$$\langle x, x_j \rangle = \left\langle \sum_{k=1}^{n} \alpha^k x_k, x_j \right\rangle = \sum_{k=1}^{n} \alpha^k \langle x_k, x_j \rangle = \alpha^j \langle x_j, x_j \rangle$$

and, hence

$$x = \sum_{k=1}^{n} \alpha^k x_k \quad \text{where } \alpha^k = \frac{\langle x, x_k \rangle}{\|x_k\|^2}.$$

If we let $u_k = x_k / \|x_k\|$, then $(u_1, \cdots, u_n)$ is an orthonormal n-tuple and

$$x = \sum_{k=1}^{n} \langle x, u_k \rangle u_k .$$

The inner product $\langle x, u_k \rangle$ is the component of x in the direction u_k and is called the Fourier coefficient of x with respect to u_k.

In general if $(u_s)_{s \in S}$ is an orthonormal family of points in an inner product space X, then $\{\langle x, u_s \rangle : s \in S\}$ is called the set of (generalized) **Fourier coefficients** of x with respect to (u_s).

4·2 Proposition. *If $(u_1, \cdots, u_n)$ is a finite orthonormal family of points in an inner product space X and $x \in X$, then $\|x - \sum_{k=1}^{n} \alpha^k u_k\|$ has its minimum value when $\alpha^k = \langle x, u_k \rangle$ and*

$$\sum_{k=1}^{n} |\langle x, u_k \rangle|^2 \leq \|x\|^2.$$

PROOF. For any numbers $\alpha^1, \cdots, \alpha^n$ in $\mathbf{F}$, we have

$$\begin{aligned} 0 \leq \left\| x - \sum_{k=1}^{n} \alpha^k u_k \right\|^2 &= \left\langle x - \sum_{k=1}^{n} \alpha^k u_k , x - \sum_{k=1}^{n} \alpha^k u_k \right\rangle \\ &= \|x\|^2 - \sum_{k=1}^{n} \alpha^k \overline{\langle x, u_k \rangle} - \sum_{k=1}^{n} \overline{\alpha^k} \langle x, u_k \rangle + \sum_{k=1}^{n} |\alpha^k|^2 \\ &= \|x\|^2 + \sum_{k=1}^{n} |\langle x, u_k \rangle - \alpha^k|^2 - \sum_{k=1}^{n} |\langle x, u_k \rangle|^2. \end{aligned}$$

Thus, $\|x - \sum_{k=1}^{n} \alpha^k u_k\|$ has its minimum value when $\alpha^k = \langle x, u_k \rangle$. If we set $\alpha^k = \langle x, u_k \rangle$ then we obtain

$$\sum_{k=1}^{n} |\langle x, u_k \rangle|^2 \leq \|x\|^2. \blacksquare$$

If Y is the linear subspace of X spanned by $\{u_1, \cdots, u_n\}$, then we see that $\sum_{k=1}^{n} \langle x, u_k \rangle u_k$ is the point in Y closest to x; that is,

$$P_Y(x) = \sum_{k=1}^{n} \langle x, u_k \rangle u_k \quad \text{for all } x \in X.$$

Therefore, $x - \sum_{k=1}^{n} \langle x, u_k \rangle u_k$ is in $Y^{\perp}$.

Now let $(u_s)_{s \in S}$ be an orthonormal family of points in X whose range is of arbitrary cardinal number. We will show that, for any point $x \in X$, the set $\{\langle x, u_s\rangle : s \in S\}$ of Fourier coefficients of x contains only a countable number of nonzero elements.

4 · 3 Proposition. *Let $(u_s)_{s \in S}$ be an orthonormal family of points in an inner product space and let x be any element in X. Then, the set $\{s \in S : \langle x, u_s\rangle \neq 0\}$ is countable.*

PROOF. Let $T = \{s \in S : \langle x, u_s\rangle \neq 0\}$ and $T_n = \{s \in S : |\langle x, u_s\rangle| > 1/n\}$. Then $T = \bigcup_{n=1}^{\infty} T_n$. If T' is a subset of T_n having m elements, then

$$\sum_{s \in T'} |\langle x, u_s\rangle|^2 > \frac{m}{n^2}.$$

Since by Proposition 4 · 2

$$\sum_{s \in T'} |\langle x, u_s\rangle|^2 \leq \|x\|^2,$$

we see that $m < n^2 \|x\|^2$. Thus, each T_n is finite and consequently T is countable. ∎

Since for a given $x \in X$ the set $\{\langle x, u_s\rangle : s \in T\}$ of nonzero Fourier coefficients is countable, we can ascribe a meaning to $\sum_{s \in S} |\langle x, u_s\rangle|^2$. If T is finite, then $\sum_{s \in S} |\langle x, u_s\rangle|^2$ is the finite sum $\sum_{s \in T} |\langle x, u_s\rangle|^2$. If T is countably infinite, take an enumeration $\{s_k : k = 1, 2, \cdots\}$ of T and let $\sum_{s \in S} |\langle x, u_s\rangle|^2$ be the series $\sum_{k=1}^{\infty} |\langle x, u_{s_k}\rangle|^2$. Since by Proposition 4 · 2 $\sum_{k=1}^{n} |\langle x, u_{s_k}\rangle|^2 \leq \|x\|^2$ for any positive integer n, this series converges and $\sum_{k=1}^{\infty} |\langle x, u_{s_k}\rangle|^2 \leq \|x\|^2$. The sum of this series does not depend on the particular enumeration of T since any other enumeration of T would yield a rearrangement of this absolutely convergent series.

In the above discussion we have proved:

4 · 4 Proposition (Bessel's Inequality). *Let $(u_s)_{s \in S}$ be an orthonormal family of points in an inner product space X. Then, for any $x \in X$,*

$$\sum_{s \in S} |\langle x, u_s\rangle|^2 \leq \|x\|^2.$$

If X is a Hilbert space, then by the same line of reasoning as above we may assign a meaning to $\sum_{s \in S}\langle x, u_s\rangle u_s$ where $(u_s)_{s \in S}$ is an orthonormal family of points in X. If, for $x \in X$, the set $\{\langle x, u_s\rangle : s \in T\}$ of nonzero Fourier coefficients is finite, then this denotes a finite sum and there is no problem. Otherwise, let $\{s_k : k = 1, 2, \cdots\}$ be an enumeration of T and define

$$\sum_{s \in S} \langle x, u_s\rangle u_s = \sum_{k=1}^{\infty} \langle x, u_{s_k}\rangle u_{s_k}.$$

To ensure that this is a meaningful definition we must show that $\sum_{k=1}^{\infty} \langle x, u_{s_k}\rangle u_{s_k}$ converges and that any rearrangement has the same sum.

Let $y_n = \sum_{k=1}^{n} \langle x, u_{s_k}\rangle u_{s_k}$. Then, for $m > n$ we have

$$\|y_m - y_n\|^2 = \left\| \sum_{k=n+1}^{m} \langle x, u_{s_k}\rangle u_{s_k} \right\|^2$$
$$= \sum_{k=n+1}^{m} |\langle x, u_{s_k}\rangle|^2.$$

Since $\sum_{k=1}^{\infty} |\langle x, u_{s_k}\rangle|^2$ converges, (y_n) is a Cauchy sequence in the Hilbert space X and, hence, converges. Let $\sum_{k=1}^{\infty} \langle x, u_{s_k}\rangle u_{s_k} = y$.

It remains to be shown that any rearrangement of $\sum_{k=1}^{\infty} \langle x, u_{s_k}\rangle u_{s_k}$ also converges to y. Take another enumeration $\{t_k : k = 1, 2, \cdots\}$ of T such that $t_k = s_{f(k)}$. Let $z_n = \sum_{k=1}^{n} \langle x, u_{t_k}\rangle u_{t_k}$. For each $\varepsilon > 0$ there exists a positive integer n_0 such that

$$\sum_{k=n_0+1}^{n_0+p} |\langle x, u_{s_k}\rangle|^2 < \varepsilon^2 \quad \text{for all } p \geq 1.$$

Take m_0 such that $\{k : k \leq n_0\} \subset \{f(k) : k \leq m_0\}$. Then $m_0 \geq n_0$. For any $m, n \geq m_0$, there exists a $p \geq 1$ such that

$$\|y_m - z_n\|^2 \leq \sum_{k=n_0+1}^{n_0+p} |\langle x, u_{s_k}\rangle|^2 < \varepsilon^2.$$

Therefore,

$$\|y - z_n\| \leq \varepsilon \quad \text{for all } n \geq m_0$$

and, hence, $\lim z_n = y$.

Thus, we have shown that if $(u_s)_{s \in S}$ is an orthonormal family of points in a Hilbert space X and x is any element of X then the series $\sum_{s \in S} \langle x, u_s\rangle u_s$ converges. This is called the **Fourier series** of x with respect to $(u_s)_{s \in S}$.

4·5 Proposition. *If $(u_s)_{s \in S}$ is an orthonormal family of points in a Hilbert space X and x is any element in X, then $x - \sum_{s \in S} \langle x, u_s\rangle u_s$ is orthogonal to u_s for all $s \in S$.*

PROOF. Let $T = \{s \in S : \langle x, u_s\rangle \neq 0\}$. Then,

$$\sum_{s \in S} \langle x, u_s\rangle u_s = \sum_{t \in T} \langle x, u_t\rangle u_t.$$

If $s \in S \sim T$, then $\langle x, u_s\rangle = 0$ and $\langle u_t, u_s\rangle = 0$ for all $t \in T$. Thus,

$$\left\langle x - \sum_{t \in T} \langle x, u_t\rangle u_t, u_s \right\rangle = \langle x, u_s\rangle - \sum_{t \in T} \langle x, u_t\rangle\langle u_t, u_s\rangle = 0.$$

If $s \in T$, then

$$\left\langle x - \sum_{t \in T} \langle x, u_t\rangle u_t, u_s \right\rangle = \langle x, u_s\rangle - \sum_{t \in T} \langle x, u_t\rangle\langle u_t, u_s\rangle$$
$$= \langle x, u_s\rangle - \langle x, u_s\rangle = 0. \blacksquare$$

If Y is the closure of the subspace of X spanned by $\{u_s : s \in S\}$, then Proposition 4 · 5 implies that the projection P_Y of X onto Y is given by the rule:

$$P_Y(x) = \sum_{s \in S} \langle x, u_s \rangle u_s;$$

that is, the Fourier series of x with respect to $(u_s)_{s \in S}$ is the projection $P_Y(x)$ of x onto Y.

PROBLEMS

1. If $(x_1, \cdots, x_n)$ is an orthogonal family of points in an inner product space X, show that

$$\left\| \sum_{k=1}^{n} x_k \right\|^2 = \sum_{k=1}^{n} \|x_k\|^2.$$

2. If $(x_k)_1{}^\infty$ is an orthogonal family of points in a Hilbert space X, show that $\sum_{k=1}^{\infty} x_k$ converges if and only if $\sum_{k=1}^{\infty} \|x_k\|^2$ converges and in case of convergence

$$\left\| \sum_{k=1}^{\infty} x_k \right\|^2 = \sum_{k=1}^{\infty} \|x_k\|^2.$$

3. Let $(e_k)_1{}^\infty$ be the orthonormal family of points $e_k = (\delta_k{}^j)$ in the complex sequence space l_2. If x is any element in l_2, determine the set of Fourier coefficients of x with respect to $(e_k)_1{}^\infty$.

4. If $(u_k)_1{}^\infty$ is an orthonormal sequence of points in an inner product space X, show that

$$\lim \langle x, u_k \rangle = 0 \quad \text{for each } x \in X.$$

5 · THE GRAM–SCHMIDT ORTHOGONALIZATION PROCESS

Let $(x_1, \cdots, x_n)$ be orthogonal in an inner product space X and let Y be the linear subspace of X spanned by $\{x_1, \cdots, x_n\}$. If $u_k = x_k / \|x_k\|$, from the discussion following Proposition 4 · 2 we know that for any $x \in X$ the point $x - \sum_{k=1}^{n} \langle x, u_k \rangle u_k$ is in $Y^\perp$. Thus,

$$\left\langle x_j, x - \sum_{k=1}^{n} \frac{\langle x, x_k \rangle}{\|x_k\|^2} x_k \right\rangle = 0 \quad \text{for } j = 1, \ldots, n. \tag{5 · 1}$$

This indicates a way of obtaining a countable orthogonal family from a given countable linearly independent family.

5 · 2 Proposition. (Gram–Schmidt Orthogonalization Process). *If (y_n) is a countable linearly independent family of points in X, we can construct an orthogonal family (x_n) with the same cardinal number as follows:*

$$x_1 = y_1$$

$$x_n = y_n - \sum_{k=1}^{n-1} \alpha^k x_k \quad \textit{where } \alpha^k = \frac{\langle y_n, x_k \rangle}{\|x_k\|^2} \quad \textit{for } n > 1.$$

For each n, $\{x_1, \cdots, x_n\}$ spans the same subspace as $\{y_1, \cdots, y_n\}$.

PROOF. If (y_n) has a finite number of terms, say m, then the above inductive definition terminates with x_m; otherwise it continues indefinitely. Note that $x_n \neq 0$. For, if it were zero, then $(y_1, \cdots, y_n)$ would be linearly dependent.

Clearly, $\{x_n\}$ has the same cardinal number as $\{y_n\}$. Also, it is clear from the definition that, for any n, $\{x_1, \cdots, x_n\}$ spans the same subspace as $\{y_1, \cdots, y_n\}$. That (x_n) is orthogonal follows from $5 \cdot 1$. ∎

If we let $u_n = \|x_n\|^{-1}x_n$, then the Gram–Schmidt process provides a method for obtaining an orthonormal family (u_n) from a given countable linearly independent family (y_n).

As an example of this process we give a few terms in the orthonormal sequence in $C_2[-1, 1]$ obtained from the linearly independent sequence $(I^n)_0^\infty$, where I is the identity function on $[-1, 1]$:

$$x_0 = 1, \qquad u_0 = \frac{1}{\sqrt{2}},$$

$$x_1 = I - \frac{1}{2}\int_{-1}^{1} I = I, \qquad u_1 = \sqrt{\frac{3}{2}}\,I,$$

$$x_2 = I^2 - \frac{1}{2}\int_{-1}^{1} I^2 - \frac{3}{2}I\int_{-1}^{1} I^3 = I^2 - \frac{1}{3}, \quad u_2 = \frac{3\sqrt{5}}{2\sqrt{2}}\left(I^2 - \frac{1}{3}\right).$$

The sequence $(u_n)_0^\infty$ obtained in this way is the sequence of **normalized Legendre Polynomials.**

Other useful orthonormal sequences $(u_n)_0^\infty$ of polynomials can be obtained by applying this process to $(I^n)_0^\infty$ in real spaces of the type $\mathscr{L}_2{}^g(a, b)$ where g is a nondecreasing function on the finite or infinite interval (a, b). In order to carry out the process, the functions I^n must belong to $\mathscr{L}_2{}^g(a, b)$ and we must have

$$x_n = I^n - \sum_{k=0}^{n-1} \langle I^n, u_k \rangle u_k \neq 0 \quad \text{in } \mathscr{L}_2{}^g(a, b).$$

We now give sufficient conditions for the process to be carried out in $\mathscr{L}_2{}^g(a, b)$.

5 · 3 Proposition. *If g has an infinite number of points of increase* in (a, b) and, for some $r > 0$, $\int_a^b e^{r|t|}dg(t)$ exists, then the Gram-Schmidt process can be applied to $(I^n)_0^\infty$ in $\mathscr{L}_2{}^g(a, b)$ to obtain an orthonormal sequence of polynomials in $\mathscr{L}_2{}^g(a, b)$.*

PROOF. First, we show that, for each n, $I^n \in \mathscr{L}_2{}^g(a, b)$. If (a, b) is finite, it is clear that $\int_a^b I^{2n}\, dg$ exists and, hence, $I^n \in \mathscr{L}_2{}^g(a, b)$. Assume now that (a, b)

*If g is a nondecreasing function on (a, b), a point $t \in (a, b)$ is a *point of increase* of g if, for any interval (c, d) such that $t \in (c, d) \subset (a, b)$, $g(d) - g(c) > 0$.

is infinite and $\int_a^b e^{r|t|}\, dg(t)$ exists. Since $e^{r|t|} = \sum_{k=0}^{\infty} r^k |t|^k / k!$, we have

$$t^{2n} \leq \frac{(2n)!}{r^{2n}} e^{r\,|t|}.$$

This shows that the Stieltjes-measurable function I^{2n} is less than or equal to a function integrable with respect to g and, hence, is itself integrable with respect to g; that is, $I^n \in \mathscr{L}_2{}^g(a, b)$.

We now show that $x_n \neq 0$ in $\mathscr{L}_2{}^g(a, b)$ where $x_n = I^n - \sum_{k=0}^{n-1} \langle I^n, u_k \rangle u_k$. Since x_n is a polynomial of degree n, it has at most n zeros in (a, b). Let the zeros of x_n in (a, b), listed in increasing order, be: $t_1, \cdots, t_m$. Then,

$$\|x_n\|^2 = \int_a^b x_n{}^2\, dg = \int_a^{t_1} x_n{}^2\, dg + \sum_{i=2}^{m} \int_{t_{i-1}}^{t_i} x_n{}^2\, dg + \int_{t_m}^b x_n{}^2\, dg.$$

Since g has an infinite number of points of increase in (a, b), at least one of these subintervals, say (t_{i-1}, t_i), must have positive Stieltjes measure. And, since $x_n{}^2$ is continuous and positive on (t_{i-1}, t_i), we have $\int_{t_{i-1}}^{t_i} x_n{}^2\, dg > 0$ and, hence, $\|x_n\| \neq 0$; that is, $x_n \neq 0$ in $\mathscr{L}_2{}^g(a, b)$. ▮

For example, if $(a, b) = (-1, 1)$ and $g(t) = t$, then the Gram–Schmidt process can be applied to $(I^n)_0{}^\infty$ on $\mathscr{L}_2(-1, 1)$ and the sequence obtained is the sequence of normalized Legendre Polynomials.

Suppose $(a, b) = (-\infty, \infty)$ and $g(t) = \int_{-\infty}^t e^{-\tau^2}\, d\tau$. Clearly, g has an infinite number of points of increase in $(-\infty, \infty)$. Consider

$$\int_{-\infty}^{\infty} e^{|t|}\, dg(t) = \int_{-\infty}^{\infty} e^{|t|} e^{-t^2}\, dt = \int_{-\infty}^{\infty} e^{|t| - t^2}\, dt.$$

Since $|t| - t^2 \leq -|t|$ for $|t| \geq 2$ and $\int_2^\infty e^{-t}\, dt$ exists, $\int_{-\infty}^{\infty} e^{|t|}\, dg(t)$ exists. Thus, Proposition 5 · 3 implies that the Gram–Schmidt process can be applied to $(I^n)_0{}^\infty$ in the space $\mathscr{L}_2{}^g(-\infty, \infty) = \mathscr{L}_2{}^g$. The orthonormal sequence so obtained is the sequence of **normalized Hermite Polynomials.** Since (Problems 3 and 4)

$$\int_{-\infty}^{\infty} e^{-t^2}\, dt = \sqrt{\pi} \quad \text{and} \quad \int_{-\infty}^{\infty} t^{2n} e^{-t^2}\, dt = \frac{1 \cdot 3 \cdot 5 \cdots (2n-1)}{2^n} \sqrt{\pi},$$

the first few terms of the sequence (h_n) of normalized Hermite Polynomials are:

$$x_0 = 1, \qquad h_0 = \pi^{-1/4},$$

$$x_1 = I - \frac{1}{\sqrt{\pi}} \int_{-\infty}^{\infty} I\, dg = I, \qquad h_1 = \pi^{-1/4} \sqrt{2}\, I,$$

$$x_2 = I^2 - \frac{1}{\sqrt{\pi}} \int_{-\infty}^{\infty} I^2\, dg - \frac{2}{\sqrt{\pi}} I \int_{-\infty}^{\infty} I^3\, dg$$

$$= I^2 - 1/2, \qquad h_2 = \pi^{-1/4} \sqrt{2}(I^2 - 1/2).$$

In the above examples and other examples of importance, the function g is absolutely continuous. In such cases the function g' is called a **weight function** and is denoted by w. Then the inner product in $\mathscr{L}_2^g(a, b)$ becomes a Lebesgue integral:

$$\langle f_1, f_2 \rangle = \int_a^b f_1 f_2 \, dg = \int_a^b w f_1 f_2 .$$

Now consider the general case of an orthogonal sequence $(p_n)_0^\infty$ of polynomials in a real space $\mathscr{L}_2^g(a, b)$ where p_n is of degree n. The following proposition shows that p_n is orthogonal to any polynomial of degree less than n.

5 · 4 Proposition. *If $(p_n)_0^\infty$ is an orthogonal sequence of polynomials in $\mathscr{L}_2^g(a, b)$ and q_m is a polynomial of degree m, then*

$$\langle p_n, q_m \rangle = 0 \quad \textit{for all } n > m.$$

PROOF. Since $(p_0, \cdots, p_{n-1})$ is orthogonal, it is linearly independent and, hence, is a Hamel basis of the n-dimensional space of polynomials of degree $\leq n - 1$. Then, since p_n is orthogonal to $\{p_0, \cdots, p_{n-1}\}$, it is orthogonal to the subspace spanned by these vectors. ▮

We can now show that the fact that $(p_n)_0^\infty$ is orthogonal in $\mathscr{L}_2^g(a, b)$ determines each p_n up to a multiplicative constant.

5 · 5 Proposition. *If $(p_n)_0^\infty$ and $(q_n)_0^\infty$ are orthogonal sequences of polynomials in $\mathscr{L}_2^g(a, b)$, then*

$$q_n = \alpha^n p_n \quad \textit{for some } \alpha^n \in \mathbf{R}.$$

PROOF. Since $(p_0, \cdots, p_n)$ is a Hamel basis of the space of polynomials of degree $\leq n$,

$$q_n = \sum_{k=0}^{n} \alpha^k p_k .$$

Then, for $k < n$,

$$\alpha^k = \|p_k\|^{-2} \langle q_n, p_k \rangle = 0$$

and, hence, $q_n = \alpha^n p_n$. ▮

Proposition 5 · 5 implies that if the orthogonal sequence $(p_n)_0^\infty$ of polynomials in $\mathscr{L}_2^g(a, b)$ is normalized by $\|p_n\| = 1$, then the sequence is uniquely determined up to sign. In many cases another type of normalization is more convenient and we will see some examples a little later.

The next result shows that the polynomial p_n, a term in an orthogonal sequence of polynomials in $\mathscr{L}_2^g(a, b)$, has n distinct zeros and they all lie in (a, b).

5 · 6 Proposition. *If* $(p_n)_0^\infty$ *is an orthogonal sequence of polynomials in* $\mathcal{L}_2{}^g(a, b)$, *then* p_n *has n distinct real zeros in* (a, b).

PROOF. For $n \geq 1$, since p_0 is a nonzero constant function,

$$\int_a^b p_n\, dg = \frac{1}{p_0}\int_a^b p_n\, p_0\, dg = 0.$$

Thus, there is at least one point in (a, b) where p_n changes sign. Suppose p_n changes sign at the points $x_1, \cdots, x_m$ in (a, b). Since $p_n(x_k) = 0$ for $k = 1, \cdots, m$ we know that $m \leq n$. Suppose $m < n$ and let

$$q(x) = \prod_{k=1}^{m}(x - x_k).$$

Then, $q(x)$ is a polynomial of degree m and, hence,

$$\int_a^b q p_n\, dg = 0.$$

However, this is not possible since qp_n is of constant sign on (a, b). Therefore, $m = n$ and p_n has n distinct zeros in (a, b). ▮

PROBLEMS

1. In $\mathbf{R}^3$ use the Gram–Schmidt orthogonalization process to obtain an orthonormal triple from ((2, 3, 5), (1, 4, 7), (0, 2, 1)).

2. Apply the Gram–Schmidt process to $(I^n)_0^\infty$ to obtain the first three terms of an orthonormal sequence of polynomials in $\mathcal{L}_2{}^g(a, b)$ where

a. $(a,b) = (-1, 1)$, $g(t) = \int_{-1}^t (d\tau/\sqrt{1-\tau^2})$, $w(t) = (1 - t^2)^{-1/2}$ (Chebyshev polynomials of the first kind)

b. $(a, b) = (-1, 1)$, $g(t) = \int_{-1}^t \sqrt{1-\tau^2}\, d\tau$, $w(t) = (1 - t^2)^{1/2}$ (Chebyshev polynomials of the second kind)

c. $(a, b) = (0, \infty)$, $g(t) = \int_0^t e^{-\tau}\, d\tau$, $w(t) = e^{-t}$ (Laguerre polynomials).

3. Show that $\int_{-\infty}^\infty e^{-t^2}\, dt = \sqrt{\pi}$.

Suggestion. Let $A = \int_{-\infty}^\infty e^{-t^2}\, dt$. Then $A^2 = \int_{-\infty}^\infty e^{-x^2}\, dx \int_{-\infty}^\infty e^{-y^2}\, dy = \int_{\mathbf{R}^2} e^{-(x^2+y^2)}\, dx\, dy$ and using polar coordinates

$$A^2 = \lim_{n\to\infty}\int_0^{2\pi}\int_0^n \rho e^{-\rho^2}\, d\rho\, d\theta.$$

4. Show that

$$\int_{-\infty}^\infty x^{2n}e^{-x^2}\, dx = \frac{1\cdot 3\cdot 5\cdots(2n-1)}{2^n}\sqrt{\pi}.$$

6 · ORTHOGONAL BASES

An orthogonal family $(x_s)_{s\in S}$ of points in an inner product space X is said to be **complete** if no orthogonal family of points in X is a proper extension of $(x_s)_{s\in S}$. Thus, if $E = \{x_s : s \in S\}$, then the orthogonal family $(x_s)_{s\in S}$ is complete if and only if $E^\perp = \{0\}$. Note that this use of the term "complete" bears no relation to its use in the context of complete metric or normed linear spaces. A complete orthogonal family of points in X is called an **orthogonal basis** of X. If an orthogonal basis is normalized so that each term has norm 1, then it is called an **orthonormal basis** of X. In many discussions it is convenient to use orthonormal bases.

An orthogonal basis is maximal with respect to orthogonality whereas a Hamel basis is maximal with respect to linear independence. As the following example shows, an orthonormal baisis need not be a Hamel basis. In the complex sequence space l_2 consider the sequence of points (e_k) where $e_k = (\delta_k{}^j)$ and $\delta_k{}^j$ is the Kronecker delta. This is not a Hamel basis for l_2 since the sequence $(1/n)$ cannot be written as a finite linear combination of terms of (e_k). However, (e_k) is an orthonormal basis for l_2. It is clearly an orthonormal sequence. To show that it is complete, take $x = (\alpha^k) \in E^\perp$ where $E = \{e_k : k = 1, 2, \cdots\}$. Then, for all k

$$0 = \langle x, e_k \rangle = \alpha^k;$$

that is, $x = 0$.

Thus, an orthonormal basis B may fail to be a Hamel basis for an inner product space X because some vector in X cannot be expressed as a finite linear combination of terms of B. This can happen only if B is infinite (Problem 1). If B is infinite, then the following theorem shows that provided X is a Hilbert space any vector $x \in X$ can be written as an "infinite linear combination" of terms of B. More precisely this theorem shows that x is the sum of its Fourier series with respect to B.

6 · 1 Theorem. *If $E = \{u_s : s \in S\}$ where $(u_s)_{s \in S}$ is an orthonormal family of points in a Hilbert space X, then the following are equivalent:*

(1) *$(u_s)_{s \in S}$ is complete*
(2) *For each $x \in X$, $x = \sum_{s \in S} \langle x, u_s \rangle u_s$*
(3) *For each $x \in X$, $\|x\|^2 = \sum_{s \in S} |\langle x, u_s \rangle|^2$ (Parseval's Identity)*
(4) *The span of E is dense in X.*

PROOF. (1) ⇒(2). Take $x \in X$. By Proposition 4 · 5, $x - \sum_{s \in S} \langle x, u_s \rangle u_s$ is orthogonal to E. Since $(u_s)_{s \in S}$ is complete, $E^\perp = \{0\}$ and, hence,

$$x - \sum_{s \in S} \langle x, u_s \rangle u_s = 0;$$

that is, $x = \sum_{s \in S} \langle x, u_s \rangle u_s$.
(2) ⇒(3). Take $x \in X$. Then $x = \sum_{s \in S} \langle x, u_s \rangle u_s$ and

$$\|x\|^2 = \langle x, x \rangle = \left\langle \sum_{s \in S} \langle x, u_s \rangle u_s, x \right\rangle = \sum_{s \in S} \langle x, u_s \rangle \langle u_s, x \rangle$$
$$= \sum_{s \in S} |\langle x, u_s \rangle|^2.$$

(3) $\Rightarrow$(4). Take $x \in X$ and let $T = \{s \in S : \langle x, u_s \rangle \neq 0\}$. Let $\{s_k\}$ be an enumeration of T. Then, for each $\varepsilon > 0$ there exists a positive integer n such that

$$\left\| x - \sum_{k=1}^{n} \langle x, u_{s_k} \rangle u_{s_k} \right\|^2 = \|x\|^2 - \sum_{k=1}^{n} |\langle x, u_{s_k} \rangle|^2 < \varepsilon.$$

(4) $\Rightarrow$(1). Take $x \in E^\perp$. Then x is orthogonal to the span of E since for any $u_{s_1}, \cdots, u_{s_n}$ in E and $\alpha^1, \cdots, \alpha^n$ in $\mathbf{F}$

$$\left\langle \sum_{k=1}^{n} \alpha^k u_{s_k}, x \right\rangle = \sum_{k=1}^{n} \alpha^k \langle u_{s_k}, x \rangle = 0.$$

Since the span of E is dense in X there is a sequence (y_n) of points in this span such that $\lim y_n = x$. Then,

$$\langle x, x \rangle = \lim \langle y_n, x \rangle = 0$$

and, hence, $x = 0$. ∎

By using Zorn's Lemma it is easy to show that any nontrivial inner product space has an orthonormal basis.

6 · 2 Theorem. *If A is an orthonormal family of points in an inner product space X, then there exists an orthonormal basis B which is an extension of A.*

Proof. Let $\mathscr{A}$ be the partially ordered set consisting of all orthonormal families of points in X which are extensions of A. The partial order is that of extension; that is, $A_1 \prec A_2$ if A_2 is an extension of A_1. For any chain $\{(x_s)_{s \in S_t} : t \in T\}$ in $\mathscr{A}$ the family $(x_s)_{s \in S}$, where $S = \bigcup_{t \in T} S_t$, is orthonormal and is an extension of A. Hence, any chain in $\mathscr{A}$ has an upper bound and, therefore, $\mathscr{A}$ has an maximal element by Zorn's Lemma. This maximal element is an orthonormal basis of X which is an extension of A. ∎

If an inner product space X has a Hamel basis $B = (u_s)_{s \in S}$ which is orthonormal, then B is an orthonormal basis of X. For if $x \in E^\perp$ where $E = \{u_s : s \in S\}$, then there exist vectors $u_{s_1}, \cdots, u_{s_n}$ in E and scalars $\alpha^1, \cdots, \alpha^n$ in $\mathbf{F}$ such that $x = \sum_{k=1}^{n} \alpha^k u_{s_k}$. Then,

$$\langle x, x \rangle = \left\langle \sum_{k=1}^{n} \alpha^k u_{s_k}, x \right\rangle = \sum_{k=1}^{n} \alpha^k \langle u_{s_k}, x \rangle = 0$$

and, hence, $x = 0$. This shows that $E^\perp = \{0\}$. In case X has a countable Hamel basis then application of the Gram–Schmidt orthogonalization process produces a countable Hamel basis which is orthonormal and, hence, is an orthonormal basis of X.

We now show that any separable inner product space has a countable orthonormal basis. Recall that a space X is separable if there is a countable set which is dense in X.

6 · 3 Proposition. *A separable inner product space X has a countable orthonormal basis.*

PROOF. Let $E = \{a_k\}$ be a countable dense set in X with $a_1 \neq 0$. Define a countable family (x_n) inductively as follows. Let $x_1 = a_1$ and assume $x_1, \cdots, x_m$ have been chosen from E so that $(x_1, \cdots, x_m)$ is linearly independent. If $\{x_1, \cdots, x_m\}$ spans E, the construction terminates. Otherwise, let x_{m+1} be the first element in E not in the subspace spanned by $\{x_1, \cdots, x_m\}$. Then, $(x_1, \cdots, x_{m+1})$ is linearly independent. Thus, we obtain a countable linearly independent sequence (x_n) such that $\{x_n\}$ spans E. By the Gram-Schmidt orthogonalization process we obtain a countable orthonormal family (u_n) such that $\{u_n\}$ spans E. To show that (u_n) is complete, suppose that $\langle x, u_n \rangle = 0$ for all n. Then x is orthogonal to the dense set E and, hence, $x = 0$ (Problem 2, p. 288). ∎

A consequence of Proposition 6 · 3 is that any separable Hilbert space X is isometrically isomorphic to l_2^n or l_2. If X has a finite orthonormal basis $(u_1, \cdots, u_n)$, then X is essentially the same as l_2^n (Problem 2). Suppose now that X has a countably infinite orthonormal basis $(u_k)_1^\infty$. Let φ be the function from X into l_2 such that

$$\varphi(x) = (\langle x, u_k \rangle)_1^\infty, \quad x \in X.$$

That $(\langle x, u_k \rangle)$ belongs to l_2 follows from Parseval's identity:

$$\|x\|^2 = \sum_{k=1}^{\infty} |\langle x, u_k \rangle|^2.$$

This also shows that φ is an isometry: $\|\varphi(x)\| = \|x\|$. The linearity of φ is easily checked. To show that φ maps X onto l_2, take $(\alpha^k) \in l_2$. Since $\sum_{k=1}^\infty |\alpha^k|^2$ converges, we see that $\sum_{k=1}^\infty \alpha^k u_k$ converges. Let $x = \sum_{k=1}^\infty \alpha^k u_k$. Then,

$$\varphi(x) = (\langle x, u_k \rangle) = (\alpha^k)$$

which shows that φ is onto. Thus, X and l_2 are isometrically isomorphic. Furthermore, the inner product is preserved under this isomorphism:

$$\begin{aligned} \langle x, y \rangle &= \left\langle \sum_{k=1}^{\infty} \langle x, u_k \rangle u_k, \sum_{j=1}^{\infty} \langle y, u_j \rangle u_j \right\rangle \\ &= \sum_{k=1}^{\infty} \langle x, u_k \rangle \overline{\langle y, u_k \rangle} = \langle \varphi(x), \varphi(y) \rangle. \end{aligned}$$

PROBLEMS

1. If $B = (u_1, \cdots, u_n)$ is a finite orthonormal basis for an inner product space X, show that B is a Hamel basis for X.

2. If a Hilbert space X has the finite orthonormal basis $(u_1, \cdots, u_n)$, show that X is isometrically isomorphic to l_2^n.

3. If a Hilbert space X has a countable orthonormal basis, show that X is separable.

4. Show that

$$\left(\frac{1}{\sqrt{2\pi}}, \frac{1}{\sqrt{\pi}}\cos t, \frac{1}{\sqrt{\pi}}\sin t, \frac{1}{\sqrt{\pi}}\cos 2t, \frac{1}{\sqrt{\pi}}\sin 2t, \cdots\right)$$

is an orthonormal basis for $C_2([-\pi, \pi], \mathbf{R})$.

5. If B is an orthonormal basis for an inner product space X and

$$\langle x, u\rangle = \langle y, u\rangle \quad \text{for all } u \in B,$$

show that $x = y$.

6. Show that in an inner product space Parseval's identity: $\|x\|^2 = \sum_{s\in S} |\langle x, u_s\rangle|^2$, where $(u_s)_{s\in S}$ is an orthonormal family, is equivalent to

$$\langle x, y\rangle = \sum_{s\in S} \langle x, u_s\rangle\overline{\langle y, u_s\rangle}.$$

7 · THE COMPLEX EXPONENTIAL AND TRIGONOMETRIC SEQUENCES

Since the complex Hilbert space $\mathscr{L}_2[-\pi, \pi]$ is separable, we know that it has a countable orthogonal basis.

7 · 1 Theorem. *If, for each integer n,*

$$E_n(t) = e^{int}, \quad t \in [-\pi, \pi],$$

then $(E_n)_{-\infty}^{\infty}$ is an orthogonal basis for $\mathscr{L}_2[-\pi, \pi]$.

Proof. Since, for $n \neq m$,

$$\langle E_n, E_m\rangle = \int_{-\pi}^{\pi} e^{int}e^{-imt}\,dt = \int_{-\pi}^{\pi} e^{i(n-m)t}\,dt = 0,$$

(E_n) is orthogonal.

We show that (E_n) is complete by showing that its span is dense in $\mathscr{L}_2[-\pi, \pi]$. Take $f \in \mathscr{L}_2[-\pi, \pi]$ and $\varepsilon > 0$. By Theorem 3 · 8 of Chapter 8 (p. 197) there exists a continuous function g such that

$$\|f - g\| < \varepsilon/3.$$

Let $t_n = \pi - 1/n^2$ and define

$$g_n(t) = \begin{cases} g(t), & \text{for } t \in [-\pi, t_n) \\ g(t_n) + (g(-\pi) - g(t_n))\dfrac{t - t_n}{\pi - t_n} & \text{for } t \in [t_n, \pi]. \end{cases}$$

Then, g_n is continuous on $[-\pi, \pi]$ and, since $g_n(\pi) = g_n(-\pi)$, g_n can be extended to a continuous function on $\mathbf{R}$ with period 2π. Also, if $M = \max\{|g(t)| : t \in [-\pi, \pi]\}$, then

$$\|g - g_n\| = \left[\int_{-\pi}^{\pi} |g(t) - g_n(t)|^2\,dt\right]^{1/2} \leq \left[\int_{t_n}^{\pi} (2M)^2\,dt\right]^{1/2} = \frac{2M}{n}.$$

Therefore, taking $n \geq 6M/\varepsilon$, we have

$$\|g - g_n\| \leq \varepsilon/3.$$

By the Weierstrass Theorem (Corollary 2 · 9, p. 191) there exists a trigonometric polynomial T, $T(t) = \sum_{k=-m}^{m} \alpha_k e^{ikt}$, such that

$$|g_n(t) - T(t)| \leq \frac{\varepsilon}{3\sqrt{2\pi}} \quad \text{for all } t \in [-\pi, \pi].$$

Then,

$$\|g_n - T\| = \left[\int_{-\pi}^{\pi} |g_n(t) - T(t)|^2 \, dt\right]^{1/2} \leq \frac{\varepsilon}{3}$$

and, therefore,

$$\|f - T\| \leq \|f - g\| + \|g - g_n\| + \|g_n - T\| < \varepsilon. \quad \blacksquare$$

Since

$$\langle E_n, E_n \rangle = \int_{-\pi}^{\pi} e^{int} e^{-int} \, dt = 2\pi,$$

if

$$u_n(t) = \frac{1}{\sqrt{2\pi}} e^{int},$$

then $(u_n)_{-\infty}^{\infty}$ is an orthonormal basis for $L_2[-\pi, \pi]$. Theorem 6 · 1 implies that the Fourier series of any function $f \in \mathscr{L}_2[-\pi, \pi]$ with respect to this basis must converge to f in $\mathscr{L}_2[-\pi, \pi]$; that is,

$$f = \sum_{n=-\infty}^{\infty} \gamma_n u_n \quad \text{where } \gamma_n = \langle f, u_n \rangle = \frac{1}{\sqrt{2\pi}} \int_{-\pi}^{\pi} f(t) e^{-int} \, dt.$$

We emphasize that the convergence of this Fourier series to f is in the norm of $\mathscr{L}_2[-\pi, \pi]$. This is not a solution to the classical problem concerning pointwise convergence of Fourier series which will be discussed in §10.

By modifying the above orthonormal basis for $\mathscr{L}_2[-\pi, \pi]$ one can obtain an orthonormal basis for $\mathscr{L}_2[a, b]$ where $[a, b]$ is any finite interval. In fact, if

$$u_n(t) = \frac{1}{\sqrt{b - a}} e^{i2\pi nt/(b-a)}$$

then $(u_n)_{-\infty}^{\infty}$ is an orthonormal basis for $\mathscr{L}_2[a, b]$.

Now consider the real space $\mathscr{L}_2[-\pi, \pi]$.

7 · 2 Theorem. *The sequence* $(1, \cos t, \sin t, \cos 2t, \sin 2t, \cdots)$ *is an orthogonal basis for the real space* $\mathscr{L}_2[-\pi, \pi]$.

PROOF. The fact that this sequence is orthogonal is easily verified. To show that the sequence is complete, take $f \in \mathscr{L}_2[-\pi, \pi]$ such that, for all $n = 1, 2, \cdots$,

$$\int_{-\pi}^{\pi} f(t)\, dt = 0, \quad \int_{-\pi}^{\pi} f(t) \cos nt\, dt = 0, \quad \int_{-\pi}^{\pi} f(t) \sin nt\, dt = 0.$$

Then, for $n = 0, 1, 2, \cdots$

$$\int_{-\pi}^{\pi} f(t)e^{int}\, dt = \int_{-\pi}^{\pi} f(t) \cos nt\, dt + i \int_{-\pi}^{\pi} f(t) \sin nt\, dt = 0$$

and also

$$\int_{-\pi}^{\pi} f(t)e^{-int}\, dt = 0.$$

Thus, $f = 0$ in $\mathscr{L}_2[-\pi, \pi]$; that is, $f = 0$ a.e. on $[-\pi, \pi]$. This shows that the sequence is complete. ∎

Since $(e^{iqnx})_{n=-\infty}^{\infty}$, where $q = 2\pi/(b-a)$ is an orthogonal basis for $\mathscr{L}_2[a, b]$, if $f \in \mathscr{L}_2[a, b]$ and

$$\int_a^b f(x)e^{-iqnx}\, dx = 0 \quad \text{for all } n$$

then $f = 0$ in $\mathscr{L}_2[a, b]$.

We now generalize this to a real space $\mathscr{L}_2^g$, where g is a bounded nondecreasing function. First, we note that Lemmas 3 · 1 – 3 · 5 and Theorem 3 · 6 of Chapter 8 can be extended to $\mathscr{L}_2^g$ where the open intervals and step functions mentioned are now taken to be admissible and the integral $\int_a^b |f - h|$ is replaced by $\int_{-\infty}^{\infty} |f - h| dg$. Also, we note that, since g is a bounded nondecreasing function on **R**, $\lim_{x \to -\infty} g(x)$ and $\lim_{x \to \infty} g(x)$ exist and, if we let $g(-\infty) = \lim_{x \to -\infty} g(x)$ and $g(\infty) = \lim_{x \to \infty} g(x)$, then $\int_{-\infty}^{\infty} dg = g(\infty) - g(-\infty)$.

7 · 3 Theorem. *If $f \in \mathscr{L}_2^g$ where g is a bounded nondecreasing function on* **R**, *and*

$$\int_{-\infty}^{\infty} f(x)e^{itx}\, dg(x) = 0 \textit{ for all } t \in \mathbf{R},$$

then $f = 0$ in $\mathscr{L}_2^g$.

PROOF. If T is a trigonometric polynomial of period p, $T(x) = \sum_{k=-m}^{m} c_k e^{i2\pi kx/p}$, then

$$\langle f, T\rangle = \int_{-\infty}^{\infty} fTdg = \sum_{k=-m}^{m} c_k \int_{-\infty}^{\infty} f(x)e^{i2\pi kx/p}\, dg(x) = 0.$$

Thus, if there is a sequence (T_n) of trigonometric polynomials such that lim $T_n = f$ in $\mathcal{L}_2^g$, then $\langle f, f\rangle = \lim \langle f, T_n\rangle = 0$ and, hence, f $= 0$ in $\mathcal{L}_2^g$.

By the extended versions of Theorem 3 · 6 and Lemma 3 · 2 of Chapter 8, for any $\varepsilon > 0$ there exist an admissible step function h_1 and a continuous function h_2 such that $\|f - h_1\| < \varepsilon$ and $\|h_1 - h_2\| < \varepsilon$. Say that $h_1(x) = 0$ for all $x \notin [c, d]$ where $[c, d]$ is an admissible interval and let $M = \max \{|h_2(x)|^2: x \in [c, d]\}$. Take $a < c$ and $b > d$ such that $g(a) - g(-\infty) < \varepsilon /M$ and $g(\infty) - g(b) < \varepsilon /M$. Also, since g is continuous at c and d, we can take $c_1 \in (a, c)$ and $d_1 \in (d, b)$ such that $g(c) - g(c_1) < \varepsilon /M$ and $g(d_1) - g(d) < \varepsilon /M$.

Define h to be the continuous function of period $p = b - a$ such that:

$$h(x) = h_2(x),\ x \in [c, d];\ h(x) = 0,\ x \in [a, c_1] \cup [d_1, b];$$

and h is linear on $[c_1, c]$ and on $[d, d_1]$. Note that $h^2(x) \leq M$ for all $x \in$ **R**. Writing the integral $\int_{-\infty}^{\infty} (h(x) - h_1(x))^2 dg(x)$ as a sum of integrals over the intervals $[-\infty, a]$, $[a, c_1]$, $[c_1, c]$, $[c, d]$ $[d, d_1]$, $[d_1, b]$, and $[b, \infty]$, we obtain:

$$\|h - h_1\|^2 < \varepsilon + 0 + \varepsilon + \varepsilon^2 + \varepsilon + 0 + \varepsilon = 4\varepsilon + \varepsilon^2.$$

Thus, h is a continuous function on **R** of period p and $\|h - h_1\| < [4\varepsilon + \varepsilon^2]^{1/2}$.

The Weierstrass Approximation result in Corollary 2 · 9 of Chapter 8 shows that there is a trigonometric polynomial T of period p such that $|h(x) - T(x)| < \varepsilon$ for all $x \in$ **R**. Then

$$\|h - T\| = \left[\int_{-\infty}^{\infty} (h(x) - T(x))^2\, dg(x)\right]^{1/2} < \varepsilon B \text{ where } B = (g(\infty) - g(-\infty))^{1/2}.$$

Letting $\varphi(\varepsilon) = \varepsilon + (4\varepsilon + \varepsilon^2)^{1/2} + \varepsilon B$, we have:

$$\|f - T\| \leq \|f - h_1\| + \|h_1 - h\| + \|h - T\| < \varphi(\varepsilon) \text{ where } \lim_{\varepsilon\to 0} \varphi(\varepsilon) = 0.$$

For $\varepsilon = 1/n$ $(n = 1, 2, \ldots)$, let T_n denote the corresponding trigonometric polynomial. We now have a sequence (T_n) such that lim $T_n = f$ in $\mathcal{L}_2^g$ ∎

PROBLEM

a. Show that if

$$u_n(t) = \frac{1}{\sqrt{2\pi}} e^{int},$$

then $(u_n)_{-\infty}^{\infty}$ is an orthonormal basis for $\mathscr{L}_2[c, c + 2\pi]$ for any real number c.

b. Show that if $[a, b]$ is a finite interval and

$$u_n(\tau) = \frac{1}{\sqrt{b - a}} e^{2\pi in\tau/(b-a)},$$

then $(u_n)_{-\infty}^{\infty}$ is an orthonormal basis for $\mathscr{L}_2[a, b]$.

8 · THE LEGENDRE POLYNOMIALS

In §5 as an example of the application of the Gram–Schmidt orthogonalization process we applied the process to the linearly independent sequence $(I^n)_0^\infty$ in $C_2[-1, 1]$ to obtain the first three normalized Legendre polynomials. The Legendre polynomials are important in many applications in mathematical physics and in this section we discuss some of their properties.

8 · 1 Definition. *The polynomials defined by*

$$P_0(x) = 1,$$

$$P_n(x) = \frac{1}{2^n n!} \frac{d^n}{dx^n} (x^2 - 1)^n \quad \textit{for } n > 0$$

are called the ***Legendre polynomials.***

From the formula of this definition, called **Rodrigues' formula**, we can write down the first few Legendre polynomials:

$$P_1(x) = x, \quad P_2(x) = \tfrac{1}{2}(3x^2 - 1), \quad P_3(x) = \tfrac{1}{2}(5x^3 - 3x).$$

Expanding $(x^2 - 1)^n$ we see that P_n is a polynomial of degree n and that the coefficient of x^n in $P_n(x)$ is $(2n)!/(2^n(n!)^2)$. Also, $P_n(x)$ has only even powers of x if n is even and only odd powers if n is odd. Thus,

$$P_n(-x) = (-1)^n P_n(x).$$

We will show that (P_n) is an orthogonal basis of the real space $\mathscr{L}_2(-1, 1)$. First we note that if $f(x) = (x^2 - 1)^n$, then

$$f^{(k)}(x) = q_k(x)(x^2 - 1)^{n-k} \quad k \le n$$

where q_k is a polynomial of degree k. This can be proved by induction. Also, if we use integration by parts, then we have

$$\int_{-1}^{1} f = \int_{-1}^{1} (x^2 - 1)^n \, dx = x(x^2 - 1)^n \Big]_{-1}^{1} - 2n \int_{-1}^{1} x^2(x^2 - 1)^{n-1} \, dx$$

$$= -2n \int_{-1}^{1} (x^2 - 1)^n \, dx - 2n \int_{-1}^{1} (x^2 - 1)^{n-1} \, dx.$$

Thus, we obtain the reduction formula

$$\int_{-1}^{1} (x^2 - 1)^n \, dx = -\frac{2n}{2n+1} \int_{-1}^{1} (x^2 - 1)^{n-1} \, dx.$$

By repeated use of this reduction formula we obtain

$$\int_{-1}^{1} f = \int_{-1}^{1} (x^2 - 1)^n \, dx = (-1)^n 2 \frac{2^n n!}{(2n+1)(2n-1) \cdots 3}.$$

8 · 2 Proposition. *(P_n) is an orthogonal sequence in $\mathscr{L}_2(-1, 1)$ and*

$$\|P_n\|^2 = \frac{2}{2n+1}.$$

Proof. If we let $f(x) = (x^2 - 1)^n$ and integrate by parts repeatedly, then for $m \leq n$ we obtain

$$2^n n! \langle I^m, P_n \rangle = \int_{-1}^{1} x^m f^{(n)}(x) \, dx = -m \int_{-1}^{1} x^{m-1} f^{(n-1)}(x) \, dx$$

$$= \cdots = (-1)^m m! \int_{-1}^{1} f^{(n-m)}(x) \, dx.$$

Thus, if $m < n$, then

$$2^n n! \langle I^m, P_n \rangle = (-1)^m m! f^{(n-m-1)}(x)]_{-1}^{1} = 0;$$

that is, $\langle I^m, P_n \rangle = 0$ and, hence, $\langle P_m, P_n \rangle = 0$. If $m = n$, then

$$\|P_n\|^2 = \langle P_n, P_n \rangle = \frac{(2n)!}{2^n (n!)^2} \langle I^n, P_n \rangle = \frac{(2n)!}{[2^n \cdot n!]^2} (-1)^n \int_{-1}^{1} f$$

$$= \frac{(2n)!}{[2^n \cdot n!]^2} 2 \frac{2^n n!}{(2n+1)(2n-1) \cdots 3} = \frac{2}{2n+1}. \blacksquare$$

Since (P_n) is an orthogonal sequence in $\mathscr{L}_2(-1, 1)$ it follows from Proposition 5 · 6 that P_n has n distinct zeros in $(-1, 1)$. We also know that $(P_0, P_1, \cdots, P_m)$ is a Hamel basis for the $(m+1)$-dimensional space of polynomials of degree $\leq m$.

8 · 3 Theorem. *The sequence (P_n) of Legendre polynomials is an orthogonal basis of $\mathscr{L}_2(-1, 1)$.*

PROOF. To show that the orthogonal sequence (P_n) is complete we will show that its span is dense in $\mathscr{L}_2(-1, 1)$. Take $f \in \mathscr{L}_2(-1, 1)$ and $\varepsilon > 0$. By Theorem 3 · 8, p. 197, there exists a continuous function g such that

$$\|f - g\| < \varepsilon/2.$$

By the Weierstrass Approximation Theorem (Cor. 2·6, p. 190) there exists a polynomial Q such that

$$|g(x) - Q(x)| < \varepsilon/4 \quad \text{for all } x \in [-1, 1].$$

Then,

$$\|g - Q\| = \left[\int_{-1}^{1} |g(x) - Q(x)|^2 \, dx\right]^{1/2} < \frac{\varepsilon}{2}$$

and, therefore,

$$\|f - Q\| \leq \|f - g\| + \|g - Q\| < \varepsilon.$$

If the polynomial Q is of degree m, then it is a linear combination of $P_0, \cdots, P_m$ and so is in the span of (P_n). ▮

We now obtain a three term recurrence relation which allows us to compute P_{n+1} in terms of P_n and P_{n-1}.

8 · 4 Proposition. $(n + 1)P_{n+1}(x) = (2n + 1)xP_n(x) - nP_{n-1}(x)$ *for* $n \geq 1$.

PROOF. Take $n \geq 1$. Since IP_n is a polynomial of degree $n + 1$ and $(P_0, \cdots, P_{n+1})$ is a Hamel basis for the linear space of polynomials of degree $\leq n + 1$, we have

$$IP_n = \sum_{k=0}^{n+1} \alpha^k P_k, \quad \text{where } \alpha^k = \frac{2k + 1}{2} \int_{-1}^{1} xP_n(x)P_k(x) \, dx.$$

If $k < n - 1$, then IP_k is of degree less than n and P_n is orthogonal to IP_k (Proposition 5 · 4); that is, $\alpha^k = 0$ for $k < n - 1$. Also, since IP_n^2 is an odd function,

$$\alpha^n = \frac{2n + 1}{2} \int_{-1}^{1} xP_n^2(x) \, dx = 0.$$

Thus,

$$xP_n(x) = \alpha^{n-1}P_{n-1}(x) + \alpha^{n+1}P_{n+1}(x).$$

Equating the coefficients of x^{n+1}, we have

$$\frac{1}{2^n n!} \cdot \frac{(2n)!}{n!} = \frac{1}{2^{n+1}(n + 1)!} \cdot \frac{(2n + 2)!}{(n + 1)!} \alpha^{n+1}$$

or

$$\alpha^{n+1} = \frac{n + 1}{2n + 1}.$$

Then

$$\int_{-1}^{1} xP_n(x)P_{n+1}(x)\,dx = \frac{2}{2n+3}\alpha^{n+1} = \frac{2(n+1)}{(2n+3)(2n+1)} \quad \text{for } n \geq 0$$

and replacing n by $n-1$, we obtain

$$\alpha^{n-1} = \frac{2n-1}{2}\int_{-1}^{1} xP_n(x)P_{n-1}(x)\,dx = \frac{2n-1}{2}\cdot\frac{2n}{(2n+1)(2n-1)}$$
$$= \frac{n}{2n+1}.$$

With these values for the coefficients α^{n-1} and α^{n+1} we have

$$(2n+1)xP_n(x) = nP_{n-1}(x) + (n+1)P_{n+1}(x). \blacksquare$$

Since we know P_1 and P_2 we can use the recurrence relation to obtain P_3 :

$$P_3(x) = \tfrac{5}{3}x\,P_2(x) - \tfrac{2}{3}P_1(x) = \tfrac{5}{3}x\tfrac{1}{2}(3x^2-1) - \tfrac{2}{3}x$$
$$= \tfrac{1}{2}(5x^3 - 3x).$$

We could now obtain P_4 and so on.

We now introduce a generating function for the Legendre polynomials. The generating function makes it easy for us to obtain some further properties of the Legendre polynomials.

8 · 5 Proposition. $(1 - 2xt + t^2)^{-1/2} = \sum_{n=0}^{\infty} P_n(x)t^n$ *for* $|2xt - t^2| < 1$.

PROOF. Let $F(x, t) = (1 - 2xt + t^2)^{-1/2}$. Then from the binomial series expansion

$$(1-u)^{-1/2} = 1 + \sum_{k=1}^{\infty} \frac{1\cdot 3\cdots(2k-1)}{2^k k!}u^k, \quad |u| < 1$$

we obtain

$$F(x,t) = 1 + \sum_{k=1}^{\infty} \frac{1\cdot 3\cdots(2k-1)}{2^k k!}(2xt - t^2)^k$$
$$= \sum_{n=0}^{\infty} g_n(x)t^n$$

where g_n is a polynomial and the series is absolutely and uniformly convergent for $|2xt - t^2| < 1$. Since

$$\frac{\partial F(x,t)}{\partial t} = (x-t)(1 - 2xt + t^2)^{-3/2} = (x-t)(1 - 2xt + t^2)^{-1}F(x,t),$$

we have

$$(1 - 2xt + t^2)\frac{\partial F(x,t)}{\partial t} + (t - x)F(x,t) = 0.$$

That is,

$$(1 - 2xt + t^2) \sum_{n=1}^{\infty} n g_n(x) t^{n-1} + (t - x) \sum_{n=0}^{\infty} g_n(x) t^n = 0$$

or

$$g_1(x) - x g_0(x) + \sum_{n=1}^{\infty} ((n+1) g_{n+1}(x) - (2n+1) x g_n(x) + n g_{n-1}(x)) t^n = 0.$$

Thus,

$$g_1(x) = x g_0(x),$$

$$(n+1) g_{n+1}(x) = (2n+1) x g_n(x) - n g_{n-1}(x), \quad n \geq 1.$$

Now $g_0(x) = F(x, 0) = 1 = P_0(x)$ and so $g_1(x) = x = P_1(x)$. Since the recurrence formula for the g_n is the same as that for the P_n (Proposition 8 · 4) and the first two terms are equal, we have $g_n = P_n$ for all $n \geq 0$. ▌

8 · 6 Corollary. *For all* $n \geq 0$, $P_n(1) = 1$.

Proof. Set $x = 1$ in the generating function and obtain

$$(1 - 2t + t^2)^{-1/2} = (1 - t)^{-1} = \sum_{n=0}^{\infty} t^n = \sum_{n=0}^{\infty} P_n(1) t^n.$$

Then $P_n(1) = 1$. ▌

We next show that P_n is a solution of a certain second order linear differential equation, known as **Legendre's equation**.

8 · 7 Proposition. $(1 - x^2) P_n''(x) - 2x P_n'(x) + n(n+1) P_n(x) = 0$.

Proof. Let $f(x) = (x^2 - 1)^n$. Then

$$(1 - x^2) f'(x) + 2nx f(x) = 0.$$

Differentiating $n + 1$ times and using Leibnitz' rule for the $(n + 1)$th derivative of a product:

$$D^{n+1}[gf] = \sum_{k=0}^{n+1} \binom{n+1}{k} D^k g D^{n+1-k} f,$$

we have

$$\binom{n+1}{0}(1 - x^2) f^{(n+2)}(x) + \binom{n+1}{1}(-2x) f^{(n+1)}(x) + \binom{n+1}{2}(-2) f^{(n)}(x)$$
$$+ 2n\left[\binom{n+1}{0} x f^{(n+1)}(x) + \binom{n+1}{1} f^{(n)}(x)\right] = 0.$$

Dividing by $2^n n!$ and collecting similar terms, we obtain

$$(1 - x^2) P_n''(x) - 2x P_n'(x) + n(n+1) P_n(x) = 0. \;▌$$

PROBLEMS

1. The representation of a function $f \in \mathscr{L}_2(-1, 1)$ with respect to the orthogonal basis (P_n) is often called the **Fourier–Legendre series** of f. Find the Fourier–Legendre series of each of the following functions:

a. $f(x) = \begin{cases} 0 & \text{for } -1 < x < 0 \\ 1 & \text{for } 0 < x < 1 \end{cases}$

b. $f(x) = |x| \quad \text{for } -1 < x < 1.$

2. Show that

$$P_n'(x) = xP_{n-1}'(x) + nP_{n-1}(x), \quad n \geq 1.$$

Suggestion. Use Rodrigues' formula for P_n to obtain

$$P_n'(x) = \frac{2n}{2^n n!} \frac{d^n}{dx^n} (x(x^2 - 1)^{n-1})$$

and then apply Leibniz' rule for the nth derivative of a product.

3. Show that

$$P_n(x) = \frac{x^2 - 1}{n} P_{n-1}'(x) + xP_{n-1}(x), \quad n \geq 1.$$

Suggestion. Apply Leibniz' rule to

$$P_n(x) = \frac{2n}{2^n n!} \frac{d^{n-1}}{dx^{n-1}} (x(x^2 - 1)^{n-1})$$

and subtract the resulting expression for $P_n(x)$ from twice the expression for $P_n(x)$ obtained by applying Leibniz' rule to

$$P_n(x) = \frac{1}{2^n n!} \frac{d^n}{dx^n} ((x^2 - 1)^{n-1}(x^2 - 1)).$$

4. Show that

a. $$\frac{1 - x^2}{n^2} (P_n'(x))^2 + (P_n(x))^2 = \frac{1 - x^2}{n^2} (P_{n-1}'(x))^2 + (P_{n-1}(x))^2,$$

$n \geq 1.$

Suggestion. Use Problems 2 and 3.

b. $$\frac{1 - x^2}{n^2} (P_n'(x))^2 + (P_n(x))^2 \leq 1, \quad |x| \leq 1, \quad n \geq 1.$$

c. $|P_n(x)| \leq 1$, $|x| \leq 1$.

5. Show that

$$P_{2n}(0) = \frac{(-1)^n (2n)!}{2^{2n} (n!)^2}.$$

6. Show that

$$\sum_{n=0}^{\infty} \frac{x^{n+1}}{n+1} P_n(x) = \frac{1}{2} \ln\left|\frac{1+x}{1-x}\right| \quad \text{for } |x| < 1.$$

7. Show that

$$P_n'(x) = \sum_{r=0}^{[(n-1)/2]} (2n - 4r - 1)P_{n-2r-1}(x).$$

Deduce that for all $x \in [-1, 1]$

$$n^{-2}|P_n'(x)| \le 1, \quad n^{-4}|P_n''(x)| \le 1, \cdots.$$

8. The **Chebyshev polynomials** of the first kind are defined by:

$$T_n(x) = \cos(n \operatorname{arc} \cos x), \quad n = 0, 1, 2, \cdots.$$

Show that:

a.

$$T_0(x) = 1,$$

$$T_n(x) = \sum_{k=0}^{[n/2]} \binom{n}{2k} x^{n-2k}(x^2 - 1)^k, \quad n = 1, 2, \cdots.$$

Suggestion. For $-1 \le x \le 1$, let $x = \cos\theta$, $\sqrt{1 - x^2} = \sin\theta$, $0 \le \theta \le \pi$. Use deMoivre's formula

$$\cos n\theta + i \sin n\theta = (\cos\theta + i \sin\theta)^n,$$

expand the right side by the Binomial Theorem, and take the real part of the result.

b.

$$T_{n+1}(x) = 2xT_n(x) - T_{n-1}(x), \quad n = 1, 2, \cdots.$$

Suggestion. Express $\cos(n+1)\theta$ and $\cos(n-1)\theta$ in terms of cosine and sine of θ and $n\theta$.

c. The leading coefficient of $T_n(x)$ is 2^{n-1} for $n = 1, 2, \cdots$, i.e.,

$$T_n(x) = 2^{n-1}x^n + \text{terms of lower degree.}$$

d. The sequence (T_n) is an orthogonal sequence in $\mathscr{L}_2{}^g(-1, 1)$ where

$$g(x) = \int_{-1}^{x} \frac{dt}{\sqrt{1 - t^2}}.$$

Conclude that T_n has n distinct zeros in $(-1, 1)$.

e. T_n has simple zeros at the points

$$x_k = \cos\left(\frac{2k-1}{2n}\pi\right), \quad k = 1, 2, \cdots, n$$

and extreme values in $[-1, 1]$ equal to $(-1)^k$ at the $n + 1$ points

$$\bar{x}_k = \cos\frac{k\pi}{n}, \quad k = 0, 1, \ldots, n.$$

f. If p_n is any polynomial of degree n with leading coefficient 1, then

$$\max_{-1 \leq x \leq 1} \left| \frac{1}{2^{n-1}} T_n(x) \right| = \frac{1}{2^{n-1}} \leq \max_{-1 \leq x \leq 1} |p_n(x)|.$$

Suggestion. Let $q_n(x) = (1/2^{n-1})T_n(x) - p_n(x)$ and assume that $\max_{-1 \leq x \leq 1} |p_n(x)| < 1/2^{n-1}$. What can be said about the sign of $q_n(\bar{x}_k)$? What is the minimum number of zeros that q_n must have? What is the maximum degree of q_n?

g. T_n satisfies the differential equation

$$(1 - x^2)y'' - xy' + n^2y = 0.$$

Suggestion. Set $x = \cos\theta$, $y = T_n(x) = \cos n\theta$ and use the fact that $dy/dx = (dy/d\theta)/(dx/d\theta)$.

9 · THE HERMITE POLYNOMIALS

9 · 1 Definition. *The polynomials defined by*

$$H_0(x) = 1,$$

$$H_n(x) = (-1)^n e^{x^2} \frac{d^n}{dx^n} e^{-x^2} \quad \textit{for } n > 0$$

*are called the **Hermite polynomials**.*

The first few Hermite polynomials are easily found to be:

$$H_0(x) = 1, \quad H_1(x) = 2x, \quad H_2(x) = 4x^2 - 2, \quad H_3(x) = 8x^3 - 12x.$$

It is clear from the definition that H_n is a polynomial of degree n with leading coefficient 2^n. Also,

$$H_n(-x) = (-1)^n H_n(x).$$

Thus, $H_n(x)$ contains only even powers of x if n is even and only odd powers of x if n is odd.

Let

$$g(x) = \int_{-\infty}^{x} e^{-t^2}\, dt.$$

We will show that (H_n) is an orthogonal basis of the real space $\mathscr{L}_2{}^g$.

9 · 2 Proposition. *(H_n) is an orthogonal sequence in $\mathscr{L}_2{}^g$ and*

$$\|H_n\|^2 = 2^n n! \sqrt{\pi}.$$

PROOF. If we integrate by parts repeatedly, then for $m \le n$ we obtain

$$\begin{aligned}\langle H_m, H_n\rangle &= (-1)^n \int_{-\infty}^{\infty} H_m(x) \frac{d^n}{dx^n} e^{-x^2}\, dx \\ &= (-1)^{n+1} \int_{-\infty}^{\infty} \frac{d}{dx} H_m(x) \frac{d^{n-1}}{dx^{n-1}} e^{-x^2}\, dx \\ &= \cdots = \int_{-\infty}^{\infty} e^{-x^2} \frac{d^n}{dx^n} H_m(x)\, dx.\end{aligned}$$

If $m < n$, then $\langle H_m, H_n\rangle = 0$ while if $m = n$, then

$$\|H_n\|^2 = \int_{-\infty}^{\infty} e^{-x^2} \frac{d^n}{dx^n} H_n(x)\, dx = 2^n n! \int_{-\infty}^{\infty} e^{-x^2}\, dx = 2^n n! \sqrt{\pi}. \blacksquare$$

9 · 3 Theorem. *(H_n) is an orthogonal basis of $\mathscr{L}_2{}^g$.*

PROOF. Take $f \in \mathscr{L}_2{}^g$ such that $\langle f, H_n\rangle = 0$ for all n. Then $\langle f, I^n\rangle = 0$ for all n. Let

$$F(t) = \int_{-\infty}^{\infty} f(x) e^{-itx}\, dg(x) \quad \text{for } t \in \mathbf{R}.$$

For each $t \in \mathbf{R}$,

$$e^{-ixt} = \sum_{k=0}^{\infty} \frac{(-i)^k}{k\,!} t^k x^k$$

where this series converges absolutely for all $x \in \mathbf{R}$. Thus,

$$F(t) = \int_{-\infty}^{\infty} \sum_{k=0}^{\infty} f(x) e^{-x^2} \frac{(-i)^k}{k!} t^k x^k\, dx.$$

Since

$$\begin{aligned}\int_{-\infty}^{\infty} \sum_{k=0}^{\infty} |f(x)|\, e^{-x^2} \frac{|t|^k}{k!} |x^k|\, dx &= \int_{-\infty}^{\infty} |f(x)|\, e^{-x^2} e^{|t|\,|x|}\, dx \\ &\le \left[\int_{-\infty}^{\infty} |f(x)|^2 e^{-x^2}\, dx\right]^{1/2} \left[\int_{-\infty}^{\infty} e^{2|t|\,|x|} e^{-x^2}\, dx\right]^{1/2}\end{aligned}$$

and $e^{2|t|\,|x| - x^2}$ is integrable, by a form of Levi's Theorem (Problem 3, p. 131) we have

$$F(t) = \sum_{k=0}^{\infty} \frac{(-i)^k}{k!} t^k \int_{-\infty}^{\infty} f(x) x^k\, dg(x) = 0.$$

By Theorem 7 · 3, $f = 0$ in $\mathscr{L}_2{}^g$. $\blacksquare$

The Hermite polynomials satisfy a three term recurrence relation.

9 · 4 Proposition. *$H_{n+1}(x) = 2xH_n(x) - 2nH_{n-1}(x)$ for $n \geq 1$.*

PROOF. Take $n \geq 1$. Since $2IH_n$ is a polynomial of degree $n+1$ and $(H_0, \cdots, H_{n+1})$ is a Hamel basis for the linear space of polynomials of degree $\leq n+1$, we have

$$2IH_n = \sum_{k=0}^{n+1} \alpha^k H_k \quad \text{where } \alpha^k = \frac{2}{\sqrt{\pi}\, k!2^k} \int_{-\infty}^{\infty} xH_n(x)H_k(x)\, dg(x).$$

If $k < n-1$, then IH_k is of degree less than n and H_n is orthogonal to IH_k; that is, $\alpha^k = 0$ for $k < n-1$. Also, since IH_n^2 is an odd function

$$\alpha^n = \frac{2}{\sqrt{\pi}\, n!2^n} \int_{-\infty}^{\infty} xH_n^2(x)e^{-x^2}\, dx = 0.$$

Thus,

$$2xH_n(x) = \alpha^{n-1}H_{n-1}(x) + \alpha^{n+1}H_{n+1}(x)$$

and, equating the coefficients of x^{n+1}, we have

$$2^{n+1} = 2^{n+1}\alpha^{n+1}.$$

Therefore,

$$1 = \alpha^{n+1} = \frac{2}{\sqrt{\pi}(n+1)!2^{n+1}} \int_{-\infty}^{\infty} xH_n(x)H_{n+1}(x)\, dg(x)$$

or

$$\int_{-\infty}^{\infty} xH_n(x)H_{n+1}(x)\, dg(x) = \sqrt{\pi}(n+1)!2^n.$$

If we replace n by $n-1$ in this last formula, we obtain

$$\alpha^{n-1} = \frac{2}{\sqrt{\pi}(n-1)!2^{n-1}} \int_{-\infty}^{\infty} xH_{n-1}(x)H_n(x)\, dg(x)$$

$$= \frac{2}{\sqrt{\pi}(n-1)!2^{n-1}} \sqrt{\pi}\, n!2^{n-1} = 2n.$$

Thus,

$$2xH_n(x) = 2nH_{n-1}(x) + H_{n+1}(x). \quad \blacksquare$$

We now introduce a generating function for the Hermite polynomials.

9 · 5 Proposition. *$e^{2xt-t^2} = \sum_{n=0}^{\infty} \frac{H_n(x)}{n!} t^n$ for $|2xt - t^2| < c$ for any number $c > 0$.*

PROOF. Let

$$F(x, t) = e^{2xt-t^2} = \sum_{n=0}^{\infty} \frac{g_n(x)}{n!} t^n,$$

where g_n is a polynomial and the series is absolutely and uniformly convergent if $|2xt - t^2| < c$ for any number $c > 0$. Since

$$\frac{\partial F(x, t)}{\partial t} = (2x - 2t)e^{2xt-t^2} = 2xF(x, t) - 2tF(x, t),$$

we obtain

$$\sum_{n=1}^{\infty} \frac{g_n(x)}{(n-1)!} t^{n-1} = \sum_{n=0}^{\infty} \frac{2xg_n(x)}{n!} t^n - \sum_{n=0}^{\infty} \frac{2g_n(x)}{n!} t^{n+1}$$

or

$$\sum_{n=0}^{\infty} \frac{g_{n+1}(x)}{n!} t^n - \sum_{n=0}^{\infty} \frac{2xg_n(x)}{n!} t^n + \sum_{n=1}^{\infty} \frac{2ng_{n-1}(x)}{n!} t^n = 0.$$

Thus,

$$g_1(x) - 2xg_0(x) = 0$$

and

$$g_{n+1}(x) - 2xg_n(x) + 2ng_{n-1}(x) = 0 \quad \text{for } n \geq 1.$$

Since

$$g_0(x) = F(x, 0) = 1 = H_0(x)$$

and

$$g_1(x) = 2xg_0(x) = 2x = H_1(x)$$

and the g's satisfy the same recurrence relation as the Hermite polynomials, $g_n = H_n$ for all n. ▮

The Hermite polynomial H_n is a solution of a second order linear differential equation, known as **Hermite's equation**.

9 · 6 Proposition. $H_n''(x) - 2xH_n'(x) + 2nH_n(x) = 0.$

PROOF. If we differentiate H_n and use the recurrence relation 9 · 4, we obtain

$$H_n'(x) = (-1)^n 2xe^{x^2} \frac{d^n}{dx^n} e^{-x^2} + (-1)^n e^{x^2} \frac{d^{n+1}}{dx^{n+1}} e^{-x^2}$$
$$= 2xH_n(x) - H_{n+1}(x) = 2nH_{n-1}(x).$$

Thus,

$$H_n''(x) = (-1)^{n-1} 4nxe^{x^2} \frac{d^{n-1}}{dx^{n-1}} e^{-x^2} + (-1)^{n-1} 2ne^{x^2} \frac{d^n}{dx^n} e^{-x^2}$$
$$= 4nxH_{n-1}(x) - 2nH_n(x) = 2xH_n'(x) - 2nH_n(x). ▮$$

PROBLEMS

1. Show that if f is even, then f' is odd and if f is odd, then f' is even. Hence, conclude that $H_k(x) = (-1)^k e^{x^2} (d^k/dx^k)e^{-x^2}$ is even for k even and is odd for k odd.

2. Use the generating function to show that

$$\left.\begin{aligned} H_{2n}(0) &= (-1)^n \frac{(2n)!}{n!} \\ H_{2n+1}(0) &= 0. \end{aligned}\right\} \quad n = 0, 1, 2, \ldots$$

3. If $f \in \mathscr{L}_2{}^g$ where $g(x) = \int_{-\infty}^{x} e^{-t^2}\, dt$, then the representation of f with respect to the orthogonal basis (H_n) is often called the **Fourier-Hermite** series of f. Show that the Fourier-Hermite series of f is $\sum_{k=0}^{\infty} c_k H_k(x)$ where

$$c_k = \frac{1}{2^k k! \sqrt{\pi}} \int_{-\infty}^{\infty} e^{-x^2} f(x) H_k(x)\, dx = \frac{(-1)^k}{2^k k! \sqrt{\pi}} \int_{-\infty}^{\infty} f(x) \frac{d^k}{dx^k} e^{-x^2}\, dx.$$

4. Show that if $f(x) = x^n$ where n is a nonnegative integer, then

$$c_k = \begin{cases} \dfrac{n!}{2^n(k!)\left(\dfrac{n-k}{2}\right)!} & \text{for } k \le n \text{ and } n + k \text{ even} \\ 0 & \text{for } k > n \text{ or } n + k \text{ odd.} \end{cases}$$

5. Find the Fourier-Hermite series of each of the following:

a. e^{-x} **b.** e^x
c. $\cosh x$ **d.** $\sinh x$
e. $\cos x$ **f.** $\sin x$

10 · POINTWISE CONVERGENCE OF FOURIER SERIES

Recall that the trigonometic sequence $(u_k)_{-\infty}^{\infty}$, where $u_k(x) = (1/\sqrt{2\pi})e^{ikx}$, is an orthonormal basis of $\mathscr{L}_2[-\pi, \pi]$ and, hence, for each $f \in \mathscr{L}_2[-\pi, \pi]$

$$f = \sum_{k=-\infty}^{\infty} \gamma_k u_k \quad \text{where } \gamma_k = \frac{1}{\sqrt{2\pi}} \int_{-\pi}^{\pi} f(t) e^{-ikt}\, dt.$$

This means that the Fourier series $\sum \gamma_k u_k$ converges to f in the $\mathscr{L}_2$ norm; that is,

$$\lim_{n\to\infty} \int_{-\pi}^{\pi} \left| f(x) - \sum_{k=-n}^{n} c_k e^{ikx} \right|^2 dx = 0 \quad \text{where } c_k = \frac{1}{2\pi} \int_{-\pi}^{\pi} f(t) e^{-ikt}\, dt.$$

We cannot conclude from this relation that, for a given $x \in [-\pi, \pi]$, the series $\sum_{k=-\infty}^{\infty} c_k e^{ikx}$ of complex numbers converges to the number $f(x)$. In this section we consider conditions under which the series of numbers $\sum c_k e^{ikx}$ does converge to the number $f(x)$.

First, note that if $f \in \mathscr{L}(-\pi, \pi]$, then $\int_{-\pi}^{\pi} f(t)e^{-ikt}\,dt$ exists and, hence, for each $x \in \mathbf{R}$, the Fourier series $\sum c_k e^{ikx}$ is well defined (but not necessarily convergent). Since each term of the Fourier series has period 2π, we can extend f to all of $\mathbf{R}$ so that it has period 2π, i.e., $f(x) = f(x - 2m\pi)$ if $x \in ((2m-1)\pi, (2m+1)\pi]$, and consider whether or not $\sum c_k e^{ikx}$ converges to $f(x)$ for any given $x \in \mathbf{R}$. If we make this extension, then

$$c_k = \frac{1}{2\pi}\int_{y-\pi}^{y+\pi} f(t)e^{-ikt}\,dt \quad \text{for each } y \in \mathbf{R};$$

that is, c_k can be obtained by integrating over any interval of length 2π.

Let

$$s_n(x) = \sum_{k=-n}^{n} c_k e^{ikx}.$$

Then,

$$(10 \cdot 1) \qquad s_n(x) = \frac{1}{2\pi}\int_{-\pi}^{\pi} f(t)\sum_{k=-n}^{n} e^{ik(x-t)}\,dt.$$

The partial sum s_n can be expressed briefly in terms of the **Dirichlet kernel** D_n defined by

$$D_n(x) = \sum_{k=-n}^{n} e^{ikx}.$$

In fact,

$$s_n(x) = \frac{1}{2\pi}\int_{-\pi}^{\pi} f(t)D_n(x-t)\,dt.$$

It is clear from the definition that D_n is an even function of period 2π and that $\int_0^{\pi} D_n = \pi$. Also, since

$$(e^{ix/2} - e^{-ix/2})D_n(x) = e^{i(n+1/2)x} - e^{-i(n+1/2)x},$$

we obtain the expression

$$D_n(x) = \frac{\sin(n+\frac{1}{2})x}{\sin\frac{1}{2}x}, \quad x \neq 2k\pi.$$

If we make the change of variable $u = t - x$ and use the facts that f and D_n both have period 2π and D_n is an even function, then we find that

$$\begin{aligned} s_n(x) &= \frac{1}{2\pi}\int_{-\pi-x}^{\pi-x} f(x+u)D_n(-u)\,du = \frac{1}{2\pi}\int_{-\pi}^{\pi} f(x+u)D_n(u)\,du \\ &= \frac{1}{2\pi}\int_0^{\pi} f(x+u)D_n(u)\,du + \frac{1}{2\pi}\int_{-\pi}^{0} f(x+u)D_n(u)\,du \\ &= \frac{1}{2\pi}\int_0^{\pi} f(x+t)D_n(t)\,dt + \frac{1}{2\pi}\int_0^{\pi} f(x-t)D_n(-t)\,dt \\ &= \frac{1}{2\pi}\int_0^{\pi} (f(x+t) + f(x-t))D_n(t)\,dt. \end{aligned}$$

Thus, $(s_n(x))$ converges to a number $s(x)$ if and only if

$$0 = \lim_{n\to\infty}(s_n(x) - s(x)) = \lim_{n\to\infty}\frac{1}{2\pi}\int_0^\pi (f(x+t) + f(x-t) - 2s(x))D_n(t)\,dt.$$

We have proved the following result.

10 · 2 Proposition. *If f is integrable on $(-\pi, \pi]$ and has period 2π, then the sequence $(s_n(x))$ converges to the number $s(x)$ if and only if*

$$\lim_{n\to\infty}\int_0^\pi g(x, t)D_n(t)\,dt = \lim_{n\to\infty}\int_0^\pi \frac{g(x, t)}{\sin \frac{1}{2}t}\sin(n + \tfrac{1}{2})t\,dt = 0$$

where $g(x, t) = f(x+t) + f(x-t) - 2s(x)$.

The basic result in our discussion of pointwise convergence of Fourier series is the Riemann–Lebesgue Theorem which we now prove.

10 · 3 Theorem (Riemann–Lebesgue). *If f is an integrable function, then*

$$\lim_{r\to\infty}\int f(t)e^{irt}\,dt = 0.$$

PROOF. First, assume that f is the characteristic function of an open interval (a, b). Then,

$$\left|\int f(t)e^{irt}\,dt\right| = \left|\int_a^b e^{irt}\,dt\right| = \left|\frac{1}{ir}(e^{irb} - e^{ira})\right| \le \frac{2}{r}$$

and, hence,

$$\lim_{r\to\infty}\int f(t)e^{irt}\,dt = 0.$$

Since the theorem holds for the characteristic function of an open interval, it holds for any step function.

Now let f be an integrable function. Then, for each $\varepsilon > 0$ there exists a step function g such that (Problem 7, p. 127)

$$\int |f - g| < \frac{\varepsilon}{2}.$$

Thus,

$$\left|\int f(t)e^{irt}\,dt\right| \le \left|\int (f(t) - g(t))e^{irt}\,dt\right| + \left|\int g(t)e^{irt}\,dt\right|$$

$$\le \int |f - g| + \left|\int g(t)e^{irt}\,dt\right|.$$

By the first part of the proof there exists an r_0 such that for $r \ge r_0$, $|\int g(t)e^{irt}\,dt| < \varepsilon/2$ and, hence,

$$\left|\int f(t)e^{irt}\,dt\right| < \varepsilon \quad \text{for } r \ge r_0. \blacksquare$$

From the Riemann–Lebesgue Theorem it follows that if f is integrable, then

$$\lim_{r\to\infty}\int f(t)\cos rt\,dt = 0 \quad\text{and}\quad \lim_{r\to\infty}\int f(t)\sin rt\,dt = 0.$$

It also follows that the Fourier coefficients approach zero:

$$\lim_{k\to\infty} c_k = \lim_{k\to\infty}\frac{1}{2\pi}\int_{-\pi}^{\pi} f(t)e^{-ikt}\,dt = \lim_{k\to\infty}\frac{1}{2\pi}\int f(t)\chi_{(-\pi,\pi]}(t)e^{-ikt}\,dt = 0.$$

With the aid of the Riemann–Lebesgue Theorem, we obtain the following modification of Proposition 10 · 2.

10 · 4 Proposition. *If f is integrable on $(-\pi, \pi]$ and has period 2π, then $(s_n(x))$ converges to $s(x)$ if and only if for some number $\delta \in (0, \pi)$*

$$\lim_{n\to\infty}\int_0^\delta \frac{g(x, t)}{\sin \frac{1}{2}t}\sin(n + \tfrac{1}{2})t\,dt = 0.$$

PROOF. If $\delta \in (0, \pi)$, then $(\sin \frac{1}{2}t)^{-1}$ is continuous on $[\delta, \pi]$. Hence, $g(x, t)\,(\sin \frac{1}{2}t)^{-1}$ is integrable on $[\delta, \pi]$ and the Riemann–Lebesgue Theorem implies that

$$\lim_{n\to\infty}\int_\delta^\pi \frac{g(x, t)}{\sin \frac{1}{2}t}\sin(n + \tfrac{1}{2})t\,dt = 0.$$

Thus, from Proposition 10 · 2 it follows that $(s_n(x))$ converges to $s(x)$ if and only if for some $\delta \in (0, \pi)$

$$\lim_{n\to\infty}\int_0^\delta \frac{g(x, t)}{\sin \frac{1}{2}t}\sin(n + \tfrac{1}{2})t\,dt = 0. \quad \blacksquare$$

Another useful result is the following:

10 · 5 Proposition. *If f is integrable on $(-\pi, \pi]$ and has period 2π, then $(s_n(x))$ converges to $s(x)$ if and only if for some $\delta \in (0, \pi)$,*

$$\lim_{n\to\infty}\int_0^\delta \frac{g(x, t)}{t}\sin(n + \tfrac{1}{2})t\,dt = 0.$$

PROOF. Using L'Hospital's rule or Taylor's formula, we see that

$$\lim_{t\to 0}\left(\frac{1}{\sin \frac{1}{2}t} - \frac{2}{t}\right) = 0.$$

Thus, if we let $h(0) = 0$ and

$$h(t) = \frac{1}{\sin \frac{1}{2}t} - \frac{2}{t} \quad \text{for } t \in (0, \pi],$$

then h is continuous on $[0, \pi]$. This implies that all of the following Lebesgue integrals exist for each $\delta \in (0, \pi)$:

$$2\int_0^\delta \frac{g(x, t)}{t} \sin(n + \tfrac{1}{2})t\, dt$$

$$= \int_0^\delta \frac{g(x, t)}{\sin \frac{1}{2}t} \sin(n + \tfrac{1}{2})t\, dt - \int_0^\delta h(t)g(x, t) \sin(n + \tfrac{1}{2})t\, dt$$

Also, since hg is integrable on $[0, \pi]$ for each fixed $x \in \mathbf{R}$, the Riemann–Lebesgue Theorem implies

$$\lim_{n\to\infty} \int_0^\delta h(t)g(x, t) \sin(n + \tfrac{1}{2})t\, dt = 0.$$

Thus, Proposition 10 · 4 implies that $(s_n(x))$ converges to $s(x)$ if and only if for some $\delta \in (0, \pi)$

$$\lim_{n\to\infty} \int_0^\delta \frac{g(x, t)}{t} \sin(n + \tfrac{1}{2})t\, dt = 0. \blacksquare$$

From Proposition 10 · 5 we can easily obtain sufficient conditions, due to Dini, for the pointwise convergence of a Fourier series.

10 · 6 Proposition (Dini). *If f is integrable on $(-\pi, \pi]$ and has period 2π, then the Fourier series of f converges at the point x to $s(x)$ if, for some $\delta \in (0, \pi)$, $g(x, t)t^{-1}$ is integrable on* $[0, \delta]$.

PROOF. By the Riemann–Lebesgue Theorem

$$\lim_{n\to\infty} \int_0^\delta \frac{g(x, t)}{t} \sin(n + \tfrac{1}{2})t\, dt = 0$$

and, hence, Proposition 10 · 5 implies that $(s_n(x))$ converges to $s(x)$. $\blacksquare$

We now discuss a type of function which will satisfy the Dini conditions. Call a function f **almost differentiable** at a point x if $f(x+), f(x-)$, and

$$\lim_{t\to 0^+} \frac{f(x + t) - f(x+)}{t} \quad \text{and} \quad \lim_{t\to 0^+} \frac{f(x - t) - f(x-)}{-t}$$

all exist; that is, at x the graph of f has nonvertical left-hand and right-hand tangents.

10 · 7 Proposition. *If f is integrable on $(-\pi, \pi]$ with period 2π and is almost differentiable at x, then the Fourier series of f converges at x to $s(x)$ where $s(x) = \frac{1}{2}(f(x+) + f(x-))$.*

PROOF. Let $g(x, t) = f(x + t) + f(x - t) - 2s(x)$. Since f is almost differentiable at x, there exist a $\delta \in (0, \pi)$ and a number M such that for all $t \in (0, \delta)$

$$\left|\frac{g(x, t)}{t}\right| \leq \left|\frac{f(x + t) - f(x+)}{t}\right| + \left|\frac{f(x - t) - f(x-)}{t}\right| \leq M.$$

Since on $(0, \delta)$ the measurable function $g(x, t)t^{-1}$ is bounded by the integrable function M, it is integrable. Then $g(x, t)t^{-1}$ is integrable on $[0, \delta]$ and Proposition 10 · 6 implies that

$$\lim s_n(x) = \tfrac{1}{2}(f(x+) + f(x-)). \quad \blacksquare$$

In particular, if the integrable function f of period 2π is differentiable at x, then the Fourier series of f converges at x to $f(x)$.

Another useful result on pointwise convergence of Fourier series is due to Dirichlet.

10 · 8 Theorem (Dirichlet). *If f is an integrable function of period 2π which is of bounded variation on $[a, b]$, then the Fourier series of f converges at x to*

$$s(x) = \tfrac{1}{2}(f(x+) + f(x-))$$

for each $x \in (a, b)$.

PROOF. Take $x \in (a, b)$ and take a positive number $\delta < \min\{x - a, b - x, \pi\}$. Then

$$g(x, t) = f(x + t) + f(x - t) - f(x+) - f(x-)$$

is of bounded variation on $[0, \delta]$ as a function of t. Therefore, for x fixed, g is the difference of two nondecreasing functions f_1 and f_2 on $[0, \delta]$:

$$g(x, t) = f_1(t) - f_2(t), \quad t \in [0, \delta].$$

Since

$$0 = g(x, 0+) = f_1(0+) - f_2(0+),$$

if we let $h_i(t) = f_i(t) - f_i(0+)$ for $i = 1, 2$, then $h_i(0+) = 0$, h_i is nondecreasing on $[0, \delta]$ and nonnegative on $(0, \delta]$, and

$$g(x, t) = h_1(t) - h_2(t), \quad t \in [0, \delta].$$

From Proposition 10 · 5 we know that $(s_n(x))$ converges to $s(x)$ if for some $\delta \in (0, \pi)$

$$\lim_{n\to\infty} \int_0^\delta \frac{g(x, t)}{t} \sin(n + \tfrac{1}{2})t \, dt = 0.$$

Thus, $\lim s_n(x) = s(x)$ if

$$\lim_{n\to\infty} \int_0^\delta \frac{h_i(t)}{t} \sin(n + \tfrac{1}{2})t \, dt = 0 \quad \text{for } i = 1, 2.$$

Take $\varepsilon > 0$. Since $h_i(0+) = 0$ there exists an $\eta \in (0, \delta)$ such that

$$h_i(t) < \varepsilon \quad \text{for } t \in (0, \eta).$$

If we use the Second Mean Value Theorem (Corollary 7 · 3, p. 271) and the problem on p. 271, then we obtain: for some $c \in [0, \eta]$,

$$\begin{aligned}\left|\int_0^\eta \frac{h_i(t)}{t}\sin(n+\tfrac{1}{2})t\,dt\right| &= \left|h_i(\eta-)\int_c^\eta \frac{\sin(n+\frac{1}{2})t}{t}\,dt\right| \\ &< \varepsilon\left|\int_{(n+1/2)c}^{(n+1/2)\eta} \frac{\sin u}{u}\,du\right| \\ &\le 4\varepsilon \quad \text{for all } n.\end{aligned}$$

By the Riemann–Lebesgue Theorem, for all n sufficiently large

$$\left|\int_\eta^\delta \frac{h_i(t)}{t}\sin(n+\tfrac{1}{2})t\,dt\right| < \varepsilon.$$

Thus, for all n sufficiently large,

$$\left|\int_0^\delta \frac{h_i(t)}{t}\sin(n+\tfrac{1}{2})t\,dt\right| < 5\varepsilon. \quad ∎$$

PROBLEMS

1. Determine the Fourier series of each of the following functions with respect to the orthogonal basis $(1, \cos nx, \sin nx)_{n=1}^\infty$ for the real space $\mathscr{L}_2[-\pi, \pi]$.

a. $f(x) = \begin{cases} -\pi/4 & \text{for } -\pi < x < 0 \\ \pi/4 & \text{for } 0 < x < \pi \end{cases}$ **b.** $f(x) = \begin{cases} 0 & \text{for } -\pi < x < 0 \\ 1 & \text{for } 0 < x < \pi \end{cases}$

c. $f(x) = |x|$ for $-\pi < x < \pi$ **d.** $f(x) = x^2$ for $-\pi < x < \pi$

e. $f(x) = \begin{cases} 0 & \text{for } -\pi < x < 0 \\ \sin x & \text{for } 0 < x < \pi \end{cases}$

f. $f(x) = \begin{cases} -\cos x & \text{for } -\pi < x < 0 \\ \cos x & \text{for } 0 < x < \pi \end{cases}$

g. $f(x) = \begin{cases} \dfrac{-\pi + x}{2} & \text{for } -\pi < x < 0 \\ \dfrac{\pi - x}{2} & \text{for } 0 < x < \pi \end{cases}$

h. $f(x) = |\sin x|$ for $-\pi < x < \pi$
i. $f(x) = \cos ax$ for $-\pi < x < \pi$ and $0 < a < 1$
j. $f(x) = \sin ax$ for $-\pi < x < \pi$ and $0 < a < 1$

2. Show that if f is an even function, then the Fourier series of f with respect to $(1, \cos nx, \sin nx)_{n=1}^\infty$ contains only a constant term and cosine terms and if f is an odd function then the Fourier series contains only sine terms.

3. If f is defined on $(0, \pi]$, then f can be defined on $(-\pi, 0]$ so that f is even on $(-\pi, \pi]$ or f can be defined on $(-\pi, 0]$ so that f is odd on $(-\pi, \pi]$. If f is defined to be even on $(-\pi, \pi]$, then the resulting expansion of f with respect to $(1, \cos nx, \sin nx)_{n=1}^{\infty}$ is called the **Fourier cosine series** of f on $(0, \pi]$ and if f is defined to be odd on $(-\pi, \pi]$, then the resulting expansion is called the **Fourier sine series** of f on $(0, \pi]$.

a. Expand x^3 on the interval $(0, \pi]$ in a Fourier cosine series.
b. Expand x^3 on the interval $(0, \pi]$ in a Fourier sine series.
c. Expand $\cos x$ on the interval $(0, \pi]$ in a Fourier sine series.
d. Expand $\sin x$ on the interval $(0, \pi]$ in a Fourier cosine series.
e. Expand $x \sin x$ on the interval $(0, \pi]$ in a Fourier cosine series.
f. Expand $x \sin x$ on the interval $(0, \pi]$ in a Fourier sine series.

4. The sequence $(1, \cos(n\pi x/l), \sin(n\pi x/l))_{n=1}^{\infty}$ is an orthogonal basis for the real space $\mathscr{L}_2(-l, l)$. Show that the Fourier series of a function $f \in \mathscr{L}_2(-l, l)$ with respect to this basis is

$$\tfrac{1}{2}a_0 + \sum_{n=1}^{\infty}\left[a_n \cos\frac{n\pi x}{l} + b_n \sin\frac{n\pi x}{l}\right]$$

where

$$a_n = \frac{1}{l}\int_{-l}^{l} f(x)\cos\frac{n\pi x}{l}\,dx, \quad n = 0, 1, 2, \ldots$$

and

$$b_n = \frac{1}{l}\int_{-l}^{l} f(x)\sin\frac{n\pi x}{l}\,dx, \quad n = 1, 2, \ldots.$$

5. Find the Fourier series of each of the following functions on the interval shown:

a. $f(x) = \begin{cases} 0 & \text{for } -2 < x < 0 \\ 1 & \text{for } 0 < x < 2 \end{cases}$
b. $f(x) = |x|$ for $-1 < x < 1$
c. $f(x) = x^2$ for $-3 < x < 3$
d. $f(x) = \cos \pi x$ for $-1 < x < 1$.

6. Let $\sum a_k$ be a series of complex numbers and let $s_j = \sum_{k=0}^{j} a_k$. The arithmetic mean of the first $n + 1$ partial sums is

$$\sigma_n = \frac{1}{n+1}\sum_{j=0}^{n} s_j.$$

We say that the series $\sum a_k$ is $(C, 1)$ or **Cesàro summable** to s if

$$\lim \sigma_n = s.$$

Show that:

a. $\sum_{k=0}^{\infty} (-1)^k$ is Cesàro summable to 1/2.

b. If $\sum a_k$ converges to s, then it is Cesàro summable to s.

7. Let f be integrable on $(-\pi, \pi]$ and be of period 2π. Let $s_j(x) = \sum_{k=-j}^{j} c_k e^{ikx}$ be the jth partial sum of the Fourier series of f. With σ_n defined as in Problem 6 show that:

a. $\sigma_n(x) = \sum_{k=-n}^{n} \left(1 - \frac{|k|}{n+1}\right) c_k e^{ikx}$

b. $\sigma_n(x) = \frac{1}{2\pi(n+1)} \int_0^{\pi} (f(x+t) + f(x-t)) \sum_{j=0}^{n} D_j(t)\, dt$

where D_j is the Dirichlet kernel $D_j(t) = \sum_{k=-j}^{j} e^{ikt}$.

8. The **Fejer kernel** K_n is defined by

$$K_n = \frac{1}{n+1} \sum_{j=0}^{n} D_j$$

so that

$$\sigma_n(x) = \frac{1}{2\pi} \int_0^{\pi} (f(x+t) + f(x-t)) K_n(t)\, dt.$$

Show that:

a. $K_n(-t) = K_n(t)$

b. $\int_0^{\pi} K_n = \pi$

c. $K_n(t) = \frac{1}{n+1} \left[\frac{\sin(n+1)t/2}{\sin t/2}\right]^2, \quad t \neq 2k\pi$

d. $K_n(t) \geq 0$

e. $K_n(t) \leq n + 1$

f. $K_n(t) \leq c/((n+1)t^2)$ for $0 < t \leq \pi$ where c is a constant independent of n.

9. Prove Fejer's Theorem: If f is integrable on $(-\pi, \pi]$ with period 2π, then the Fourier series of f is Cesàro summable to

$$s(x) = \tfrac{1}{2}(f(x+) + f(x-))$$

at each point x at which $f(x+)$ and $f(x-)$ exist. In particular, the Fourier series of f is Cesàro summable to $f(x)$ at each point x at which f is continuous.

Suggestion. If $f(x+)$ and $f(x-)$ exist, let

$$g(x, t) = f(x+t) + f(x-t) - 2s(x).$$

For $\varepsilon > 0$ there exists a $\delta > 0$ such that $|g(x, t)| < \varepsilon$ whenever $0 \le t < \delta$. Then consider

$$|\sigma_n(x) - s(x)| \le \frac{1}{2\pi} \int_0^\delta |g(x, t)| K_n(t)\, dt + \frac{1}{2\pi} \int_\delta^\pi |g(x, t)| K_n(t)\, dt.$$

10. If, in Problem 9, f is continuous on $[-\pi, \pi]$, show that (σ_n) converges to f uniformly on $\mathbf{R}$ and thus obtain another proof of the Second Weierstrass Approximation Theorem (Corollary 2 · 9, p. 191).

11. By carrying out the following steps obtain a proof of the Fejer-Lebesgue Theorem: If f is integrable on $(-\pi, \pi)$, then the Fourier series of f is Cesàro summable to $f(x)$ at almost all $x \in (-\pi, \pi)$.

a. (Problem 5, p. 238). Take $x \in (-\pi, \pi)$ in the Lebesgue set of f and let $g(x, t) = f(x + t) + f(x - t) - 2f(x)$ and $G(x, t) = \int_0^t |g(x, u)|\, du$. For each $\varepsilon > 0$ there exists a $\delta > 0$ such that $\delta < \min\{\pi - x, x + \pi\}$ and,

$$G(x, h) \le \varepsilon h \quad \text{for } 0 \le h \le \delta.$$

b. (Problem 8c). Extend f to $[-2\pi, 2\pi]$ by defining $f(x) = 0$ for $x \in [-2\pi, 2\pi] \sim (-\pi, \pi)$. There exists an $n_0 > 1/\delta$ such that for $n \ge n_0$

$$K_n(t) < \frac{\varepsilon}{G(x, \pi) + 1} \qquad \text{for } \delta \le t \le \pi.$$

c. For $n \ge n_0$

$$2\pi |\sigma_n(x) - f(x)| \le \int_0^{1/n} |g(x, t)| K_n(t)\, dt + \int_{1/n}^\delta |g(x, t)| K_n(t)\, dt + \int_\delta^\pi |g(x, t)| K_n(t)\, dt.$$

d. (Part a and Problem 8e)

$$\int_0^{1/n} |g(x, t)| K_n(t)\, dt \le 2\varepsilon.$$

e. (Integration by parts)

$$\int_{1/n}^\delta \frac{|g(x, t)|}{t^2}\, dt = \frac{1}{\delta^2} G(x, \delta) - n^2 G\left(x, \frac{1}{n}\right) + 2 \int_{1/n}^\delta \frac{G(x, t)}{t^3}\, dt.$$

f. (Problem 8f, Parts e and a)

$$\int_{1/n}^\delta |g(x, t)| K_n(t)\, dt \le \frac{c}{n + 1} \int_{1/n}^\delta \frac{|g(x, t)|}{t^2}\, dt < 3c\varepsilon.$$

g. (Part b)

$$\int_\delta^\pi |g(x, t)| K_n(t)\, dt < \varepsilon.$$

h. (Parts c, d, f, g)

$$|\sigma_n(x) - f(x)| < \frac{3}{2\pi}(1 + c)\varepsilon.$$

11 · THE FOURIER INTEGRAL

In this section we discuss a representation of a nonperiodic integrable function on the whole line which is the counterpart of the representation of a periodic function by its Fourier series. We begin by reexamining the Fourier series representation of a periodic function. Let $f \in \mathscr{L}(a, b]$ and extend f to all of $\mathbf{R}$ so that it has period $b - a = 2r$. If we let $f_1(u) = f(ru/\pi)$, then f_1 is an integrable function of period 2π and its Fourier series representation is

$$\sum_{k=-\infty}^{\infty} c_k e^{iku} \quad \text{where } c_k = \frac{1}{2\pi}\int_{-\pi}^{\pi} f_1(t)e^{-ikt}\,dt.$$

If we let $x = ru/\pi$ and $s = rt/\pi$, we see that the Fourier series of f is

$$(11 \cdot 1) \qquad \sum_{k=-\infty}^{\infty} c_k e^{ik\pi x/r} \quad \text{where } c_k = \frac{1}{2r}\int_{-r}^{r} f(s)e^{-ik\pi s/r}\,ds.$$

It is clear that with minor modifications the results of the preceding section apply to this series.

Now consider the case where $f \in \mathscr{L}(-\infty, \infty)$, usually written simply $f \in \mathscr{L}$. Then, for all $r > 0$, $f \in \mathscr{L}(-r, r]$ and, hence, f has the Fourier series representation 11 · 1 on $[-r, r]$. If we let

$$f_r(u) = \frac{1}{\sqrt{2\pi}}\int_{-r}^{r} f(t)e^{-iut}\,dt,$$

then

$$c_k = \frac{1}{\sqrt{2\pi}} f_r\left(\frac{k\pi}{r}\right)\frac{\pi}{r}$$

and the Fourier series 11 · 1 becomes

$$(11 \cdot 2) \qquad \frac{1}{\sqrt{2\pi}}\sum_{k=-\infty}^{\infty} f_r\left(\frac{k\pi}{r}\right)e^{ik\pi x/r}\,\frac{\pi}{r}.$$

This suggests that as $r \to \infty$ the Fourier series 11 · 2 should be replaced by the Fourier integral

$$\frac{1}{\sqrt{2\pi}}\int_{-\infty}^{\infty} \hat{f}(u)e^{iux}\,du \quad \text{where } \hat{f}(u) = \frac{1}{\sqrt{2\pi}}\int_{-\infty}^{\infty} f(t)e^{-iut}\,dt.$$

11 · 3 Definition. *If $f \in \mathscr{L}$, then the* ***Fourier transform*** $\hat{f}$ *of f is defined by*

$$\hat{f}(y) = \frac{1}{\sqrt{2\pi}}\int_{-\infty}^{\infty} f(x)e^{-iyx}\,dx, \quad y \in \mathbf{R}.$$

Since $f \in \mathscr{L}$ it is clear that $\hat{f}(y)$ exists for each $y \in \mathbf{R}$. The Riemann-Lebesgue Theorem implies that

$$\lim_{y \to \pm\infty} \hat{f}(y) = 0.$$

11 · 4 Proposition. *If $f \in \mathscr{L}$, then $\hat{f}$ is continuous.*

PROOF. Take $y \in \mathbf{R}$. If the sequence $(\hat{f}(y_n))$ converges to $\hat{f}(y)$ for each sequence of points (y_n) in $\mathbf{R}$ which converges to y, then $\hat{f}$ is continuous at y.

Let $g_n(x) = f(x)e^{-iy_nx}$; then (g_n) is a sequence of functions in $\mathscr{L}$ such that

$$\lim_{n \to \infty} g_n(x) = f(x)e^{-iyx} \quad \text{for all } x \in \mathbf{R}$$

and $|g_n| \le |f|$. Therefore, the Lebesgue Dominated Convergence Theorem implies that

$$\lim \int g_n = \int f(x)e^{-iyx}\,dx$$

and, hence,

$$\lim \hat{f}(y_n) = \hat{f}(y). \quad \blacksquare$$

If we define

$$\mathscr{F}f = \hat{f} \quad \text{for } f \in \mathscr{L},$$

then $\mathscr{F}$ maps $\mathscr{L}_1$ into $C(\mathbf{R}, \mathbf{C})$. It is easy to see that $\mathscr{F}$ is linear. Also, $\mathscr{F}$ is a bounded linear transformation:

$$\|\mathscr{F}f\| = \sup_{y \in \mathbf{R}} \left| \frac{1}{\sqrt{2\pi}} \int_{-\infty}^{\infty} f(x)e^{-iyx}\,dx \right| \le \frac{1}{\sqrt{2\pi}} \int_{-\infty}^{\infty} |f| = \frac{1}{\sqrt{2\pi}} \|f\|.$$

We now consider pointwise convergence of the Fourier integral

$$\lim_{r \to \infty} \int_{-r}^{r} \hat{f}(u)e^{iux}\,du.$$

If $\hat{f} \in \mathscr{L}$, then this limit exists and is $\int_{-\infty}^{\infty} \hat{f}(u)e^{iux}\,dx$. If $\hat{f} \notin \mathscr{L}$, then the limit may still exist and our objective is to give sufficient conditions for the convergence at a point x which are comparable to the conditions of Theorem 10 · 8. For this we need the following result.

11 · 5 Proposition.

$$\lim_{b \to \infty} \int_0^b \frac{\sin x}{x}\,dx = \frac{\pi}{2}.$$

PROOF. It is easily verified that the alternating series $\sum_{k=1}^{\infty} \int_{(k-1)\pi}^{k\pi} (\sin x/x)\,dx$ converges. If we let

$$s_n = \sum_{k=1}^{n} \int_{(k-1)\pi}^{k\pi} \frac{\sin x}{x}\,dx,$$

then for $b \geq \pi$ there exists a positive integer n such that $n\pi \leq b < (n+1)\pi$. Then $\int_0^b (\sin x/x)\,dx$ lies between s_n and s_{n+1} and, hence, $\lim_{b\to\infty} \int_0^b (\sin x/x)\,dx$ exists.

To evaluate this limit we recall that

$$\int_0^\pi \frac{\sin(n+\frac{1}{2})t}{\sin\frac{1}{2}t}\,dt = \int_0^\pi D_n = \pi.$$

Also, if $h(0) = 0$ and

$$h(t) = \frac{1}{\sin\frac{1}{2}t} - \frac{2}{t} \quad \text{for } t \in (0, \pi],$$

then h is continuous on $[0, \pi]$. Since

$$2\int_0^\pi \frac{\sin(n+\frac{1}{2})t}{t}\,dt = \int_0^\pi \frac{\sin(n+\frac{1}{2})t}{\sin\frac{1}{2}t}\,dt - \int_0^\pi h(t)\sin(n+\tfrac{1}{2})t\,dt,$$

using the Riemann-Lebesgue Theorem we obtain

$$\lim_{n\to\infty} 2\int_0^\pi \frac{\sin(n+\frac{1}{2})t}{t}\,dt = \lim_{n\to\infty}\int_0^\pi \frac{\sin(n+\frac{1}{2})t}{\sin\frac{1}{2}t}\,dt = \pi.$$

Therefore,

$$\lim_{n\to\infty}\int_0^\pi \frac{\sin(n+\frac{1}{2})t}{t}\,dt = \lim_{n\to\infty}\int_0^{(n+1/2)\pi} \frac{\sin x}{x}\,dx = \frac{\pi}{2}$$

and, hence,

$$\lim_{b\to\infty}\int_0^b \frac{\sin x}{x}\,dx = \frac{\pi}{2}. \quad ∎$$

Proposition 11 · 5 shows that $\int_0^\infty (\sin x/x)\,dx$ exists as an improper Riemann integral although $\sin/I \notin \mathscr{L}(0, \infty)$.

Now suppose $f \in \mathscr{L}$ and for $r > 0$ let

$$s_r(x) = \frac{1}{\sqrt{2\pi}}\int_{-r}^r \hat{f}(u)e^{iux}\,du.$$

Using the Tonelli and Fubini Theorems, we obtain

$$\begin{aligned}
s_r(x) &= \frac{1}{2\pi}\int_{-r}^r \int_{-\infty}^\infty f(t)e^{iu(x-t)}\,dt\,du \\
&= \frac{1}{2\pi}\int_{-\infty}^\infty \left(\int_{-r}^r e^{iu(x-t)}\,du\right) f(t)\,dt \\
&= \frac{1}{\pi}\int_{-\infty}^\infty f(t)\,\frac{\sin r(x-t)}{x-t}\,dt = \frac{1}{\pi}\int_{-\infty}^\infty f(x+t)\,\frac{\sin rt}{t}\,dt \\
&= \frac{1}{\pi}\int_0^\infty (f(x+t)+f(x-t))\,\frac{\sin rt}{t}\,dt.
\end{aligned}$$

Since, for each $\delta > 0$, $[f(x+t)+f(x-t)]t^{-1}$ is integrable with respect to t on $[\delta, \infty)$, the Riemann-Lebesgue Theorem implies that

$$\lim_{r\to\infty} s_r(x) = \lim_{r\to\infty} \frac{1}{\pi}\int_0^\delta \frac{f(x+t)+f(x-t)}{t}\sin rt\,dt.$$

Also, Proposition 11 · 5 implies that

$$\lim_{r\to\infty} \frac{2}{\pi}\int_0^\delta \frac{\sin rt}{t}\,dt = \lim_{r\to\infty} \frac{2}{\pi}\int_0^{r\delta} \frac{\sin u}{u}\,du = 1.$$

Thus, $\lim_{r\to\infty} s_r(x) = s(x)$ if for some $\delta > 0$

(11.6) $$\lim_{r\to\infty} \frac{1}{\pi}\int_0^\delta (f(x+t)+f(x-t)-2s(x))\frac{\sin rt}{t}\,dt = 0.$$

11 · 7 Theorem (Jordan). *If $f \in \mathscr{L}$ and if f is of bounded variation on $[a, b]$, then for each $x \in (a, b)$*

$$\lim_{r\to\infty} \frac{1}{\sqrt{2\pi}}\int_{-r}^{r} \hat{f}(u)e^{iux}\,du = \tfrac{1}{2}(f(x+)+f(x-)).$$

PROOF. Take $x \in (a, b)$ and take $\delta > 0$ such that $\delta < \min\{x-a, b-x\}$. If we let $s(x) = \frac{1}{2}(f(x+)+f(x-))$ and $g(x, t) = f(x+t)+f(x-t)-2s(x)$, then g is of bounded variation with respect to t on $[0, \delta]$. Therefore g is the difference of two nondecreasing functions f_1 and f_2 on $[0, \delta]$:

$$g(x, t) = f_1(t) - f_2(t), \quad t \in [0, \delta].$$

Since

$$0 = g(x, 0+) = f_1(0+) - f_2(0+),$$

if we let $h_i(t) = f_i(t) - f_i(0+)$ for $i = 1, 2$, then $h_i(0+) = 0$, h_i is nondecreasing on $[0, \delta]$ and nonnegative on $(0, \delta]$, and

$$g(x, t) = h_1(t) - h_2(t), \quad t \in [0, \delta].$$

From 11 · 6 we know that $\lim_{r\to\infty} s_r(x) = s(x)$ if for some $\delta > 0$

$$\lim_{r\to\infty} \int_0^\delta \frac{g(x, t)}{t}\sin rt\,dt = 0.$$

Thus, $\lim_{r\to\infty} s_r(x) = s(x)$ if

$$\lim_{r\to\infty} \int_0^\delta \frac{h_i(t)}{t}\sin rt\,dt = 0, \quad i = 1, 2.$$

Take $\varepsilon > 0$. Since $h_i(0+) = 0$ there exists an $\eta \in (0, \delta)$ such that $h_i(t) < \varepsilon$ for $t \in (0, \eta)$. If we use the Second Mean Value Theorem and the problem on p. 271, we obtain: for some $c \in [0, \eta]$

$$\left|\int_0^\eta \frac{h_i(t)}{t} \sin rt\, dt\right| = \left|h_i(\eta-) \int_c^\eta \frac{\sin rt}{t}\, dt\right|$$

$$< \varepsilon \left|\int_{rc}^{r\eta} \frac{\sin u}{u}\, du\right|$$

$$\leq 4\varepsilon \quad \text{for all } r > 0.$$

By the Riemann-Lebesgue Theorem for all r sufficiently large

$$\left|\int_\eta^\delta \frac{h_i(t)}{t} \sin rt\, dt\right| < \varepsilon.$$

Thus, for all r sufficiently large

$$\left|\int_0^\delta \frac{h_i(t)}{t} \sin rt\, dt\right| < 5\varepsilon. \quad \blacksquare$$

PROBLEMS

1. If $f(x) = e^{-x^2/2}$, determine $\hat{f}$.

2. If $f \in \mathscr{L}$ and $g(x) = f(x + c)$, show that $\hat{g}(y) = \hat{f}(y)e^{icy}$.

3. If f and g are in $\mathscr{L}$, let the **convolution** of f and g, denoted by $f * g$, be defined by

$$(f * g)(x) = \frac{1}{\sqrt{2\pi}} \int_{-\infty}^{\infty} f(t)g(x - t)\, dt.$$

Show that:

a. $f * g = g * f$.

b. If $h = f * g$, then $\hat{h} = \hat{f}\hat{g}$.

Suggestion. Refer to Problem 3, p. 246.

4. If, for all $r > 0$, $f \in \mathscr{L}(-r, r)$ define the $(C, 1)$ or **Cesàro means** of $\int_{-\infty}^{\infty} f$ by

$$\sigma_r = \frac{1}{r} \int_0^r \int_{-s}^{s} f(t)\, dt\, ds.$$

We say that $\int_{-\infty}^{\infty} f$ is **Cesàro summable** to s if $\lim_{r\to\infty} \sigma_r = s$. Show that:

a. $\sigma_r = \displaystyle\int_{-r}^{r} \left(1 - \frac{|t|}{r}\right) f(t)\, dt$

b. If $\displaystyle\int_{-\infty}^{\infty} f = s$, then $\displaystyle\lim_{r\to\infty} \sigma_r = s$.

Suggestion. Let $f_r(t) = (1 - |t|/r) f(t)\chi_{[-r,r]}(t)$ and use Problem 9, p. 131.

5. Let $f \in \mathscr{L}$. The Cesàro means of the Fourier integral

$$\frac{1}{\sqrt{2\pi}}\int_{-\infty}^{\infty} \hat{f}(u)e^{iux}\,du$$

are

$$\sigma_r(x) = \frac{1}{\sqrt{2\pi}}\int_{-r}^{r}\left(1 - \frac{|u|}{r}\right)\hat{f}(u)e^{iux}\,du.$$

Show that

$$\sigma_r(x) = \frac{1}{2\pi}\int_{-\infty}^{\infty} f(t)\int_{-r}^{r}\left(1 - \frac{|u|}{r}\right)e^{iu(x-t)}\,du\,dt.$$

6. The **Fejer kernel** K_r is defined by

$$K_r(s) = \int_{-r}^{r}\left(1 - \frac{|u|}{r}\right)e^{ius}\,du$$

so that

$$\sigma_r(x) = \frac{1}{2\pi}\int_{-\infty}^{\infty} f(t)K_r(x-t)\,dt.$$

Show that:

a. $K_r(s) = \dfrac{2}{rs^2}(1 - \cos rs) \le \dfrac{4}{rs^2}, \quad s \neq 0$

b. $K_r(-s) = K_r(s)$

c. $0 \le K_r(s) \le r$

d. $\displaystyle\int_0^{\infty} K_r(s)\,ds = \pi$

e. $\displaystyle\sigma_r(x) = \frac{1}{2\pi}\int_0^{\infty}(f(x+t) + f(x-t))K_r(t)\,dt.$

7. Prove Fejer's Theorem for the Fourier integral: If $f \in \mathscr{L}$, then the Fourier integral of f is Cesàro summable to

$$s(x) = \tfrac{1}{2}(f(x+) + f(x-))$$

at each point x at which $f(x+)$ and $f(x-)$ exist. In particular, the Fourier integral of f is Cesàro summable to $f(x)$ at each point x at which f is continuous.

Suggestion. Refer to Problem 9 of §10.

8. Prove the Fejer-Lebesgue Theorem for the Fourier integral: If $f \in \mathscr{L}$, then the Fourier integral of f is Cesàro summable to $f(x)$ for all x in the Lebesgue set of f; that is,

$$\lim_{r\to\infty} \frac{1}{\sqrt{2\pi}}\int_{-r}^{r}\left(1 - \frac{|u|}{r}\right)\hat{f}(u)e^{iux}\,du = f(x) \quad \text{for a.a.}x.$$

Suggestion. Refer to Problem 11 of §10.

9. If both $f \in \mathscr{L}$ and $\hat{f} \in \mathscr{L}$, show that for all x in the Lebesgue set of f

$$\frac{1}{\sqrt{2\pi}} \int_{-\infty}^{\infty} \hat{f}(u) e^{iux}\, du = f(x).$$

Suggestion. Let $h_r(u) = (1 - |u|/r)\hat{f}(u)e^{iux}\, \chi_{[-r,\, r]}(u)$. Use the form of the Lebesgue Dominated Convergence Theorem given in Problem 9, p. 131, and Problem 8, above.

10. If $f \in \mathscr{L}$ and $\hat{f} = 0$, show that $f = 0$ a.e.

Bibliography

1. Bachman, G. and Narici, L. *Functional Analysis.* Academic Press, New York, 1966.
2. Banach, S. *Théorie des Opérations Linéaires.* Chelsea Publishing Co., New York, 1955.
3. Bartle, R. G. *The Elements of Integration.* John Wiley & Sons, Inc., New York, 1966.
4. Berberian, S. K. *Measure and Integration.* The Macmillan Company, New York, 1965.
5. Boas, R. P., Jr. *A Primer of Real Functions.* Carus Mathematical Monograph number thirteen. The Mathematical Association of America. Distributed by John Wiley and Sons, Inc., New York, 1960.
6. Bohnenblust, H. F. and A. Sobczyk. Extensions of functionals on complex linear spaces, *Bulletin Amer. Math. Soc.,* **44**(1938), 91–93.
7. Buck, R. C., Editor. *Studies in Mathematics, Vol. 1, Studies in Modern Analysis.* The Mathematical Association of America. Distributed by Prentice-Hall, Inc., Englewood Cliffs, New Jersey, 1968.
8. Collatz, L. *Functional Analysis and Numerical Mathematics.* Translated from the German by H. Oser. Academic Press, New York, 1966.
9. Coppel, W. A. J. B. Fourier. On the occasion of his Two Hundredth Birthday, *The American Math. Monthly,* **76**(1969), p. 468–483.
10. Dieudonné, J. *Treatise on Analysis.* Vol. I. *Foundations of Modern Analysis,* Enlarged and Corrected Printing, 1969. Vol. II, Translated by I. G. MacDonald, 1970. Vols. III & IV, in preparation. Academic Press, New York.
11. du Bois-Reymond, P. Versuch einer Classification der willkürlichen Functionen reeller Argumente, *Journal für Math.,* **79**(1875), 21–37.
12. Dunford, N. and Schwartz, J. T. *Linear Operators, Vol. 1. General Theory.* Interscience Publishers, Inc., New York, 1958.
13. Faddeev, D. K. and Faddeeva, V. N. *Computational Methods of Linear Algebra.* W. H. Freeman and Company, San Francisco, 1963.
14. Faddeeva, V. N. *Computational Methods of Linear Algebra.* Dover Publications, Inc., New York, 1959.
15. Graves, L. M. *The Theory of Functions of Real Variables,* Second Edition. McGraw-Hill, New York, 1956.
16. Haaser, N. B., LaSalle, J. P., and Sullivan, J. A. *A Course in Mathematical Analysis, Vol. I, Introduction to Analysis.* Blaisdell Publishing Company, Waltham, Mass., 1959.
17. Haaser, N. B., LaSalle, J. P., and Sullivan, J. A. *A Course in Mathematical Analysis, Vol. II, Intermediate Analysis.* Blaisdell Publishing Company, Waltham, Mass., 1964.
18. Halmos, P. R. *Measure Theory.* D. Van Nostrand Company, Princeton, New Jersey, 1950.
19. Halmos, P. R. *Naïve Set Theory.* D. Van Nostrand Company, Princeton, New Jersey, 1960.

20. Hewitt, E. and Stromberg, K. *Real and Abstract Analysis*. Springer-Verlag New York, Inc., New York, 1965.

21. Kamke, E. *Theory of Sets*. Translated from the German by F. Bagemihl. Dover Publications, Inc., New York, 1950.

22. Kelley, J. L. *General Topology*. D. Van Nostrand Company, Princeton, New Jersey, 1955.

23. Kolmogorov, A. N. and Fomin, S. V. *Elements of the Theory of Functions and Functional Analysis, Vol. 1 Metric and Normed Spaces*. Translated from the Russian by L. F. Boron. Graylock Press, Rochester, New York, 1957.

24. Kolmogorov, A. N. and Fomin, S. V. *Elements of the Theory of Functions and Functional Analysis, Vol. 2 Measure. The Lebesgue Integral. Hilbert Space*. Translated from the Russian by H. Kamel and H. Komm. Graylock Press, Albany, New York, 1961.

25. Landau, E. *Foundations of Analysis*. Translated from the German by F. Steinhardt. Chelsea Publishing Company, New York, 1951.

26. Lebesgue, H. *Leçons sur l'intégration et la recherche des fonctions primitives*. Paris, 1904; 2nd ed. 1928.

27. Loomis, L. H. *An Introduction to Abstract Harmonic Analysis*. D. Van Nostrand Company, Princeton, New Jersey, 1953.

28. McShane, E. J. and Botts, T. A. *Real Analysis*. D. Van Nostrand Company, Princeton, New Jersey, 1959.

29. Natanson, I. P. *Theory of Functions of a Real Variable*. 2 Vols. Frederick Ungar Publishing Co., New York, 1955, 1960.

30. Patterson, E. M. *Topology*. Oliver and Boyd, Edinburgh, 1959.

31. Riesz, F. Sur les opérations fonctionnelles linéaires, Comptes Rendus Acad. Sci. Paris, **149**(1909), p. 974–977.

32. Riesz, F. and Sz.-Nagy, B. *Functional Analysis*. Translated from the French by L. F. Boron. Frederick Ungar Publishing Co., New York, 1955.

33. Rogosinski, W. W. *Volume and Integral*, Oliver and Boyd, Edinburgh, 1952.

34. Royden, H. L. *Real Analysis, Second Edition*. The Macmillan Company, New York, 1968.

35. Rudin, W. *Principles of Mathematical Analysis, Second Edition*. McGraw-Hill Book Company, New York, 1964.

36. Simmons, G. F. *Introduction to Topology and Modern Analysis*. McGraw-Hill Book Company, 1963.

37. Suppes, P. C. *Axiomatic Set Theory*. D. Van Nostrand Company, Princeton, New Jersey, 1960. Corrected and enlarged edition published by Dover Publications, Inc., New York, 1972.

38. Taylor, A. E. *Introduction to Functional Analysis*. John Wiley & Sons, New York, 1958.

39. Taylor, A. E. *General Theory of Functions and Integration*. Blaisdell Publishing Company, Waltham, Mass., 1965.

40. Varga, R. S. *Matrix Iterative Analysis*. Prentice-Hall, Inc., Englewood Cliffs, New Jersey, 1962.

41. Waerden, B. L. van der. Ein einfaches Beispiel einer nichtdifferenzierbaren stetigen Funktion, *Math. Zeitschr.*, **32**(1930), 474–475.

42. Yosida, K. *Functional Analysis*. Springer-Verlag, Berlin, 1965.

Index of Symbols

Index

A CATALOG OF SELECTED

DOVER BOOKS

IN SCIENCE AND MATHEMATICS

A CATALOG OF SELECTED

DOVER BOOKS

IN SCIENCE AND MATHEMATICS

Astronomy

BURNHAM'S CELESTIAL HANDBOOK, Robert Burnham, Jr. Thorough guide to the stars beyond our solar system. Exhaustive treatment. Alphabetical by constellation: Andromeda to Cetus in Vol. 1; Chamaeleon to Orion in Vol. 2; and Pavo to Vulpecula in Vol. 3. Hundreds of illustrations. Index in Vol. 3. 2,000pp. 6⅛ x 9¼. 23567-X, 23568-8, 23673-0 Three-vol. set

THE EXTRATERRESTRIAL LIFE DEBATE, 1750–1900, Michael J. Crowe. First detailed, scholarly study in English of the many ideas that developed from 1750 to 1900 regarding the existence of intelligent extraterrestrial life. Examines ideas of Kant, Herschel, Voltaire, Percival Lowell, many other scientists and thinkers. 16 illustrations. 704pp. 5⅜ x 8½. 40675-X

A HISTORY OF ASTRONOMY, A. Pannekoek. Well-balanced, carefully reasoned study covers such topics as Ptolemaic theory, work of Copernicus, Kepler, Newton, Eddington's work on stars, much more. Illustrated. References. 521pp. 5⅜ x 8½. 65994-1

AMATEUR ASTRONOMER'S HANDBOOK, J. B. Sidgwick. Timeless, comprehensive coverage of telescopes, mirrors, lenses, mountings, telescope drives, micrometers, spectroscopes, more. 189 illustrations. 576pp. 5⅝ x 8¼. (Available in U.S. only.) 24034-7

STARS AND RELATIVITY, Ya. B. Zel'dovich and I. D. Novikov. Vol. 1 of *Relativistic Astrophysics* by famed Russian scientists. General relativity, properties of matter under astrophysical conditions, stars, and stellar systems. Deep physical insights, clear presentation. 1971 edition. References. 544pp. 5⅝ x 8¼. 69424-0

Chemistry

CHEMICAL MAGIC, Leonard A. Ford. Second Edition, Revised by E. Winston Grundmeier. Over 100 unusual stunts demonstrating cold fire, dust explosions, much more. Text explains scientific principles and stresses safety precautions. 128pp. 5⅜ x 8½. 67628-5

THE DEVELOPMENT OF MODERN CHEMISTRY, Aaron J. Ihde. Authoritative history of chemistry from ancient Greek theory to 20th-century innovation. Covers major chemists and their discoveries. 209 illustrations. 14 tables. Bibliographies. Indices. Appendices. 851pp. 5⅜ x 8½. 64235-6

CATALYSIS IN CHEMISTRY AND ENZYMOLOGY, William P. Jencks. Exceptionally clear coverage of mechanisms for catalysis, forces in aqueous solution, carbonyl- and acyl-group reactions, practical kinetics, more. 864pp. 5⅜ x 8½. 65460-5

THE HISTORICAL BACKGROUND OF CHEMISTRY, Henry M. Leicester. Evolution of ideas, not individual biography. Concentrates on formulation of a coherent set of chemical laws. 260pp. 5⅜ x 8½. 61053-5

A SHORT HISTORY OF CHEMISTRY, J. R. Partington. Classic exposition explores origins of chemistry, alchemy, early medical chemistry, nature of atmosphere, theory of valency, laws and structure of atomic theory, much more. 428pp. 5⅜ x 8½. (Available in U.S. only.) 65977-1

GENERAL CHEMISTRY, Linus Pauling. Revised 3rd edition of classic first-year text by Nobel laureate. Atomic and molecular structure, quantum mechanics, statistical mechanics, thermodynamics correlated with descriptive chemistry. Problems. 992pp. 5⅜ x 8½. 65622-5

Engineering

DE RE METALLICA, Georgius Agricola. The famous Hoover translation of greatest treatise on technological chemistry, engineering, geology, mining of early modern times (1556). All 289 original woodcuts. 638pp. 6¾ x 11. 60006-8

FUNDAMENTALS OF ASTRODYNAMICS, Roger Bate et al. Modern approach developed by U.S. Air Force Academy. Designed as a first course. Problems, exercises. Numerous illustrations. 455pp. 5⅜ x 8½. 60061-0

DYNAMICS OF FLUIDS IN POROUS MEDIA, Jacob Bear. For advanced students of ground water hydrology, soil mechanics and physics, drainage and irrigation engineering and more. 335 illustrations. Exercises, with answers. 784pp. 6⅛ x 9¼. 65675-6

ANALYTICAL MECHANICS OF GEARS, Earle Buckingham. Indispensable reference for modern gear manufacture covers conjugate gear-tooth action, gear-tooth profiles of various gears, many other topics. 263 figures. 102 tables. 546pp. 5⅜ x 8½. 65712-4

MECHANICS, J. P. Den Hartog. A classic introductory text or refresher. Hundreds of applications and design problems illuminate fundamentals of trusses, loaded beams and cables, etc. 334 answered problems. 462pp. 5⅜ x 8½. 60754-2

MECHANICAL VIBRATIONS, J. P. Den Hartog. Classic textbook offers lucid explanations and illustrative models, applying theories of vibrations to a variety of practical industrial engineering problems. Numerous figures. 233 problems, solutions. Appendix. Index. Preface. 436pp. 5⅜ x 8½. 64785-4

STRENGTH OF MATERIALS, J. P. Den Hartog. Full, clear treatment of basic material (tension, torsion, bending, etc.) plus advanced material on engineering methods, applications. 350 answered problems. 323pp. 5⅜ x 8½. 60755-0

A HISTORY OF MECHANICS, René Dugas. Monumental study of mechanical principles from antiquity to quantum mechanics. Contributions of ancient Greeks, Galileo, Leonardo, Kepler, Lagrange, many others. 671pp. 5⅜ x 8½. 65632-2

FUNCTIONAL ANALYSIS (Second Corrected Edition), George Bachman and Lawrence Narici. Excellent treatment of subject geared toward students with background in linear algebra, advanced calculus, physics and engineering. Text covers introduction to inner-product spaces, normed, metric spaces, and topological spaces; complete orthonormal sets, the Hahn-Banach Theorem and its consequences, and many other related subjects. 1966 ed. 544pp. 6⅛ x 9¼. 40251-7

ASYMPTOTIC EXPANSIONS OF INTEGRALS, Norman Bleistein & Richard A. Handelsman. Best introduction to important field with applications in a variety of scientific disciplines. New preface. Problems. Diagrams. Tables. Bibliography. Index. 448pp. 5⅜ x 8½. 65082-0

FAMOUS PROBLEMS OF GEOMETRY AND HOW TO SOLVE THEM, Benjamin Bold. Squaring the circle, trisecting the angle, duplicating the cube: learn their history, why they are impossible to solve, then solve them yourself. 128pp. 5⅜ x 8½. 24297-8

VECTOR AND TENSOR ANALYSIS WITH APPLICATIONS, A. I. Borisenko and I. E. Tarapov. Concise introduction. Worked-out problems, solutions, exercises. 257pp. 5⅝ x 8¼. 63833-2

THE ABSOLUTE DIFFERENTIAL CALCULUS (CALCULUS OF TENSORS), Tullio Levi-Civita. Great 20th-century mathematician's classic work on material necessary for mathematical grasp of theory of relativity. 452pp. 5⅝ x 8¼. 63401-9

AN INTRODUCTION TO ORDINARY DIFFERENTIAL EQUATIONS, Earl A. Coddington. A thorough and systematic first course in elementary differential equations for undergraduates in mathematics and science, with many exercises and problems (with answers). Index. 304pp. 5⅜ x 8½. 65942-9

FOURIER SERIES AND ORTHOGONAL FUNCTIONS, Harry F. Davis. An incisive text combining theory and practical example to introduce Fourier series, orthogonal functions and applications of the Fourier method to boundary-value problems. 570 exercises. Answers and notes. 416pp. 5⅜ x 8½. 65973-9

COMPUTABILITY AND UNSOLVABILITY, Martin Davis. Classic graduate-level introduction to theory of computability, usually referred to as theory of recurrent functions. New preface and appendix. 288pp. 5⅜ x 8½. 61471-9

ASYMPTOTIC METHODS IN ANALYSIS, N. G. de Bruijn. An inexpensive, comprehensive guide to asymptotic methods–the pioneering work that teaches by explaining worked examples in detail. Index. 224pp. 5⅜ x 8½ 64221-6

ESSAYS ON THE THEORY OF NUMBERS, Richard Dedekind. Two classic essays by great German mathematician: on the theory of irrational numbers; and on transfinite numbers and properties of natural numbers. 115pp. 5⅜ x 8½. 21010-3

POPULAR LECTURES ON MATHEMATICAL LOGIC, Hao Wang. Noted logician's lucid treatment of historical developments, set theory, model theory, recursion theory and constructivism, proof theory, more. 3 appendixes. Bibliography. 1981 edition. ix + 283pp. 5⅜ x 8½. 67632-3

CALCULUS OF VARIATIONS, Robert Weinstock. Basic introduction covering isoperimetric problems, theory of elasticity, quantum mechanics, electrostatics, etc. Exercises throughout. 326pp. 5⅜ x 8½. 63069-2

THE CONTINUUM: A Critical Examination of the Foundation of Analysis, Hermann Weyl. Classic of 20th-century foundational research deals with the conceptual problem posed by the continuum. 156pp. 5⅜ x 8½. 67982-9

CHALLENGING MATHEMATICAL PROBLEMS WITH ELEMENTARY SOLUTIONS, A. M. Yaglom and I. M. Yaglom. Over 170 challenging problems on probability theory, combinatorial analysis, points and lines, topology, convex polygons, many other topics. Solutions. Total of 445pp. 5⅜ x 8½. Two-vol. set.
Vol. I: 65536-9 Vol. II: 65537-7

A SURVEY OF NUMERICAL MATHEMATICS, David M. Young and Robert Todd Gregory. Broad self-contained coverage of computer-oriented numerical algorithms for solving various types of mathematical problems in linear algebra, ordinary and partial, differential equations, much more. Exercises. Total of 1,248pp. 5⅜ x 8½. Two volumes. Vol. I: 65691-8 Vol. II: 65692-6

INTRODUCTION TO PARTIAL DIFFERENTIAL EQUATIONS WITH APPLICATIONS, E. C. Zachmanoglou and Dale W. Thoe. Essentials of partial differential equations applied to common problems in engineering and the physical sciences. Problems and answers. 416pp. 5⅜ x 8½. 65251-3

THE THEORY OF GROUPS, Hans J. Zassenhaus. Well-written graduate-level text acquaints reader with group-theoretic methods and demonstrates their usefulness in mathematics. Axioms, the calculus of complexes, homomorphic mapping, p-group theory, more. Many proofs shorter and more transparent than older ones. 276pp. 5⅜ x 8½. 40922-8

DISTRIBUTION THEORY AND TRANSFORM ANALYSIS: An Introduction to Generalized Functions, with Applications, A. H. Zemanian. Provides basics of distribution theory, describes generalized Fourier and Laplace transformations. Numerous problems. 384pp. 5⅜ x 8½. 65479-6

Math–Decision Theory, Statistics, Probability

ELEMENTARY DECISION THEORY, Herman Chernoff and Lincoln E. Moses. Clear introduction to statistics and statistical theory covers data processing, probability and random variables, testing hypotheses, much more. Exercises. 364pp. 5⅜ x 8½. 65218-1

Math–Geometry and Topology

ELEMENTARY CONCEPTS OF TOPOLOGY, Paul Alexandroff. Elegant, intuitive approach to topology from set-theoretic topology to Betti groups; how concepts of topology are useful in math and physics. 25 figures. 57pp. 5⅜ x 8½. 60747-X

COMBINATORIAL TOPOLOGY, P. S. Alexandrov. Clearly written, well-organized, three-part text begins by dealing with certain classic problems without using the formal techniques of homology theory and advances to the central concept, the Betti groups. Numerous detailed examples. 654pp. 5⅜ x 8½. 40179-0

EXPERIMENTS IN TOPOLOGY, Stephen Barr. Classic, lively explanation of one of the byways of mathematics. Klein bottles, Moebius strips, projective planes, map coloring, problem of the Koenigsberg bridges, much more, described with clarity and wit. 43 figures. 210pp. 5⅜ x 8½. 25933-1

CONFORMAL MAPPING ON RIEMANN SURFACES, Harvey Cohn. Lucid, insightful book presents ideal coverage of subject. 334 exercises make book perfect for self-study. 55 figures. 352pp. 5⅝ x 8¼. 64025-6

THE GEOMETRY OF RENÉ DESCARTES, René Descartes. The great work founded analytical geometry. Original French text, Descartes's own diagrams, together with definitive Smith-Latham translation. 244pp. 5⅜ x 8½. 60068-8

THE THIRTEEN BOOKS OF EUCLID'S ELEMENTS, translated with introduction and commentary by Sir Thomas L. Heath. Definitive edition. Textual and linguistic notes, mathematical analysis. 2,500 years of critical commentary. Unabridged. 1,414pp. 5⅜ x 8½. Three-vol. set.
Vol. I: 60088-2 Vol. II: 60089-0 Vol. III: 60090-4

GEOMETRY OF COMPLEX NUMBERS, Hans Schwerdtfeger. Illuminating, widely praised book on analytic geometry of circles, the Moebius transformation, and two-dimensional non-Euclidean geometries. 200pp. 5⅝ x 8¼. 63830-8

DIFFERENTIAL GEOMETRY, Heinrich W. Guggenheimer. Local differential geometry as an application of advanced calculus and linear algebra. Curvature, transformation groups, surfaces, more. Exercises. 62 figures. 378pp. 5⅜ x 8½. 63433-7

CURVATURE AND HOMOLOGY: Enlarged Edition, Samuel I. Goldberg. Revised edition examines topology of differentiable manifolds; curvature, homology of Riemannian manifolds; compact Lie groups; complex manifolds; curvature, homology of Kaehler manifolds. New Preface. Four new appendixes. 416pp. 5⅜ x 8½. 40207-X

TOPOLOGY, John G. Hocking and Gail S. Young. Superb one-year course in classical topology. Topological spaces and functions, point-set topology, much more. Examples and problems. Bibliography. Index. 384pp. 5⅝ x 8¼. 65676-4

LECTURES ON CLASSICAL DIFFERENTIAL GEOMETRY, Second Edition, Dirk J. Struik. Excellent brief introduction covers curves, theory of surfaces, fundamental equations, geometry on a surface, conformal mapping, other topics. Problems. 240pp. 5⅜ x 8½. 65609-8

Math–History of

A SHORT ACCOUNT OF THE HISTORY OF MATHEMATICS, W. W. Rouse Ball. One of clearest, most authoritative surveys from the Egyptians and Phoenicians through 19th-century figures such as Grassman, Galois, Riemann. Fourth edition. 522pp. 5⅜ x 8½. 20630-0

THE HISTORY OF THE CALCULUS AND ITS CONCEPTUAL DEVELOPMENT, Carl B. Boyer. Origins in antiquity, medieval contributions, work of Newton, Leibniz, rigorous formulation. Treatment is verbal. 346pp. 5⅜ x 8½. 60509-4

THE HISTORICAL ROOTS OF ELEMENTARY MATHEMATICS, Lucas N. H. Bunt, Phillip S. Jones, and Jack D. Bedient. Fundamental underpinnings of modern arithmetic, algebra, geometry and number systems derived from ancient civilizations. 320pp. 5⅜ x 8½. 25563-8

A HISTORY OF MATHEMATICAL NOTATIONS, Florian Cajori. This classic study notes the first appearance of a mathematical symbol and its origin, the competition it encountered, its spread among writers in different countries, its rise to popularity, its eventual decline or ultimate survival. Original 1929 two-volume edition presented here in one volume. xxviii+820pp. 5⅜ x 8½. 67766-4

GAMES, GODS & GAMBLING: A History of Probability and Statistical Ideas, F. N. David. Episodes from the lives of Galileo, Fermat, Pascal, and others illustrate this fascinating account of the roots of mathematics. Features thought-provoking references to classics, archaeology, biography, poetry. 1962 edition. 304pp. 5⅜ x 8½. (Available in U.S. only.) 40023-9

OF MEN AND NUMBERS: The Story of the Great Mathematicians, Jane Muir. Fascinating accounts of the lives and accomplishments of history's greatest mathematical minds–Pythagoras, Descartes, Euler, Pascal, Cantor, many more. Anecdotal, illuminating. 30 diagrams. Bibliography. 256pp. 5⅜ x 8½. 28973-7

HISTORY OF MATHEMATICS, David E. Smith. Nontechnical survey from ancient Greece and Orient to late 19th century; evolution of arithmetic, geometry, trigonometry, calculating devices, algebra, the calculus. 362 illustrations. 1,355pp. 5⅜ x 8½. Two-vol. set. Vol. I: 20429-4 Vol. II: 20430-8

A CONCISE HISTORY OF MATHEMATICS, Dirk J. Struik. The best brief history of mathematics. Stresses origins and covers every major figure from ancient Near East to 19th century. 41 illustrations. 195pp. 5⅜ x 8½. 60255-9

Physics

OPTICAL RESONANCE AND TWO-LEVEL ATOMS, L. Allen and J. H. Eberly. Clear, comprehensive introduction to basic principles behind all quantum optical resonance phenomena. 53 illustrations. Preface. Index. 256pp. 5⅜ x 8½. 65533-4

ULTRASONIC ABSORPTION: An Introduction to the Theory of Sound Absorption and Dispersion in Gases, Liquids and Solids, A. B. Bhatia. Standard reference in the field provides a clear, systematically organized introductory review of fundamental concepts for advanced graduate students, research workers. Numerous diagrams. Bibliography. 440pp. 5⅜ x 8½. 64917-2

QUANTUM THEORY, David Bohm. This advanced undergraduate-level text presents the quantum theory in terms of qualitative and imaginative concepts, followed by specific applications worked out in mathematical detail. Preface. Index. 655pp. 5⅜ x 8½. 65969-0

ATOMIC PHYSICS (8th edition), Max Born. Nobel laureate's lucid treatment of kinetic theory of gases, elementary particles, nuclear atom, wave-corpuscles, atomic structure and spectral lines, much more. Over 40 appendices, bibliography. 495pp. 5⅜ x 8½. 65984-4

AN INTRODUCTION TO HAMILTONIAN OPTICS, H. A. Buchdahl. Detailed account of the Hamiltonian treatment of aberration theory in geometrical optics. Many classes of optical systems defined in terms of the symmetries they possess. Problems with detailed solutions. 1970 edition. xv + 360pp. 5⅜ x 8½. 67597-1

THIRTY YEARS THAT SHOOK PHYSICS: The Story of Quantum Theory, George Gamow. Lucid, accessible introduction to influential theory of energy and matter. Careful explanations of Dirac's anti-particles, Bohr's model of the atom, much more. 12 plates. Numerous drawings. 240pp. 5⅜ x 8½. 24895-X

ELECTRONIC STRUCTURE AND THE PROPERTIES OF SOLIDS: The Physics of the Chemical Bond, Walter A. Harrison. Innovative text offers basic understanding of the electronic structure of covalent and ionic solids, simple metals, transition metals and their compounds. Problems. 1980 edition. 582pp. 6⅛ x 9¼. 66021-4

HYDRODYNAMIC AND HYDROMAGNETIC STABILITY, S. Chandrasekhar. Lucid examination of the Rayleigh-Benard problem; clear coverage of the theory of instabilities causing convection. 704pp. 5⅜ x 8¼. 64071-X

INVESTIGATIONS ON THE THEORY OF THE BROWNIAN MOVEMENT, Albert Einstein. Five papers (1905–8) investigating dynamics of Brownian motion and evolving elementary theory. Notes by R. Fürth. 122pp. 5⅜ x 8½. 60304-0

THE PHYSICS OF WAVES, William C. Elmore and Mark A. Heald. Unique overview of classical wave theory. Acoustics, optics, electromagnetic radiation, more. Ideal as classroom text or for self-study. Problems. 477pp. 5⅜ x 8½. 64926-1

CATALOG OF DOVER BOOKS

METHODS OF THERMODYNAMICS, Howard Reiss. Outstanding text focuses on physical technique of thermodynamics, typical problem areas of understanding, and significance and use of thermodynamic potential. 1965 edition. 238pp. 5⅜ x 8½. 69445-3

TENSOR ANALYSIS FOR PHYSICISTS, J. A. Schouten. Concise exposition of the mathematical basis of tensor analysis, integrated with well-chosen physical examples of the theory. Exercises. Index. Bibliography. 289pp. 5⅜ x 8½. 65582-2

RELATIVITY IN ILLUSTRATIONS, Jacob T. Schwartz. Clear nontechnical treatment makes relativity more accessible than ever before. Over 60 drawings illustrate concepts more clearly than text alone. Only high school geometry needed. Bibliography. 128pp. 6⅛ x 9¼. 25965-X

THE ELECTROMAGNETIC FIELD, Albert Shadowitz. Comprehensive undergraduate text covers basics of electric and magnetic fields, builds up to electromagnetic theory. Also related topics, including relativity. Over 900 problems. 768pp. 5⅝ x 8¼. 65660-8

GREAT EXPERIMENTS IN PHYSICS: Firsthand Accounts from Galileo to Einstein, edited by Morris H. Shamos. 25 crucial discoveries: Newton's laws of motion, Chadwick's study of the neutron, Hertz on electromagnetic waves, more. Original accounts clearly annotated. 370pp. 5⅜ x 8½. 25346-5

RELATIVITY, THERMODYNAMICS AND COSMOLOGY, Richard C. Tolman. Landmark study extends thermodynamics to special, general relativity; also applications of relativistic mechanics, thermodynamics to cosmological models. 501pp. 5⅜ x 8½. 65383-8

LIGHT SCATTERING BY SMALL PARTICLES, H. C. van de Hulst. Comprehensive treatment including full range of useful approximation methods for researchers in chemistry, meteorology and astronomy. 44 illustrations. 470pp. 5⅜ x 8½. 64228-3

STATISTICAL PHYSICS, Gregory H. Wannier. Classic text combines thermodynamics, statistical mechanics and kinetic theory in one unified presentation of thermal physics. Problems with solutions. Bibliography. 532pp. 5⅜ x 8½. 65401-X

Paperbound unless otherwise indicated. Available at your book dealer, online at **www.doverpublications.com**, or by writing to Dept. GI, Dover Publications, Inc., 31 East 2nd Street, Mineola, NY 11501. For current price information or for free catalogues (please indicate field of interest), write to Dover Publications or log on to **www.doverpublications.com** and see every Dover book in print. Dover publishes more than 500 books each year on science, elementary and advanced mathematics, biology, music, art, literary history, social sciences, and other areas.